Animal Cell Electroporation and Electrofusion Protocols

Methods in Molecular Biology™ Series
John M. Walker, SERIES EDITOR

55. **Plant Cell Electroporation and Electrofusion Protocols,** edited by *Jac A. Nickoloff, 1995*
54. **YAC Protocols,** edited by *David Markie, 1995*
53. **Yeast Protocols:** *Methods in Cell and Molecular Biology,* edited by *Ivor H. Evans, 1995*
52. **Capillary Electrophoresis:** *Principles, Instrumentation, and Applications,* edited by *Kevin D. Altria, 1995*
51. **Antibody Engineering Protocols,** edited by *Sudhir Paul, 1995*
50. **Species Diagnostics Protocols:** *PCR and Other Nucleic Acid Methods,* edited by *Justin P. Clapp, 1995*
49. **Plant Gene Transfer and Expression Protocols,** edited by *Heddwyn Jones, 1995*
48. **Animal Cell Electroporation and Electrofusion Protocols,** edited by *Jac A. Nickoloff, 1995*
47. **Electroporation Protocols for Microorganisms,** edited by *Jac A. Nickoloff, 1995*
46. **Diagnostic Bacteriology Protocols,** edited by *Jenny Howard and David M. Whitcombe, 1995*
45. **Monoclonal Antibody Protocols,** edited by *William C. Davis, 1995*
44. ***Agrobacterium* Protocols,** edited by *Kevan M. A. Gartland and Michael R. Davey, 1995*
43. **In Vitro Toxicity Testing Protocols,** edited by *Sheila O'Hare and Chris K. Atterwill, 1995*
42. **ELISA:** *Theory and Practice,* by *John R. Crowther, 1995*
41. **Signal Transduction Protocols,** edited by *David A. Kendall and Stephen J. Hill, 1995*
40. **Protein Stability and Folding:** *Theory and Practice,* edited by *Bret A. Shirley, 1995*
39. **Baculovirus Expression Protocols,** edited by *Christopher D. Richardson, 1995*
38. **Cryopreservation and Freeze-Drying Protocols,** edited by *John G. Day and Mark R. McLellan, 1995*
37. **In Vitro Transcription and Translation Protocols,** edited by *Martin J. Tymms, 1995*
36. **Peptide Analysis Protocols,** edited by *Ben M. Dunn and Michael W. Pennington, 1994*
35. **Peptide Synthesis Protocols,** edited by *Michael W. Pennington and Ben M. Dunn, 1994*
34. **Immunocytochemical Methods and Protocols,** edited by *Lorette C. Javois, 1994*
33. ***In Situ* Hybridization Protocols,** edited by *K. H. Andy Choo, 1994*
32. **Basic Protein and Peptide Protocols,** edited by *John M. Walker, 1994*
31. **Protocols for Gene Analysis,** edited by *Adrian J. Harwood, 1994*
30. **DNA–Protein Interactions,** edited by *G. Geoff Kneale, 1994*
29. **Chromosome Analysis Protocols,** edited by *John R. Gosden, 1994*
28. **Protocols for Nucleic Acid Analysis by Nonradioactive Probes,** edited by *Peter G. Isaac, 1994*
27. **Biomembrane Protocols:** *II. Architecture and Function,* edited by *John M. Graham and Joan A. Higgins, 1994*
26. **Protocols for Oligonucleotide Conjugates:** *Synthesis and Analytical Techniques,* edited by *Sudhir Agrawal, 1994*
25. **Computer Analysis of Sequence Data:** *Part II,* edited by *Annette M. Griffin and Hugh G. Griffin, 1994*
24. **Computer Analysis of Sequence Data:** *Part I,* edited by *Annette M. Griffin and Hugh G. Griffin, 1994*
23. **DNA Sequencing Protocols,** edited by *Hugh G. Griffin and Annette M. Griffin, 1993*
22. **Microscopy, Optical Spectroscopy, and Macroscopic Techniques,** edited by *Christopher Jones, Barbara Mulloy, and Adrian H. Thomas, 1993*
21. **Protocols in Molecular Parasitology,** edited by *John E. Hyde, 1993*
20. **Protocols for Oligonucleotides and Analogs:** *Synthesis and Properties,* edited by *Sudhir Agrawal, 1993*
19. **Biomembrane Protocols:** *I. Isolation and Analysis,* edited by *John M. Graham and Joan A. Higgins, 1993*
18. **Transgenesis Techniques:** *Principles and Protocols,* edited by *David Murphy and David A. Carter, 1993*
17. **Spectroscopic Methods and Analyses:** *NMR, Mass Spectrometry, and Metalloprotein Techniques,* edited by *Christopher Jones, Barbara Mulloy, and Adrian H. Thomas, 1993*
16. **Enzymes of Molecular Biology,** edited by *Michael M. Burrell, 1993*
15. **PCR Protocols:** *Current Methods and Applications,* edited by *Bruce A. White, 1993*
14. **Glycoprotein Analysis in Biomedicine,** edited by *Elizabeth F. Hounsell, 1993*
13. **Protocols in Molecular Neurobiology,** edited by *Alan Longstaff and Patricia Revest, 1992*
12. **Pulsed-Field Gel Electrophoresis:** *Protocols, Methods, and Theories,* edited by *Margit Burmeister and Levy Ulanovsky, 1992*
11. **Practical Protein Chromatography,** edited by *Andrew Kenney and Susan Fowell, 1992*
10. **Immunochemical Protocols,** edited by *Margaret M. Manson, 1992*
9. **Protocols in Human Molecular Genetics,** edited by *Christopher G. Mathew, 1991*
8. **Practical Molecular Virology:** *Viral Vectors for Gene Expression,* edited by *Mary K. L. Collins, 1991*
7. **Gene Transfer and Expression Protocols,** edited by *Edward J. Murray, 1991*
6. **Plant Cell and Tissue Culture,** edited by *Jeffrey W. Pollard and John M. Walker, 1990*
5. **Animal Cell Culture,** edited by *Jeffrey W. Pollard and John M. Walker, 1990*

Methods in Molecular Biology™ • 48

Animal Cell Electroporation and Electrofusion Protocols

Edited by

Jac A. Nickoloff

Harvard University, Boston, MA

Springer Science+Business Media, LLC

© 1995 Springer Science+Business Media New York
Originally published by Humana Press Inc. in 1995

All rights reserved.

No part of this book may be reproduced, stored in a retrieval system, or transmitted in any form or by any means, electronic, mechanical, photocopying, microfilming, recording, or otherwise without written permission from the Publisher. Methods in Molecular Biology™ is a trademark of
Springer Science+Business Media, LLC.

All authored papers, comments, opinions, conclusions, or recommendations are those of the author(s) and do not necessarily reflect the views of the publisher.

This publication is printed on acid-free paper. ∞
ANSI Z39.48-1984 (American National Standards Institute)
Permanence of Paper for Printed Library Materials

Photocopy Authorization Policy:
Authorization to photocopy items for internal or personal use, or the internal or personal use of specific clients, is granted by Springer Science+Business Media, LLC.,
provided that the base fee of US $4.00 per copy, plus US $00.20
per page, is paid directly to the Copyright Clearance Center at 222 Rosewood Drive, Danvers, MA 01923. For those organizations that have been granted a photocopy license from the CCC, a separate system of payment has been arranged and is acceptable to Springer Science+Business Media, LLC.
The fee code for users of the Transactional Reporting Service is: [0-89603-304-X/95 $4.00 + $00.20].

Library of Congress Cataloging-in-Publication Data

Animal cell electroporation and electrofusion protocols/edited by Jac A. Nickoloff.
    p.   cm.—(Methods in molecular biology™;48)
    Includes index.
    ISBN 978-1-4899-4061-2   ISBN 978-1-59259-535-8 (eBook)
    DOI 10.1007/978-1-59259-535-8
    1. Electroporation.   2. Electrofusion.   3. Animal cell biotechnology.   I. Nickoloff, Jac A.
II.   Series: Methods in molecular biology™ (Totowa, NJ); 48.
QH585.5.E48A55 1995
591.87'028—dc20   95-30825
                                                                         CIP

Preface

The ability to introduce macromolecules into animal cells, including DNA, RNA, proteins, and other bioactive compounds has facilitated a broad range of biological studies, from biochemistry and biophysics to molecular biology, cell biology, and whole animal studies. Gene transfer technology in particular will continue to play an essential role in studies aimed at improving our understanding of the relationships between the gene structure and function, and it has important practical applications in both biotechnology and biomedicine, as evidenced by the current intense interest in gene therapy. Although DNA and other macromolecules may be introduced into cells by a variety of methods, including chemical treatments and microinjection, electroporation has proven to be simpler to perform, more efficient, and effective with a wider variety of cell types than other techniques. The early and broad success of electric field-mediated DNA transfer soon prompted researchers to investigate electroporation for transferring other types of molecules into cells, including RNA, enzymes, antibodies, and analytic dyes.

Animal Cell Electroporation and Electrofusion Protocols begins with three chapters that describe the theoretical and practical aspects of electroporation, including a review of the commercially available instrumentation. These introductory chapters will be of particular interest to those new to electric field technologies and to those developing protocols for as yet untested species or cell types. Nineteen chapters follow that present well-tested protocols for electroporation of proteins and DNA into insect, fish, and mammalian cells. These chapters also include discussions of the research questions currently being addressed with particular model systems, highlighting their advantages and limitations, and each presents typical electroporation results. Protocols are given for transfecting common cell types, such as Chinese hamster ovary cells and normal human cells, as well as for more specialized applications, such as the electroporation of adherent cells *in situ*, and mouse embryonic stem cells for producing transgenic mice. Two chapters describe how specific DNA sequences and electroporation buffer components can affect gene transfer efficiency, and another describes techniques for quantitation of transient gene expression from reporter genes. The final two electroporation chapters describe clinically relevant protocols, including electroporation of cardiac cells, which is thought to occur during electric shock treatment of cardiac arrhythmia, and the use of electroporation for gene therapy.

The parameters used to electroporate different cell types are often similar. However, when high efficiencies are desired, conditions used for cell growth and preparation, electrical parameters, and post-electroporation handling and selection must be carefully controlled. Therefore, each electroporation chapter describes a detailed protocol for the transfection of a particular cell type. These chapters also include discussions of problems that may be encountered, valuable troubleshooting advice, and recommendations for optimizing a protocol. By including protocols from a diverse set of species and cell types, this volume will facilitate cross-fertilization of ideas among researchers working on different systems. By comparing different procedures, one can gain additional problem-solving insight, as well as learn how others use electroporation to achieve various experimental goals. Although it was not possible to include protocols for all cell types that have been successfully electroporated, many of the chapters include references to protocols for additional cell types.

Cell fusion technologies also have played important roles in diverse areas of biological research, including cancer biology, immunology, and developmental biology. For many years, cell fusion was achieved by treating cells with chemical agents, such as polyethylene glycol, or biological agents, such as Sendai virus, but these procedures are not effective with all cell types. As with electroporation, the membrane effects of electric fields are highly general, and electrofusion has an equally widespread applicability. Five chapters describe electrofusion protocols useful for studies of somatic cell genetics and development, and for generating lines that produce monoclonal antibodies. The final two chapters describe spectrofluorometric and flow cytometry assays useful for monitoring the efficiency of cell-cell and cell-tissue fusion and are therefore of general interest to those employing electrofusion in their research.

I want to thank all of the contributors for their timely, high quality submissions, and for their valuable suggestions, with special thanks to Dr. Richard Heller for his many recommendations of potential contributors. I also want to thank series editor John Walker for his valuable advice.

Jac A. Nickoloff

Contents

Preface v
Contents for the Companion Volumes xiii
Contributors xvii

PART I. THEORY AND INSTRUMENTATION 1
CH. 1. Electroporation Theory: *Concepts and Mechanisms,*
James C. Weaver 3
CH. 2. Effects of Pulse Length and Strength on Electroporation Efficiency,
Sek Wen Hui 29
CH. 3. Instrumentation,
Gunter A. Hofmann 41

PART II. ELECTROPORATION PROTOCOLS 61
CH. 4. The Introduction of Proteins into Mammalian Cells by Electroporation,
William F. Morgan and Joseph P. Day 63
CH. 5. Electroporation of Antigen-Presenting Cells for T-Cell Recognition and Cytotoxic T-Lymphocyte Priming,
Weisan Chen and James McCluskey 73
CH. 6. Electroporation of Antibodies into Mammalian Cells,
Paul L. Campbell, James McCluskey, Jing Ping Yeo, and Ban-Hock Toh 83
CH. 7. Electroporation of Adherent Cells *In Situ* for the Introduction of Nonpermeant Molecules,
Leda H. Raptis, Kevin L. Firth, Heather L. Brownell, Andrea Todd, W. Craig Simon, Brian M. Bennett, Leslie W. MacKenzie, and Maria Zannis-Hadjopoulos 93
CH. 8. Electrotransformation of Chinese Hamster Ovary Cells,
Danielle Gioioso Taghian and Jac A. Nickoloff 115
CH. 9. Electroporation of Rat Pituitary Cells,
Ruth H. Paulssen, Eyvind J. Paulssen, and Kaare M. Gautvik 123
CH. 10. Electroporation of Plasmid DNA into Normal Human Fibroblasts,
F. Andrew Ray 133
CH. 11. Electroporation-Mediated Gene Transfer into Hepatocytes,
Alphonse Le Cam 141

Сн. 12.	Electroporation of Human Lymphoblastoid Cells, *Fen Xia and Howard L. Liber*	151
Сн. 13.	The Use of Electroporated Bovine Spermatozoa to Transfer Foreign DNA into Oocytes, *Marc Gagné, François Pothier, and Marc-André Sirard*	161
Сн. 14.	Electroporation of Embryonic Stem Cells for Generating Transgenic Mice and Studying In Vitro Differentiation, *John S. Mudgett and Thomas J. Livelli*	167
Сн. 15.	Electrotransfection with "Intracellular" Buffer, *Maurice J. B. van den Hoff, Vincent M. Christoffels, Wil T. Labruyère, Antoon F. M. Moorman, and Wouter H. Lamers*	185
Сн. 16.	Effect of Cis-Located Human Satellite DNA on Electroporation Efficiency, *Djenann Saint-Dic and Michael S. DuBow*	199
Сн. 17.	Quantitation of Transient Gene Expression, *Michael K. Showe and Louise C. Showe*	211
Сн. 18.	Stable Integration of Vectors at High Copy Number for High-Level Expression in Animal Cells, *James Barsoum*	225
Сн. 19.	Electroporation of *Drosophila* Embryos, *K. Puloma Kamdar, Thao N. Wagner, and Victoria Finnerty*	239
Сн. 20.	Transformation of Fish Cells and Embryos, *Koji Inoue, Jun-ichiro Hata, and Shinya Yamashita*	245
Сн. 21.	Electroporation of Cardiac Cells, *Leslie Tung*	253
Сн. 22.	Electroporation for Gene Therapy, *Kathryn E. Matthews, Sukhendu B. Dev, Frances Toneguzzo, and Armand Keating*	273
PART III.	ELECTROFUSION PROTOCOLS	281
Сн. 23.	Electrofusion of Mammalian Cells, *Kenneth L. White*	283
Сн. 24.	Stabilizing Antibody Secretion of Human Epstein Barr Virus-Activated B-Lymphocytes with Hybridoma Formation by Electrofusion, *Susan Perkins and Steven K. H. Foung*	295
Сн. 25.	Electrofusion of Mammalian Oocytes and Embryonic Cells, *Josef Fulka, Jr., Robert M. Moor, and Josef Fulka*	309
Сн. 26.	Nuclear Transfer in Bovine Embryos, *Akira Iritani and Tasuku Mitani*	317
Сн. 27.	Electrofusion of Mouse Embryos to Produce Tetraploids, *Ulrich Petzoldt*	331

Contents

CH. 28. Spectrofluorometric Assay for Cell-Tissue Electrofusion,
Richard Heller *341*

CH. 29. Cytometric Detection and Quantitation of Cell–Cell Electrofusion Products,
Mark J. Jaroszeski, Richard Gilbert, and Richard Heller *355*

Index *365*

CONTENTS FOR THE COMPANION VOLUME

Electroporation Protocols for Microorganisms

- CH. 1. Electroporation Theory: *Concepts and Mechanisms*,
  James C. Weaver
- CH. 2. Instrumentation,
  Gunter A. Hofmann
- CH. 3. Direct Plasmid Transfer Between Bacterial Species and Electrocuring,
  Helen L. Withers
- CH. 4. Transfer of Episomal and Integrated Plasmids from *Saccharomyces cerevisiae* to *Escherichia coli* by Electroporation,
  Laura Gunn, Jennifer Whelden, and Jac A. Nickoloff
- CH. 5. Production of cDNA Libraries by Electroporation,
  Christian E. Gruber
- CH. 6. Electroporation of RNA into *Saccharomyces cerevisiae*,
  Daniel R. Gallie
- CH. 7. Electrofusion of Yeast Protoplasts,
  Herbert Weber and Hermann Berg
- CH. 8. *Escherichia coli* Electrotransformation,
  Elizabeth M. Miller and Jac A. Nickoloff
- CH. 9. Electrotransformation in *Salmonella*,
  Kenneth E. Sanderson, P. Ronald MacLachlan, and Andrew Hessel
- CH. 10. Electrotransformation of *Pseudomonas*,
  Jonathan J. Dennis and Pamela A. Sokol
- CH. 11. Electroporation of *Xanthomonas*,
  Teresa J. White and Carlos F. Gonzalez
- CH. 12. Transformation of *Brucella* Species with Suicide and Broad Host-Range Plasmids,
  John R. McQuiston, Gerhardt G. Schurig, Nammalwar Sriranganathan, and Stephen M. Boyle
- CH. 13. Electroporation of *Francisella tularensis*,
  Gerald S. Baron, Svetlana V. Myltseva, and Francis E. Nano
- CH. 14. A Simple and Rapid Method for Transformation of *Vibrio* Species by Electroporation,
  Hajime Hamashima, Makoto Iwasaki, and Taketoshi Arai
- CH. 15. Genetic Transformation of *Bacteroides* spp. Using Electroporation,
  C. Jeffrey Smith
- CH. 16. Electrotransformation of *Agrobacterium*,
  Jhy-Jhu Lin

Ch. 17. Electroporation of *Helicobacter pylori*,
   Ellyn D. Segal
Ch. 18. Electrotransformation of *Streptococci*,
   Robert E. McLaughlin and Joseph J. Ferretti
Ch. 19. Transformation of *Lactococcus* by Electroporation,
   Helge Holo and Ingolf F. Nes
Ch. 20. Transformation of *Lactobacillus* by Electroporation,
   Thea W. Aukrust, May B. Brurberg, and Ingolf F. Nes
Ch. 21. Electrotransformation of Staphlococci,
   Jean C. Lee
Ch. 22. Electroporation and Efficient Transformation of *Enterococcus faecalis* Grown in High Concentrations of Glycine,
   Brett D. Shepard and Michael S. Gilmore
Ch. 23. Introduction of Recombinant DNA into *Clostridium* spp.,
   Mary K. Phillips-Jones
Ch. 24. Electroporation of Mycobacteria,
   T. Parish and N. G. Stoker
Ch. 25. Electrotransformation of the Spirochete *Borrelia burgdorferi*,
   D. Scott Samuels
Ch. 26. Yeast Transformation and the Preparation of Frozen Spheroplasts for Electroporation,
   Lisa Stowers, James Gautsch, Richard Dana, and Merl F. Hoekstra
Ch. 27. Ten-Minute Electrotransformation of *Saccharomyces cerevisiae*,
   Martin Grey and Martin Brendel
Ch. 28. Electroporation of *Schizosaccharomyces pombe*,
   Mark T. Hood and C. S. Stachow
Ch. 29. Gene Transfer by Electroporation of Filamentous Fungi,
   M. Kapoor
Ch. 30. Transformation of *Candida maltosa* by Electroporation,
   Dietmar Becher and Stephen G. Oliver
Ch. 31. Electroporation of *Physarum polycephalum*,
   Timothy G. Burland and Juliet Bailey
Ch. 32. Electroporation of *Dictyostelium discoideum*,
   David Knecht and Ka Ming Pang
Ch. 33. Gene Transfer by Electroporation of *Tetrahymena*,
   Jacek Gaertig and Martin A. Gorovsky
Ch. 34. Transfection of the African and American Trypanosomes,
   John M. Kelly, Martin C. Taylor, Gloria Rudenko, and Pat A. Blundell
Ch. 35. Electroporation in *Giardia lamblia*,
   A. L. Wang, Tiina Sepp, and C. C. Wang

Index

CONTENTS FOR THE COMPANION VOLUME

Plant Cell Electroporation and Electrofusion Protocols

PART I. THEORY AND INSTRUMENTATION

CH. 1. Electroporation Theory: *Concepts and Mechanisms*,
  James C. Weaver
CH. 2. Effects of Pulse Length and Strength on Electroporation Efficiency,
  Sek Wen Hui
CH. 3. Instrumentation,
  Gunter A. Hofmann

PART II. ELECTROPORATION PROTOCOLS

CH. 4. Electroporation of *Agrobacterium tumefaciens*,
  Amke den Dulk-Ras and Paul J. J. Hooykaas
CH. 5. Electroporation of DNA into the Unicellular Green Alga *Chlamydomonas reinhardtii*,
  Laura R. Keller
CH. 6. Pollen Electrotransformation in Tobacco,
  James A. Saunders and Benjamin F. Matthews
CH. 7. Electroporation of Tobacco Leaf Protoplasts Using Plasmid DNA or Total Genomic DNA,
  Patrick Gallois, Keith Lindsey, and Renee Malone
CH. 8. Electroporation of *Brassica*,
  Frank Siegemund and Klaus Eimert
CH. 9. Transformation of Maize by Electroporation of Embryos,
  Carol A. Rhodes, Kathleen A. Marrs, and Lynn E. Murry
CH. 10. Transient Gene Expression Analysis in Electroporated Maize Protoplasts,
  Kathleen A. Marrs and J. C. Carle Urioste
CH. 11. Reporter Genes and Transient Assays for Plants,
  Benjamin F. Matthews, James A. Saunders, Joan S. Gebhardt, Jhy-Jhu Lin, and Susan M. Koehler

PART III. ELECTROFUSION PROTOCOLS

CH. 12. Electrofusion of Plant Protoplasts: *Selection and Screening for Somatic Hybrids of* Nicotiana,
  Harold N. Trick and George W. Bates

CH. 13. Protoplast Electrofusion and Regeneration in Potato,
   Jianping Cheng and James A. Saunders
CH. 14. Polymer-Supported Electrofusion of Protoplasts: *A Novel Method and a Synergistic Effect*,
   Lei Zhang

Index

Contributors

JAMES BARSOUM • *Biogen Inc., Cambridge, MA*
BRIAN M. BENNETT • *Department of Pharmacology and Toxicology, Queen's University, Kingston, Canada*
HEATHER L. BROWNELL • *Department of Microbiology and Immunology, Queen's University, Kingston, Canada*
PAUL L. CAMPBELL • *Centre for Transfusion Medicine and Immunology, Flinders Medical Center, Bedford Park, Australia*
WEISAN CHEN • *Centre for Transfusion Medicine and Immunology, Flinders Medical Center, Bedford Park, Australia*
VINCENT M. CHRISTOFFELS • *Department of Anatomy and Embryology, University of Amsterdam, Academic Medical Center, Amsterdam, The Netherlands*
JOSEPH P. DAY • *Laboratory of Radiobiology and Environmental Health, University of California, San Francisco, CA*
SUKHENDU B. DEV • *Genetronics Inc., San Diego, CA*
MICHAEL S. DUBOW • *Department of Microbiology and Immunology, McGill University, Montreal, Canada*
VICTORIA FINNERTY • *Department of Biology, Emory University, Atlanta, GA*
KEVIN L. FIRTH • *Ask Science Products, Inc., Kingston, Canada*
STEVEN K. H. FOUNG • *Department of Pathology, Stanford University School of Medicine, Stanford, CA*
JOSEF FULKA • *Institute of Animal Production, Prague, Czech Republic*
JOSEF FULKA, JR. • *Institute of Animal Production, Prague, Czech Republic*
MARC GAGNÉ • *Unit of Reproductive Ontogenetic Research, CHUL Research Center, and Department of Animal Science, University of Laval, Quebec, Canada*

KAARE M. GAUTVIK • *Department of Medical Biochemistry, Institute of Basic Medical Sciences, University of Oslo, Norway*
RICHARD GILBERT • *Department of Chemical Engineering, College of Engineering, University of South Florida, Tampa, FL*
JUN-ICHIRO HATA • *Central Research Laboratory, Nippon Suisan Kaisha, Ltd., Tokyo, Japan*
RICHARD HELLER • *Department of Chemical Engineering, College of Engineering, and Department of Surgery, College of Medicine, University of South Florida, Tampa, FL*
GUNTER A. HOFMANN • *Genetronics Inc., San Diego, CA*
SEK WEN HUI • *Membrane Biophysics Laboratory, Biophysics Department, Roswell Park Cancer Institute, Buffalo, NY*
KOJI INOUE • *Central Resarch Laboratory, Nippon Suisan Kaisha, Ltd., Tokyo, Japan*
AKIRA IRITANI • *Department of Biotechnology, College of Biology-Oriented Science and Technology, Kinki University, Wakayama, Japan*
MARK J. JAROSZESKI • *Department of Chemical Engineering, College of Engineering, University of South Florida, Tampa, FL*
K. PULOMA KAMDAR • *Department of Biology, Emory University, Atlanta, GA*
ARMAND KEATING • *Oncology Research Program, Toronto Hospital Research Institute, Toronto, Canada*
WIL T. LABRUYÈRE • *Department of Anatomy and Embryology, University of Amsterdam, Academic Medical Center, Amsterdam, The Netherlands*
WOUTER H. LAMERS • *Department of Anatomy and Embryology, University of Amsterdam, Academic Medical Center, Amsterdam, The Netherlands*
ALPHONSE LE CAM • *INSERM U-3476, Hôpital Arnaud de Villeneuve, Montpellier, France*
HOWARD L. LIBER • *Department of Cancer Biology, Harvard University School of Public Health, Boston, MA*
THOMAS J. LIVELLI • *Howard Hughes Medical Institute, Columbia University College of Physicians and Surgeons, New York, NY*

Contributors

LESLIE W. MACKENZIE • *Department of Anatomy and Cell Biology, Queen's University, Kingston, Canada*
KATHRYN E. MATTHEWS • *Oncology Research Program, Toronto Hospital Research Institute, Toronto, Canada*
JAMES MCCLUSKEY • *Centre for Transfusion Medicine and Immunology, Flinders Medical Center, Bedford Park, Australia*
TASUKU MITANI • *Laboratory of Developmental Biology, Meiji Institute of Health Science, Naruda, Japan*
ROBERT M. MOOR • *Department of Development and Signalling, AFRC Babraham Institute, Cambridge, England*
ANTOON F. M. MOORMAN • *Department of Anatomy and Embryology, University of Amsterdam, Academic Medical Center, Amsterdam, The Netherlands*
WILLIAM F. MORGAN • *Laboratory of Radiobiology and Environmental Health and Department of Radiation Oncology, University of California, San Francisco, CA*
JOHN S. MUDGETT • *Department of Molecular Immunology, Merck Research Laboratories, Rahway, NJ*
JAC A. NICKOLOFF • *Department of Cancer Biology, Harvard University School of Public Health, Boston, MA*
EYVIND J. PAULSSEN • *Department of Internal Medicine, University of Texas Southwestern Medical Center, Dallas, TX*
RUTH H. PAULSSEN • *Department of Pharmacology, University of Texas Southwestern Medical Center, Dallas, TX*
SUSAN PERKINS • *Department of Pathology, Stanford University School of Medicine, Stanford, CA*
ULRICH PETZOLDT • *Fachbereich Biologie, Philipps-Universität Marburg, Germany*
FRANÇOIS POTHIER • *Unit of Reproductive Ontogenetic Research, CHUL Research Center, and Department of Animal Science, University of Laval, Quebec, Canada*
LEDA H. RAPTIS • *Department of Microbiology and Immunology, Queen's University, Kingston, Canada*
F. ANDREW RAY • *Department of Microbiology, Immunology, and Molecular Genetics, The Albany Medical College, Albany, NY*

DJENANN SAINT-DIC • *Department of Microbiology and Immunology, McGill University, Montreal, Canada*
LOUISE C. SHOWE • *The Wistar Institute, Philadelphia, PA*
MICHAEL K. SHOWE • *The Wistar Institute, Philadelphia, PA*
W. CRAIG SIMON • *Department of Pharmacology and Toxicology, Queen's University, Kingston, Canada*
MARC-ANDRÉ SIRARD • *Department of Animal Science, University of Laval, Quebec, Canada*
DANIELLE GIOIOSO TAGHIAN • *Department of Cancer Biology, Harvard University School of Public Health, Boston, MA*
ANDREA TODD • *McGill University Cancer Center, Montreal, Canada*
BAN-HOCK TOH • *Centre for Transfusion Medicine and Immunology, Flinders Medical Center, Bedford Park, Australia*
FRANCES TONEGUZZO • *The Office for Biotechnology and Trade Mark Licensing, Harvard University, Boston, MA*
LESLIE TUNG • *Department of Biomedical Engineering, Johns Hopkins University, Baltimore, MD*
MAURICE J. B. VAN DEN HOFF • *Department of Anatomy and Embryology, Academic Medical Center, University of Amsterdam, Amsterdam, The Netherlands*
THAO N. WAGNER • *Department of Biology, Emory University, Atlanta, GA*
JAMES C. WEAVER • *Harvard–MIT Division of Health Sciences and Technology, Massachusetts Institute of Technology, Cambridge, MA*
KENNETH L. WHITE • *Department of Animal, Dairy, and Veterinary Sciences, Biotechnology Center, Utah State University, Logan, UT*
FEN XIA • *Department of Cancer Biology, Harvard University School of Public Health, Boston, MA*
SHINYA YAMASHITA • *Central Research Laboratory, Nippon Suisan Kaisha, Ltd., Tokyo, Japan*
JING PING YEO • *Centre for Transfusion Medicine and Immunology, Flinders Medical Centre, Bedford Park, Australia*
MARIA ZANNIS-HADJOPOULOS • *McGill University Cancer Center, Montreal, Canada*

PART I

THEORY AND INSTRUMENTATION

CHAPTER 1

Electroporation Theory

Concepts and Mechanisms

James C. Weaver

1. Introduction

Application of strong electric field pulses to cells and tissue is known to cause some type of structural rearrangement of the cell membrane. Significant progress has been made by adopting the hypothesis that some of these rearrangements consist of temporary aqueous pathways ("pores"), with the electric field playing the dual role of causing pore formation and providing a local driving force for ionic and molecular transport through the pores. Introduction of DNA into cells in vitro is now the most common application. With imagination, however, many other uses seem likely. For example, in vitro electroporation has been used to introduce into cells enzymes, antibodies, and other biochemical reagents for intracellular assays; to load larger cells preferentially with molecules in the presence of many smaller cells; to introduce particles into cells, including viruses; to kill cells purposefully under otherwise mild conditions; and to insert membrane macromolecules into the cell membrane itself. Only recently has the exploration of in vivo electroporation for use with intact tissue begun. Several possible applications have been identified, viz. combined electroporation and anticancer drugs for improved solid tumor chemotherapy, localized gene therapy, transdermal drug delivery, and noninvasive extraction of analytes for biochemical assays.

The present view is that electroporation is a universal bilayer membrane phenomenon *(1–7)*. Short (μs to ms) electric field pulses that cause the transmembrane voltage, $U(t)$, to rise to about 0.5–1.0 V cause electroporation. For isolated cells, the necessary single electric field pulse amplitude is in the range of 10^3–10^4 V/cm, with the value depending on cell size. Reversible electrical breakdown (REB) then occurs and is accompanied by greatly enhanced transport of molecules across the membrane. REB also results in a rapid membrane discharge, with $U(t)$ returning to small values after the pulse ends. Membrane recovery is often orders of magnitude slower. Cell stress probably occurs because of relatively nonspecific chemical exchange with the extracellular environment. Whether or not the cell survives probably depends on the cell type, the extracellular medium composition, and the ratio of intra- to extracellular volume. Progress toward a mechanistic understanding has been based mainly on theoretical models involving transient aqueous pores. An electric field pulse in the extracellular medium causes the transmembrane voltage, $U(t)$, to rise rapidly. The resulting increase in electric field energy within the membrane and ever-present thermal fluctuations combine to create and expand a heterogeneous population of pores. Scientific understanding of electroporation at the molecular level is based on the hypothesis that pores are microscopic membrane perforations, which allow hindered transport of ions and molecules across the membrane.

These pores are presently believed to be responsible for the following reasons:

1. Dramatic electrical behavior, particularly REB, during which the membrane rapidly discharges by conducting small ions (mainly Na^+ and Cl^-) through the transient pores. In this way, the membrane protects itself from destructive processes;
2. Mechanical behavior, such as rupture, a destructive phenomenon in which pulses too small or too short cause REB and lead to one or more supracritical pores, and these expand so as to remove a portion of the cell membrane; and
3. Molecular transport behavior, especially the uptake of polar molecules into the cell interior.

Both the transient pore population, and possibly a small number of metastable pores, may contribute. In the case of cells, relatively nonspecific molecular exchange between the intra- and extracellular volumes probably occurs, and can lead to chemical imbalances. Depending on the ratio of intra- and extracellular volume, the composition of the extracellular medium, and the cell type, the cell may not recover from the associated stress and will therefore die.

2. Basis of the Cell Bilayer Membrane Barrier Function

It is widely appreciated that cells have membranes in order to separate the intra- and extracellular compartments, but what does this really mean? Some molecules utilized by cells have specific transmembrane transport mechanisms, but these are not of interest here. Instead, we consider the relatively nonspecific transport governed by diffusive permeation. In this case, the permeability of the membrane to a molecule of type "s" is $P_{m,s}$, which is governed by the relative solubility (partition coefficient), $g_{m,s}$, and the diffusion constant, $D_{m,s}$, within the membrane. In the simple case of steady-state transport, the rate of diffusive, nonspecific molecular transport, N_s, is:

$$N_s = A_m P_{m,s} \Delta C_s = A_m [g_{m,s} D_{m,s}/d] \Delta C_s \tag{1}$$

where N_s, is the number of molecules of type "s" per unit time transported, ΔC_s is the concentration difference across the membrane, d ≈ 6 nm is the bilayer membrane thickness, and A_m is the area of the bilayer portion of the cell membrane. As discussed below, for charged species, the small value of $g_{m,s}$ is the main source of the large barrier imposed by a bilayer membrane.

Once a molecule dissolves in the membrane, its diffusive transport is proportional to Δc_s and $D_{m,s}$. The dependence on $D_{m,s}$ gives a significant, but not tremendously rapid, decrease in molecular transport as size is increased. The key parameter is $g_{m,s}$, which governs entry of the molecule into the membrane. For electrically neutral molecules, $g_{m,s}$ decreases with molecular size, but not dramatically. In the case of charged molecules, however, entry is drastically reduced as charge is increased. The essential features of a greatly reduced $g_{m,s}$ can be understood in terms of electrostatic energy considerations.

The essence of the cell membrane is a thin (≈6 nm) region of low dielectric constant (K_m ≈ 2–3) lipid, within which many important proteins reside. Fundamental physical considerations show that a thin sheet of low dielectric constant material should exclude ions and charged molecules. This exclusion is owing to a "Born energy" barrier, i.e., a significant cost in energy that accompanies movement of charge from a high dielectric medium, such as water (dielectric constant K_w ≈ 80), into a low dielectric medium, such as the lipid interior of a bilayer membrane (dielectric constant K_m ≈ 2) *(8)*.

The Born energy associated with a particular system of dielectrics and charges, W_{Born}, is the electrostatic energy needed to assemble that system of dielectric materials and electric charge. W_{Born} can be computed by specifying the distribution of electrical potential and the distribution of charge, or it can be computed by specifying the electric field, E, and the permittivity $\varepsilon = K\varepsilon_0$ (K is the dielectric constant and $\varepsilon_0 = 8.85 \times 10^{-12}$ F/m) *(9)*. Using the second approach:

$$W_{Born} \equiv \int_{\substack{\text{all space} \\ \text{except ion}}} 1/2 \, \varepsilon E^2 dV \tag{2}$$

The energy cost for insertion of a small ion into a membrane can now be understood by estimating the maximum change in Born energy, $\Delta W_{Born,max}$, as the ion is moved from water into the lipid interior of the membrane. It turns out that W_{Born} rises rapidly as the ion enters the membrane, and that much of the change occurs once the ion is slightly inside the low dielectric region. This means that it is reasonable to make an estimate based on treating the ion as a charged sphere of radius r_s and charge $q = ze$ with $z = \pm 1$ where $e = 1.6 \times 10^{-19}$ C. The sphere is envisioned as surrounded by water when it is located far from the membrane, and this gives ($W_{Born,i}$). When it is then moved to the center of the membrane, there is a new electrostatic energy, ($W_{Born,f}$). The difference in these two energies gives the barrier height, $\Delta W_{Born} \equiv W_{Born,f} - W_{Born,i}$. Even for small ions, such as Na$^+$ and Cl$^-$, this barrier is substantial (Fig. 1). More detailed, numerical computations confirm that ΔW_{Born} depends on both the membrane thickness, d, and ion radius, r_s.

Here we present a simple estimate of ΔW_{Born}. It is based on the recognition that if the ion diameter is small, $2r_s \approx 0.4$ nm, compared to the membrane thickness, $d \approx 3$–6 nm, then ΔW_{Born} can be estimated by neglecting the finite size of the membrane. This is reasonable, because the largest electric field occurs near the ion, and this in turn means that the details of the membrane can be replaced with bulk lipid. The resulting estimate is:

$$\Delta W_{Born} \approx e^2/8\pi\varepsilon_0 r_s [1/K_m - 1/K_w] \approx 65 \, kT \tag{3}$$

where T = 37°C = 310 K. A complex numerical computation for a thin low dielectric constant sheet immersed in water confirms this simple estimate (Fig. 1). This barrier is so large that spontaneous ion transport

Electroporation Theory

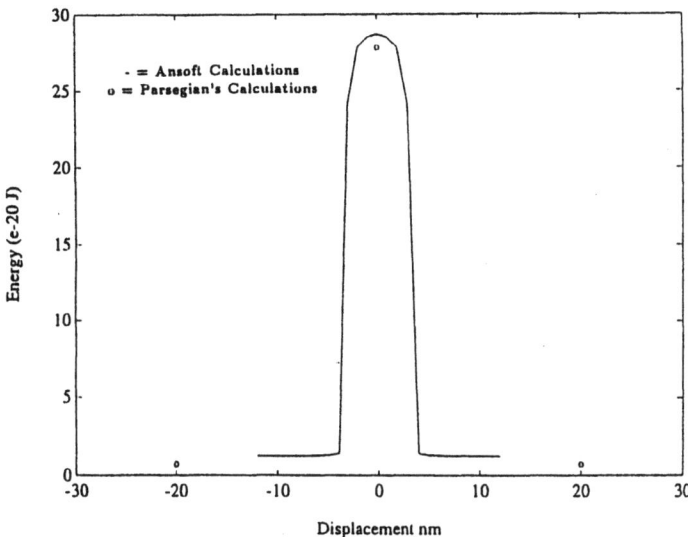

Fig. 1. Numerical calculation of the Born energy barrier for transport of a charged sphere across a membrane (thickness $d = 4$ nm). The numerical solution was obtained by using commercially available software (Ansoft, Inc., Pittsburgh, PA) to solve Poisson's equation for a continuum model consisting of a circular patch of a flow dielectric constant material ($K_m = 2$) immersed in water ($K_w = 80$). The ion was represented by a charged sphere of radius ($r_s = 0.2$ nm), and positioned at a number of different displacements on the axis of rotation of the disk. No pore was present. The electric field and the corresponding electrostatic energy were computed for each case to obtain the values plotted here as a solid line ("- Ansoft Calculations"). The single value denoted by o ("Parsegian's Calculations;" 8) is just under the Ansoft peak. As suggested by the simple estimate of Eq. (2), the barrier is large, viz. $\Delta W \approx 2.8 \times 10^{-19}$ J ≈ 65 kT. As is well appreciated, this effectively rules out significant spontaneous ion transport. The appearance of aqueous pathways ("pores"; Fig. 2) provides a large reduction in this barrier. Reproduced with permission (47).

resulting from thermal fluctuations is negligible. For example, a large transmembrane voltage, U_{direct}, would be needed to force an ion directly across the membrane. The estimated value is $U_{direct} \approx 65kT/e = 1.7$ V for $z = \pm 1$. However, 1.7 V is considerably larger than the usual "resting values" of the transmembrane voltage (about 0.1 ± 0.05 V). The scientific literature on electroporation is consistent with the idea that some sort of membrane structural rearrangement occurs at a smaller voltage.

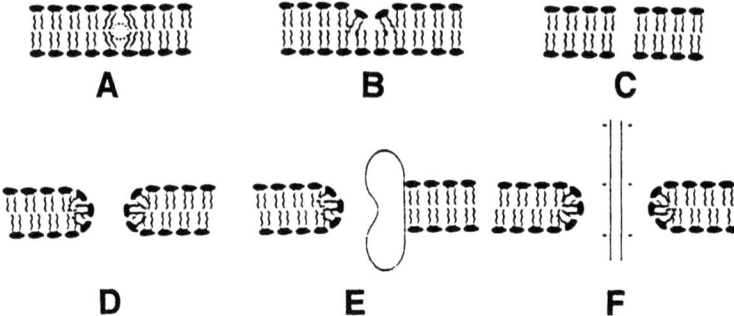

Fig. 2. Illustrations of hypothetical structures of both transient and metastable membrane conformations that may be involved in electroporation *(4)*. (**A**) Membrane-free volume fluctuation *(62)*, (**B**) Aqueous protrusion into the membrane ("dimple") *(12,63)*, (**C**) Hydrophobic pore first proposed as an immediate precursor to hydrophilic pores *(10)*, (**D**) Hydrophilic pore *(10,17,18)*; that is generally regarded as the "primary pore" through which ions and molecules pass, (**E**) Composite pore with one or more proteins at the pore's inner edge *(20)*, and (**F**) Composite pore with "foot-in-the-door" charged macromolecule inserted into a hydrophilic pore *(31)*. Although the actual transitions are not known, the transient aqueous pore model assumes that transitions from A → B → C or D occur with increasing frequency as U is increased. Type E may form by entry of a tethered macromolecule during the time that U is significantly elevated, and then persist after U has decayed to a small value because of pore conduction. These hypothetical structures have not been directly observed. Instead, evidence for them comes from interpretation of a variety of experiments involving electrical, optical, mechanical, and molecular transport behavior. Reproduced with permission *(4)*.

3. Aqueous Pathways ("Pores") Reduce the Membrane Barrier

A significant reduction in ΔW_{Born} occurs if the ion (1) is placed into a (mobile) aqueous cavity or (2) can pass through an aqueous channel *(8)*. Both types of structural changes have transport function based on a local aqueous environment, and can therefore be regarded as aqueous pathways. Both allow charged species to cross the membrane much more readily. Although both aqueous configurations lower ΔW_{Born}, the greater reduction is achieved by the pore *(8)*, and is the basis of the "transient aqueous pore" theory of electroporation.

Why should the hypothesis of pore formation be taken seriously? As shown in Fig. 2, it is imagined that some types of prepore structural changes can occur in a microscopic, fluctuating system, such as the bilayer membrane. Although the particular structures presented there are plausible, there is no direct evidence for them. In fact, it is unlikely that transient pores can be visualized by any present form of microscopy, because of the small size, short lifetime, and lack of a contrast-forming interaction. Instead, information regarding pores will probably be entirely indirect, mainly through their involvement in ionic and molecular transport *(4)*. Without pores, a still larger voltage would be needed to move multivalent ions directly across the membrane. For example, if $z = \pm 2$, then $U_{direct} \approx 7$ V, which for a cell membrane is huge.

Qualitatively, formation of aqueous pores is a plausible mechanism for transporting charged molecules across the bilayer membrane portion of cell membranes. The question of how pores form in a highly interactive way with the instantaneous transmembrane voltage has been one of the basic challenges in understanding electroporation.

4. Large U(t) Simultaneously Causes Increased Permeability and a Local Driving Force

Electroporation is more than an increase in membrane permeability to water-soluble species owing to the presence of pores. The temporary existence of a relatively large electric field within the pores also provides an important, local driving force for ionic and molecular transport. This is emphasized below, where it is argued that massive ionic conduction through the transient aqueous pores leads to a highly interactive membrane response. Such an approach provides an explanation of how a planar membrane can rupture at small voltages, but exhibits a protective REB at large voltages. At first this seems paradoxical, but the transient aqueous pore theory predicts that the membrane is actually protected by the rapid achievement of a large conductance. The large conductance limits the transmembrane voltage, rapidly discharges the membrane after a pulse, and thereby saves the membrane from irreversible breakdown (rupture). The local driving force is also essential to the prediction of an approximate plateau in the transport of charged molecules.

5. Membrane-Level and Cell-Level Phenomena

For applications, electroporation should be considered at two levels: (1) the membrane level, which allows consideration of both artificial and cell membranes, and (2) the cellular level, which leads to consideration of secondary processes that affect the cell. The distinction of these two levels is particularly important to the present concepts of reversible and irreversible electroporation. A key concept at the membrane level is that molecular transport occurs through a dynamic pore population. A related hypothesis is that electroporation itself can be reversible at the membrane level, but that large molecular transport can lead to significant chemical stress of a cell, and it is this secondary, cell-level event that leads to irreversible cell electroporation. This will be brought out in part of the presentation that follows.

6. Reversible and Irreversible Electroporation at the Membrane Level

Put simply, reversible electroporation involves creation of a dynamic pore population that eventually collapses, returning the membrane to its initial state of a very few pores. As will be discussed, reversible electroporation generally involves REB, which is actually a temporary high conductance state. Both artificial planar bilayer membranes and cell membranes are presently believed capable of experiencing reversible electroporation. In contrast, the question of how irreversible electroporation occurs is reasonably well understood for artificial planar bilayer membranes, but significantly more complicated for cells.

7. Electroporation in Artificial Planar and in Cell Membranes

Artificial planar bilayer membrane studies led to the first proposals of a theoretical mechanism for electroporation (10–16). However, not all aspects of planar membrane electroporation are directly relevant to cell membrane electroporation. Specifically, quantitative understanding of the stochastic rupture ("irreversible breakdown") in planar membranes was the first major accomplishment of the pore hypothesis. Although cell membranes can also be damaged by electroporation, there are two possible mechanisms. The first possibility is lysis resulting from a secondary result of reversible electroporation of the cell membrane. According to this hypothesis, even though the membrane recovers (the dynamic pore population returns to the initial state), there can be so much molecular transport that the cell is chemically or osmotically stressed, and this secondary event leads to cell destruction through lysis. The second possibility is that rupture of an isolated portion of a cell membrane occurs, because one or more bounded portions of the membrane behave like small planar membranes. If this is the case, the mechanistic understanding of planar membrane rupture is relevant to cells.

8. Energy Cost to Create a Pore at Zero Transmembrane Voltage ($U = 0$)

The first published descriptions of pore formation in bilayer membranes were based on the idea that spontaneous (thermal fluctuation driven) structural changes in the membrane could create pores. A basic premise was that the large pores could destroy a membrane by rupture, which was suggested to occur as a purely mechanical event, i.e., without electrical assistance *(17,18)*. The energy needed to make a pore was considered to involve two contributions. The first is the "edge energy," which relates to the creation of a stressed pore edge, of length $2\pi r$, so that if the "edge energy" (energy cost per length) was γ, then the cost to make the pore's edge was $2\pi r\gamma$. The second is the "area energy" change associated with removal of a circular patch of membrane, $-\pi r^2 \Gamma$. Here Γ is the energy per area (both sides of the membrane) of a flat membrane.

Put simply, this process is a "cookie cutter" model for a pore creation. The free energy change, $\Delta W_p(r)$, is based on a gain in edge energy and a simultaneous reduction in area energy. The interpretation is simple: a pore-free membrane is envisioned, then a circular region is cut out of the membrane, and the difference in energy between these two states calculated, and identified as ΔW_p. The corresponding equation for the pore energy is:

$$\Delta W_p(r) = 2\pi\gamma r - \pi\Gamma r^2 \text{ at } U = 0 \qquad (4)$$

A basic consequence of this model is that $\Delta W_p(r)$ describes a parabolic barrier for pores. In its simplest form, one can imagine that pores might be first made, but then expanded at the cost of additional energy. If the barrier peak is reached, however, then pores moving over the barrier can expand indefinitely, leading to membrane rupture. In the initial models (which did not include the effect of the transmembrane voltage), spontaneous thermal fluctuations were hypothesized to create pores, but the probability of surmounting the parabolic barrier was thought to be small. For this reason, it was concluded that spontaneous rupture of a red blood cell membrane by spontaneous pore formation and expansion was concluded to be negligible *(17)*. At essentially the same time, it was independently suggested that pores might provide sites in the membrane where spontaneous translocation of membrane lipid molecules ("flip flop") should preferentially occur *(18)*.

9. Energy Cost to Create a Pore at U > 0

In order to represent the electrical interaction, a pore is regarded as having an energy associated with the change of its specific capacitance, C_p. This was first presented in a series of seven back-to-back papers *(10–16)*. Early on, it was recognized that it was unfavorable for ions to enter small pores because of the Born energy change discussed previously. For this reason, a relatively small number of ions will be available within small pores to contribute to the electrical conductance of the pore. With this justification, a pore is represented by a water-filled, rather than electrolyte-filled, capacitor. However, for small hydrophilic pores, even if bulk electrolyte exists within the pores, the permittivity would be $\varepsilon \approx 70\varepsilon_0$, only about 10% different from that of pure water.

In this case, the pore resistance is still large, $R_p = \rho_e h/\pi r^2$, and is also large in comparison to the spreading resistance discussed below. If so, the voltage across the pore is approximately U. With this in mind, in the presence of a transmembrane electric field, the free energy of pore formation should be *(10)*:

$$\Delta W_p(r,U) = 2\pi\gamma r - \pi\Gamma r^2 - 0.5 C_p U^2 r^2 \tag{5}$$

Here U is the transmembrane voltage spatially averaged over the membrane. A basic feature is already apparent in the above equation: as U increases, the pore energy, ΔW_p, decreases, and it becomes much more favorable to create pores. In later versions of the transient aqueous pore model, the smaller, local transmembrane voltage, U_p, for a conducting pore is used. As water replaces lipid to make a pore, the capacitance of the membrane increases slightly.

10. Heterogeneous Distribution of Pore Sizes

A spread in pore sizes is fundamentally expected *(19–22)*. The origin of this size heterogeneity is the participation of thermal fluctuations along with electric field energy within the membrane in making pores. The basic idea is that these fluctuations spread out the pore population as pores expand against the barrier described by $\Delta W_p(r,U)$. Two extreme cases illustrate this point: (1) occasional escape of large pores over the barrier described by $\Delta W_p(r,U)$ leads to rupture, and (2) the rapid creation of many small pores ($r \approx r_{min}$) causes the large conductance that is responsible for REB. In this sense, rupture is a large-pore phenomenon, and REB is a small-pore phenomenon. The moderate value of U(t) associated with rupture leads to only a modest conductance, so that there is ample time for the pore population to evolve such that one or a small number of large pores appear and diffusively pass over the barrier, which is still fairly large. The pore population associated with REB is quite different; at larger voltages, a great many more small pores appear, and these discharge the membrane before the pore population evolves any large "critical" pores that lead to rupture.

11. Quantitative Explanation of Rupture

As the transmembrane voltage increases, the barrier $\Delta W_p(r, U)$ changes its height, ΔW_{max} and the location of its peak. The latter is associated with a critical pore radius, r_c, such that pores with $r > r_c$ tend to expand without limit. A property of $\Delta W_p(r, U)$ is that both ΔW_{max} and r_c decrease as U increases. This provides a readily visualized explanation of planar membrane rupture: as U increases, the barrier height decreases, and this increases the probability of the membrane acquiring one or more pores with $r > r(U)_c$. The appearance of even one supracritical pore is, however, sufficient to rupture the membrane. Any pore with $r > r_c$ tends to expand until it reaches the macroscopic aperture that defines the planar membrane. When this occurs, the membrane material has all collected at the aperture, and it makes no sense to talk about a membrane being present. In this case, the membrane is destroyed.

The critical pore radius, r_c, associated with the barrier maximum, $\Delta W_{p,max} = \Delta W_p(r_c, U)$, is *(10)*:

$$r_c = (\gamma/\Gamma + 0.5C_pU^2) \text{ and } \Delta W_{p,max} = \pi\gamma^2/\Gamma + 0.5C_pU^2) \qquad (6)$$

The associated pore energy, $\Delta W_{p,max}$ also decreases. Overcoming energy barriers generally depends nonlinearly on parameters, such as U, because Boltzmann factors are involved. For this reason, a nonlinear dependence on U was expected.

The electrical conductance of the membrane increases tremendously because of the appearance of pores, but the pores, particularly the many small ones, are not very good conductors. The reason for this relatively poor conduction of ions by small pores is again the Born energy change; conduction within a pore can be suppressed over bulk electrolyte conduction because of Born energy exclusion owing to the nearby low dielectric constant lipid. The motion of ions through a pore only somewhat larger than the ion itself can be sterically hindered. This has been accounted for by using the Renkin equation to describe the essential features of hindrance *(23)*. This function provides for reduced transport of a spherical ion or molecule of radius r_s through cylindrical pathway of radius r (representing a pore) *(20,21,24)*.

12. Planar Membrane Destruction by Emergence of Even One "Critical Pore"

As a striking example of the significance of heterogeneity within the pore population, it has been shown that one or a small number of large pores can destroy the membrane by causing rupture *(11)*. The original approach treated the diffusive escape of pores over an energy barrier. Later, an alternative, simpler approach for theoretically estimating the average membrane lifetime against rupture, $\bar{\tau}$, was proposed *(25)*. This approach used an absolute rate estimate for critical pore appearance in which a Boltzmann factor containing $\Delta W_p/kT$ and an order of magnitude estimate for the prefactor was used. The resulting estimate for the rate of critical pore appearance is:

$$\bar{\tau} \approx (1/\nu_0 V_m) \exp(+\Delta W_{p,c}/kT) \qquad (7)$$

This estimate used an attempt rate density, ν_0, which is based on a collision frequency density within the fluid bilayer membrane. The order of magnitude of ν_0 was obtained by estimating the volume density of collisions per time in the fluid membrane. The factor $V_m = hA_m$ is the total volume of the membrane. By choosing a plausible value (e.g., 1 s), the value of $\Delta W_{p,c}$, and hence of U_c, can be found. This is interpreted as the critical voltage for rupture. Because of the strong nonlinear behavior of Eq. (7), using values, such as 0.1 or 10 s, results in only small differences in the predicted $U_c \approx 0.3$–0.5 V.

13. Behavior of the Transmembrane Voltage During Rupture

Using this approach, reasonable (but not perfect) agreement for the behavior of $U(t)$ was found. Both the experimental and theoretical behaviors of U exhibit a sigmoidal decay during rupture, but the duration of the decay phase is longer for the experimental values. Both are much longer than the rapid discharge found for REB. Many experiments have shown that both artificial planar bilayer membranes and cell membranes exhibit REB, and its occurrence coincides with tremendously enhanced molecular transport across cell membranes. However, the term "breakdown" is misleading, because REB is now believed to be a protective behavior, in which the membrane acquires a very large conductance in the form of pores. In planar membranes challenged by short pulses (the "charge injection" method mentioned above), a characteristic of REB is the progressively faster membrane discharge as larger and larger pulses are used *(26)*.

14. Reversible Electroporation

Unlike reversible electroporation (rupture) of planar membranes, in which the role of one or a small number of critical pores is dominant, reversible electroporation is believed to involve the rapid creation of so many small pores that membrane discharge occurs before any critical pores can evolve from the small pores. The transition in a planar membrane from rupture to REB can be qualitatively understood in terms of a competition between the kinetics of pore creation and of pore expansion. If only a few pores are present owing to a modest voltage pulse, the membrane discharges very slowly (e.g., ms) and there is time for evolution of critical pores. If a very large number of pores are present because of a large pulse, then the high conductance of these pores discharges the membrane rapidly, before rupture can occur. One basic challenge in a mechanistic understanding is to find a quantitative description of the transition from rupture to REB, i.e., to show that a planar membrane can experience rupture for modest pulses, but makes a transition to REB as the pulse amplitude is increased *(19–22)*. This requires a physical model for both pore creation and destruction, and also the behavior of a dynamic, heterogeneous pore population.

15. Conducting Pores Slow Their Growth

An important aspect of the interaction of conducting pores with the changing transmembrane voltage is that pores experience a progressively smaller expanding force as they expand *(21,27)*. This occurs because there are inhomogeneous electric fields (and an associated "spreading resistance") just outside a pore's entrance and exit, such that as the pore grows, a progressively greater fraction of U appears across this spreading resistance. This means that less voltage appears across the pore itself, and therefore, the electrical expanding pressure is less. For this reason, pores tend to slow their growth as they expand. The resistance of the internal portion of the pore is also important, and as already mentioned, has a reduced internal resistance because $\sigma_p < \sigma_e$ because of Born energy "repulsion." The voltage divider effect means simply that the voltage across the pore is reduced to:

$$U_p = U [R_p/(R_p + R_s)] \leq U \qquad (8)$$

Here R_p is the electrical resistance associated with the pore interior, and R_s is the resistance associated with the external inhomogeneous electric field near the entrance and exit to the pore. The fact that U_p becomes less than U means that the electrical expanding force owing to the gradient of ΔW_p in pore radius space is reduced. In turn, this means that pores grow more slowly as they become larger, a basic pore response that contributes to reversibility *(21,27)*.

16. Reversible Electroporation and "Reversible Electrical Breakdown"

For planar membranes, the transition from irreversible behavior ("rupture") to reversible behavior ("REB" or incomplete reversible electrical breakdown) can be explained by the evolution of a dynamic, heterogeneous pore population *(20–22,24)*. One prediction of the transient aqueous pore model is that a planar membrane should also exhibit incomplete reversible electrical breakdown, i.e., a rapid discharge that does not bring U down to zero. Indeed, this is predicted to occur for somewhat smaller pulses than those that produce REB. Qualitatively, the following is believed to occur. During the initial rapid discharge, pores rapidly shrink and some disappear. As a result, the membrane conductance, G(t), rapidly reaches such a small value that further discharge occurs very slowly. On the time scale (µs) of the experiment, discharge appears to stop, and the membrane has a small transmembrane voltage, e.g., $U \approx 50$ mV.

Although irreversible electroporation of planar membranes now seems to be reasonably accounted for by a transient aqueous pore theory, the case of irreversibility in cells is more complicated and still not fully understood. The rupture of planar membranes is explained by recognizing that expansion of one or more supracritical pores can destroy the membrane. When it is created, the planar membrane covers a macroscopic aperture, but also connects to a meniscus at the edge of the aperture. This meniscus also contains phospholipids, and can be thought of as a reservoir that can exchange phospholipid molecules with the thinner bilayer membrane. As a result of this connection to the meniscus, the bilayer membrane has a total surface tension (both sides of the membrane), Γ, which favors expansion of pores. Thus, during rupture, the membrane material is carried by pore expansion into the meniscus, and the membrane itself vanishes.

However, there is no corresponding reservoir of membrane molecules in the case of the closed membrane of a vesicle or cell. For this reason, if the osmotic pressure difference across the cell membrane is zero, the cell membrane effectively has $\Gamma = 0$. For this reason, a simple vesicle cannot rupture *(28)*. Although a cell membrane has the same topology as a vesicle, the cell membrane is much more complicated, and usually contains other, membrane-connecting structures. With this in mind, suppose that a portion of a cell membrane is bounded by the cytoskeleton or some other cellular structure, such that membrane molecules can accumulate there if pores are created (Fig. 2). If so, these bounded portions of the cell membrane may be able to rupture, since a portion of the cell membrane would behave like a microscopic planar bilayer membrane. This localized but limited rupture would create an essentially permanent hole in the cell membrane, and would lead to cell death. Another possibility is that reversible electroporation occurs, with REB and a large, relatively non-specific molecular transport *(see* Section 21.) across the cell membrane.

17. Tremendous Increase in Membrane Conductance, G(t) During REB

Creation of aqueous pathways across the membrane is, of course, the phenomenon of interest. This is represented by the total membrane conductance, $G(t) = 1/R(t)$. As pores appear during reversible electroporation, R changes by orders of magnitude. A series of electrical experiments using a planar bilayer membrane provided conditions and results that motivated the choice of particular parameters, including the use of a very short (0.4 µs) square pulse *(26)*. In these experiments, a current pulse of amplitude I_i passes through R_N, thereby creating a voltage pulse, V_0 (Fig. 2). For $0 < t < t_{pulse}$ current flows into and/or across the membrane, and at $t = t_{pulse}$, the pulse is terminated by opening the switch. Because the generator is then electronically disconnected, membrane discharge can occur only through the membrane for a planar membrane (not true for a cell). Predictions of electroporation behavior were obtained by generating self-consistent numerical solutions to these equations.

18. Evidence for Metastable Pores

Pores do not necessarily disappear when U returns to small values. For example, electrical experiments with artificial planar bilayer membranes have shown that small pores remain after U is decreased. Other experiments with cells have examined the response of cells to dyes supplied after electrical pulsing, and find that a subpopulation of cells takes up these molecules *(29,30)*. Although not yet understood quantitatively in terms of an underlying mechanism, it is qualitatively plausible that some type of complex, metastable pores can form. Such pores may involve other components of a cell, e.g., the cytoskeleton or tethered cytoplasmic molecules (Fig. 2), that lead to metastable pores. For example, entry of a portion of a tethered, charged molecule should lead to a "foot-in-the-door" mechanism in which the pore cannot close *(31)*. However, pore destruction is not well understood. Initial theories assumed that pore disappearance occurs independently of other pores. This is plausible, since pores are widely spaced even when the total (aqueous) area is maximum *(22)*. Although this approximate treatment has contributed to reasonable theoretical descriptions of some experimental behavior, a complete, detailed treatment of pore disappearance remains an unsolved problem.

19. Interaction of the Membrane with the External Environment

It is not sufficient to describe only the membrane. Instead, an attempt to describe an experiment should include that part of the experimental apparatus that directly interacts with the membrane. Specifically, the electrical properties of the bathing electrolyte, electrodes, and output characteristics of the pulse generator should be included. Otherwise, there is no possibility for including the limiting effects of this part of the experiment. Clearly there is a pathway by which current flows in order to cause interfacial polarization, and thereby increase $U(t)$.

An initial attempt to include membrane–environment interactions used a simple circuit model to represent the most important aspects of the membrane and the external environment, which shows the relationship among the pulse generator, the charging pathway resistance, and the membrane *(19,21)*. The membrane is represented as the membrane capacitance, C, connected in parallel with the membrane resistance, $R(t)$. As pores begin to appear in the membrane, the membrane conductance $G(t) = 1/R(t)$ starts to increase, and therefore $R(t)$ drops. The membrane does not experience the applied pulse immediately, however, since the membrane capacitance has to charge through the external resistance of the electrolyte, which baths the membrane, the electrode resistance, and the output resistance of the pulse generator. This limitation is represented by a single resistor, R_E. This explicit, but approximate, treatment of the membrane's environment provides a reasonable approach to achieving theoretical descriptions of measurable quantities that can be compared to experimental results.

20. Fractional Aqueous Area of the Membrane During Electroporation

The membrane capacitance is treated as being constant, which is consistent with experimental data (32). It is also consistent with the theoretical model, as shown by computer simulations that use the model to predict correctly basic features of the transmembrane voltage, $U(t)$. The simulation allows the slight change in C to be predicted simultaneously, and finds that only a small fraction ($F_{w,max} \approx 5 \times 10^{-4}$) of the membrane becomes aqueous through the appearance of pores. The additional capacitance owing to this small amount of water leads to a slight (on the order of 1%) change in the capacitance (22), which is consistent with experimental results (32).

The fractional aqueous area, $F_w(t)$, changes rapidly with time as pores appear, but is predicted to be less than about 0.1% of the membrane, even though tremendous increases in ionic conduction and molecular transport take place. This is in reasonable agreement with experimental findings. According to present understanding, the minimum pore size is $r_{min} \approx 1$ nm, which means that the small ions that comprise physiologic saline can be conducted. For larger or more charged species, however, the available fractional aqueous area, $F_{w,s}$, is expected to decrease. This is a consequence of a heterogeneous pore population. With increasing molecular size and/or charge, fewer and fewer pores should participate, and this means that $F_{w,s}$ should decrease as the size and charge of "s" increase.

21. Molecular Transport Owing to Reversible Electroporation

Tremendously increased molecular transport (33,34) is probably the most important result of electroporation for biological research (Table 1). Although clearly only partially understood, much of the evidence to date supports the view that electrophoretic transport through pores is the major mechanism for transport of charged molecules (20,24,35,36).

Table 1
Candidate Mechanisms for Molecular Transport Through Pores (20)[a]

Mechanism	Molecular basis
Drift	Velocity in response to a local physical (e.g., electrical) field
Diffusion	Microscopic random walk
Convection	Fluid flow carrying dissolved molecules

[a]The dynamic pore population of electroporation is expected to provide aqueous pathways for molecular transport. Water-soluble molecules should be transported through the pores that are large enough to accommodate them, but with some hindrance. Although not yet well established, electrical drift may be the primary mechanism for charged molecules (20–35).

One surprising observation is the molecular transport caused by a single exponential pulse can exhibit a plateau, i.e., transport becomes independent of field pulse magnitude, even though the net molecular transport results in uptake that is far below the equilibrium value $N_s = V_{cell} c_{ext}$ (37–40). Here N_s is the number of molecules taken up by a single cell, V_{cell} is the cell volume, and c_{ext} is the extracellular concentration in a large volume of pulsing solution.

A plateauing of uptake that is independent of equilibrium uptake ($\bar{n}_s = V_{cell} c_{s,ext}$) may be a fundamental attribute of electroporation. Initial results from a transient aqueous pore model show that the transmembrane voltage achieves an almost constant value for much of the time during an exponential pulse. If the local driving force is therefore almost constant, the transport of small charged molecules through the pores may account for an approximate plateau (24). Transport of larger molecules may require deformation of the pores, but the approximate constancy of $U(t)$ should still occur, since the electrical behavior is dominated by the many smaller pores. These partial successes of a transient aqueous pore theory are encouraging, but a full understanding of electroporative molecular transport is still to be achieved.

22. Terminology and Concepts: Breakdown and Electropermeabilization

Based on the success of the transient aqueous pore models in providing reasonably good quantitative descriptions of several key features of electroporation, the existence of pores should be regarded as an attractive hypothesis (Table 2). With this in mind, two widely used terms, "breakdown" and "electropermeabilization," should be re-examined. First, "breakdown" in the sense of classic dielectric breakdown is mis-

Table 2
Successes of the Transient Aqueous Pore Model[a]

Behavior	Pore theory accomplishment
Stochastic nature of rupture	Explained by diffusive escape of very large pores *(10)*
Rupture voltage, U_c	Average value reasonably predicted *(10,25,64)*
Reversible electrical breakdown	Transition from rupture to REB correctly predicted *(21)*
Fractional aqueous area	$F_{w,ions} \leq 10^{-3}$ predicted; membrane conductance agrees *(22)*
Small change in capacitance	Predicted to be <2% for reversible electroporation *(22)*
Plateau in charged molecule transport	Approximate plateau predicted for exponential pulses *(24)*

[a]Successful predictions of the transient aqueous pore model for electroporation at the present time. These more specific descriptions are not accounted for simply by an increased permeability or an ionizing type of dielectric breakdown. The initial, combined theoretical and experimental studies convincingly showed that irreversible breakdown ("rupture") was not the result of a deterministic mechanism, such as compression of the entire membrane, but could instead be quantitatively accounted for by transient aqueous pores *(10)*. Recent observations of charged molecule uptake by cells that exhibits a plateau, but is far below the equilibrium value cannot readily be accounted for by any simple, long-lasting membrane permeability increase, but is predicted by the transient aqueous pore model.

leading. After all, the maximum energy available to a monovalent ion or molecule for $U \approx 0.5–1$ V is only about one-half to 1 ev. This is too small to ionize most molecules, and therefore cannot lead to conventional avalanche breakdown in which ion pairs are formed *(41)*. Instead, a better term would be "high conductance state," since it is the rapid membrane rearrangement to form conducting aqueous pathways that discharges the membrane under biochemically mild conditions *(42)*. Second, in the case of electropermeabilization, "permeabilization" implies only that a state of increased permeability has been obtained. This phenomenological term is directly relevant only to transport. It does not lead to the concept of a stochastic membrane destruction, the idea of "reversible electrical breakdown" as a protective process in the transition from rupture to REB, or the plateau in molecular transport for small charged molecules. Thus, although electroporation clearly causes an increase in permeability, electroporation is much more, and the abovementioned additional features cannot be explained solely by an increase in permeability.

23. Membrane Recovery

Recovery of the membrane after pulsing is clearly essential to achieving reversible behavior. Presently, however, relatively little is known about the kinetics of membrane recovery after the membrane has been discharged by REB. Some studies have used "delayed addition" of molecules to determine the integrity of cell membranes at different times after pulsing. Such experiments suggest that a subpopulation of cells occurs that has delayed membrane recovery, as these cells are able to take up molecules after the pulse. In addition to "natural recovery" of cell membranes, the introduction of certain surfactants has been found to accelerate membrane recovery, or at least re-establishment of the barrier function of the membrane *(43)*. Accelerated membrane recovery may have implications for medical therapies for electrical shock injury, and may also help us to understand the mechanism by which membranes recover.

24. Cell Stress and Viability

Complete cell viability, not just membrane recovery, is usually important to biological applications of electroporation, but in the case of electroporation, determination of cell death following electroporation is nontrivial. After all, by definition, electroporation alters the permeability of the membrane. This means that membrane-based short-term tests (vital stains, membrane exclusion probes) are therefore not necessarily valid *(29)*. If, however, the cells in question can be cultured, assays based on clonal growth should provide the most stringent test, and this can be carried out relatively rapidly if microcolony (2–8 cells) formation is assessed *(44)*. This was done using microencapsulated cells. The cells are initially incorporated into agarose gel microdrops (GMDs), electrically pulsed to cause electroporation, cultured while in the microscopic (e.g., 40–100 µm diameter) GMDs, and then analyzed by flow cytometry so that the subpopulation of viable cells can be determined *(45,46)*.

Cellular stress caused by electroporation may also lead to cell death without irreversible electroporation itself having occurred. According to our present understanding of electroporation itself, both reversible and irreversible electroporation result in transient openings (pores) of the membrane. These pores are often large enough that molecular transport is expected to be relatively nonspecific. As already noted, for irreversible electroporation, it is plausible that a portion of the cell membrane behaves much like a small planar membrane, and therefore can undergo rupture. In the case of reversible electroporation, significant molecular transport between the intra- and extracellular volumes may lead to a significant chemical imbalance. If this imbalance is too large, recovery may not occur, with cell death being the result. Here it is hypothesized that the volumetric ratio:

$$R_{vol} \equiv (V_{extracellular}/V_{intracellular}) \tag{9}$$

may correlate with cell death or survival *(47)*. According to this hypothesis, for a given cell type and extracellular medium composition, $R_{vol} \gg 1$ (typical of in vitro conditions, such as cell suspensions and anchorage-dependent cell culture) should favor cell death, whereas the other extreme $R_{vol} \ll 1$ (typical of in vivo tissue conditions) should favor cell survival. If correct, for the same degree of electroporation, significantly less damage may occur in tissue than in body fluids or under most in vitro conditions.

25. Tissue Electroporation

Tissue electroporation is a relatively new extension of single-cell electroporation under in vitro conditions, and is of interest because of possible medical applications, such as cancer tumor therapy *(48–50)*, transdermal drug delivery *(51,52)*, noninvasive transdermal chemical sensing *(4)*, and localized gene therapy *(53,54)*. It is also of interest because of its role in electrical injury *(43,55,56)*. The interest in tissue electroporation is growing rapidly, and may lead to many new medical applications. The basic concept is that application of electric field pulses to tissue generally results in a localized, large electric field developing across the lipid-based barriers within the tissue. This can result in the creation of new aqueous pathways across the barrier, just where they are needed in order to achieve local drug delivery. Relevant barriers are not only the single bilayer membranes of cells, but one or more tissue monolayers in which cells are connected by tight junctions (essentially two bilayers in series per monolayer), and the stratum corneum of the skin, which can be regarded very approximately as about 100 bilayer membranes in series. In such cases, it is envisioned that electroporation is to be used with living human subjects. With this in mind, it is significant that several studies support the view that electroporation conditions can be found that result in negligible damage, both in isolated cells *(57–59)* and in intact tissue in vivo *(60,61)*. Increased use of electroporation for drug delivery implies that a much better mechanistic understanding of electroporation will be needed to secure both scientific and regulatory acceptance.

26. Summary

The basic features of electrical and mechanical behavior of electroporated cell membranes are reasonably well established experimentally. Overall, the electrical and mechanical features of electroporation are consistent with a transient aqueous pore hypothesis, and several features, such as membrane rupture and reversible electrical breakdown, are reasonably well described quantitatively. This gives confidence that "electroporation" is an attractive hypothesis, and that the appearance of temporary pores owing to the simultaneous contributions of thermal fluctuations ("kT energy") and an elevated transmembrane voltage ("electric field energy") is the microscopic basis of electroporation.

Acknowledgments

I thank J. Zahn, T. E. Vaughan, M. A. Wang, R. M. Prausnitz, R. O. Potts, U. Pliquett, J. Lin, R. Langer, L. Hui, E. A. Gift, S. A. Freeman, Y. Chizmadzhev, and V. G. Bose for many stimulating and critical discussions. This work supported by NIH Grant GM34077, Army Research Office Grant No. DAAL03-90-G-0218, NIH Grant ES06010, and a computer equipment grant from Stadwerke Düsseldorf, Düsseldorf, Germany.

References

1. Neumann, E., Sowers, A., and Jordan, C. (eds.) (1989) *Electroporation and Electrofusion in Cell Biology.* Plenum, New York.
2. Tsong, T. Y. (1991) Electroporation of cell membranes. *Biophys. J.* **60,** 297–306.
3. Chang, D. C., Chassy, B. M., Saunders, J. A., and Sowers, A. E. (eds.) (1992) *Guide to Electroporation and Electrofusion.* Academic.
4. Weaver, J. C. (1993) Electroporation: a general phenomenon for manipulating cells and tissue. *J. Cell. Biochem.* **51,** 426–435.
5. Orlowski, S. and Mir, L. M. (1993) Cell electropermeabilization: a new tool for biochemical and pharmacological studies. *Biochim. Biophys. Acta* **1154,** 51–63.
6. Weaver, J. C. (1994) Electroporation in cells and tissues: a biophysical phenomenon due to electromagnetic fields. *Radio Sci.* (in press).
7. Weaver, J. C. and Chizmadzhev, Y. A. Electroporation, in *CRC Handbook of Biological Effects of Electromagnetic Fields,* 2nd ed. (Polk, C. and Postow, E., eds.), CRC, Boca Raton (submitted).
8. Parsegian, V. A. (1969) Energy of an ion crossing a low dielectric membrane: solutions to four relevant electrostatic problems. *Nature* **221,** 844–846.
9. Zahn, M. (1979) *Electromagnetic Field Theory: A Problems Solving Approach,* Wiley, New York.
10. Abidor, I. G., Arakelyan, V. B., Chernomordik, L. V., Chizmadzhev, Yu. A., Pastushenko, V. F., and Tarasevich, M. R. (1979) Electric breakdown of bilayer

membranes: I. The main experimental facts and their qualitative discussion. *Bioelectrochem. Bioenerg.* **6,** 37–52.
11. Pastushenko, V. F., Chizmadzhev, Yu. A., and Arakelyan, V. B. (1979) Electric breakdown of bilayer membranes: II. Calculation of the membrane lifetime in the steady-state diffusion approximation. *Bioelectrochem. Bioenerg.* **6,** 53–62.
12. Chizmadzhev, Yu. A., Arakelyan, V. B., and Pastushenko, V. F. (1979) Electric breakdown of bilayer membranes: III. Analysis of possible mechanisms of defect origin. *Bioelectrochem. Bioenerg.* **6,** 63–70.
13. Pastushenko, V. F., Chizmadzhev, Yu. A., and Arakelyan, V. B. (1979) Electric breakdown of bilayer membranes: IV. Consideration of the kinetic stage in the case of the single-defect membrane. *Bioelectrochem. Bioenerg.* **6,** 71–79.
14. Arakelyan, V. B., Chizmadzhev, Yu. A., and Pastushenko, V. F. (1979) Electric breakdown of bilayer membranes: V. Consideration of the kinetic stage in the case of the membrane containing an arbitrary number of defects. *Bioelectrochem. Bioenerg.* **6,** 81–87.
15. Pastushenko, V. F., Arakelyan, V. B., and Chizmadzhev, Yu. A. (1979) Electric breakdown of bilayer membranes: VI. A stochastic theory taking into account the processes of defect formation and death: membrane lifetime distribution function. *Bioelectrochem. Bioenerg.* **6,** 89–95.
16. Pastushenko, V. F., Arakelyan, V. B., and Chizmadzhev, Yu. A. (1979) Electric breakdown of bilayer membranes: VII. A stochastic theory taking into account the processes of defect formation and death: statistical properties. *Bioelectrochem. Bioenerg.* **6,** 97–104.
17. Litster, J. D. (1975) Stability of lipid bilayers and red blood cell membranes. *Phys. Lett.* **53A,** 193,194.
18. Taupin, C., Dvolaitzky, M., and Sauterey, C. (1975) Osmotic pressure induced pores in phospholipid vesicles. *Biochemistry* **14,** 4771–4775.
19. Powell, K. T., Derrick, E. G., and Weaver, J. C. (1986) A quantitative theory of reversible electrical breakdown. *Bioelectrochem. Bioelectroenerg.* **15,** 243–255.
20. Weaver, J. C. and Barnett, A. (1992) Progress towards a theoretical model of electroporation mechanism: membrane electrical behavior and molecular transport, in *Guide to Electroporation and Electrofusion* (Chang, D. C., Chassy, B. M., Saunders, J. A., and Sowers, A. E., eds.), Academic.
21. Barnett, A. and Weaver, J. C. (1991) Electroporation: a unified, quantitative theory of reversible electrical breakdown and rupture. *Bioelectrochem. Bioenerg.* **25,** 163–182.
22. Freeman, S. A., Wang, M. A., and Weaver, J. C. (1994) Theory of electroporation for a planar bilayer membrane: predictions of the fractional aqueous area, change in capacitance and pore-pore separation. *Biophysical J.* **67,** 42–56.
23. Renkin, E. M. (1954) Filtration, diffusion and molecular sieving through porous cellulose membranes. *J. Gen. Physiol.* **38,** 225–243.
24. Wang, M. A., Freeman, S. A., Bose, V. G., Dyer, S., and Weaver, J. C. (1993) Theoretical modelling of electroporation: electrical behavior and molecular transport, in *Electricity and Magnetism in Biology and Medicine* (Blank, M., ed.), San Francisco, pp. 138–140.

25. Weaver, J. C. and Mintzer, R. A. (1981) Decreased bilayer stability due to transmembrane potentials. *Phys. Lett.* **86A,** 57–59.
26. Benz, R., Beckers, F., and Zimmermann, U. (1979) Reversible electrical breakdown of lipid bilayer membranes: a charge-pulse relaxation study. *J. Membrane Biol.* **48,** 181–204.
27. Pastushenko, V. F. and Chizmadzhev, Yu. A. (1982) Stabilization of conducting pores in BLM by electric current. *Gen. Physiol. Biophys.* **1,** 43–52.
28. Sugar, I. P. and Neumann, E. (1984) Stochastic model for electric field-induced membrane pores: electroporation. *Biophys. Chemistry* **19,** 211–225.
29. Weaver, J. C., Harrison, G. I., Bliss, J. G., Mourant, J. R., and Powell, K. T. (1988) Electroporation: high frequency of occurrence of the transient high permeability state in red blood cells and intact yeast. *FEBS Lett.* **229,** 30–34.
30. Tsoneva, I., Tomov, T., Panova, I., and Strahilov, D. (1990) Effective production by electrofusion of hybridomas secreting monodonal antibodies against Hc-antigen of *Salmonella. Bioelectrochem. Bioenerg.* **24,** 41–49.
31. Weaver, J. C. (1993) Electroporation: a dramatic, nonthermal electric field phenomenon, in *Electricity and Magnetism in Biology and Medicine* (Blank, M., ed.), San Francisco, pp. 95–100.
32. Chernomordik, L. V., Sukharev, S. I., Abidor, I. G., and Chizmadzhev, Yu. A. (1982) The study of the BLM reversible electrical breakdown mechanism in the presence of UO_2^{2+}. *Bioelectrochem. Bioenerg.* **9,** 149–155.
33. Neumann, E. and Rosenheck, K. (1972) Permeability changes induced by electric impulses in vesicular membranes. *J. Membrane Biol.* **10,** 279–290.
34. Kinosita, K. Jr. and Tsong, T. Y. (1978) Survival of sucrose-loaded erythrocytes in circulation. *Nature* **272,** 258–260.
35. Klenchin, V. A., Sukharev, S. I., Serov, S. M., Chernomordik, L. V., and Chizmadzhev, Yu. A. (1991) Electrically induced DNA uptake by cells is a fast process involving DNA electrophoresis. *Biophys. J.* **60,** 804–811.
36. Sukharev, S. I., Klenchin, V. A., Serov, S. M., Chernomordik, L. V., and Chizmadzhev, Y. A. (1992) Electroporation and electrophoretic DNA transfer into cells. *Biophys. J.* **63,** 1320–1327.
37. Prausnitz, M. R., Lau, B. S., Milano, C. D., Conner, S., Langer, R., and Weaver, J. C. (1993) A quantitative study of electroporation showing a plateau in net molecular transport. *Biophys. J.* **65,** 414–422.
38. Prausnitz, M. R., Milano, C. D., Gimm, J. A., Langer, R., and Weaver, J. C. (1994) Quantitative study of molecular transport due to electroporation: uptake of bovine serum albumin by human red blood cell ghosts. *Biophys. J.* **66,** 1522–1530.
39. Gift, E. A. and Weaver, J. C. (1995) Observation of extremely heterogeneous electroporative uptake which changes with electric field pulse amplitude in *Saccharomyces cerevisiae. Biochim. Biophys. Acta* **1234(1),** 52–62.
40. Hui, L., Gift, E. A., and Weaver, J. C. Uptake of Bovine Serum Albumin by Yeast due to Electroporation: Existence of a Plateau as Pulse Amplitude is Increased (in preparation).
41. Lillie (1958) Glass, in *Handbook of Physics* (Condon, E. U. and Odishaw, H., eds.), McGraw-Hill, New York, pp. 8–83, 8–107.

42. Neumann, E., Sprafke, A., Boldt, E., and Wolf, H. (1992) Biophysical digression on membrane electroporation, in *Guide to Electroporation and Electrofusion* (Chang, D. C., Chassy, B. M., Saunders, J. A., and Sowers, A. E., eds.), Academic.
43. Lee, R. C., River, L. P., Pan, F.-S., Ji, L., and Wollmann, R. L. (1992) Surfactant induced sealing of electropermeabilized skeletal muscle membranes *in vivo. Proc. Natl. Acad. Sci. USA* **89,** 4524–4528.
44. Gift, E. A. and Weaver, J. C. (1993) Cell survival following electroporation: quantitative assessment using large numbers of microcolonies, in *Electricity and Magnetism in Biology and Medicine* (Blank, M., ed.), San Francisco, pp. 147–150.
45. Weaver, J. C., Bliss, J. G., Powell, K. T., Harrison, G. I., and Williams, G. B. (1991) Rapid clonal growth measurements at the single-cell level: gel microdroplets and flow cytometry. *Bio/Technology* **9,** 873–877.
46. Weaver, J. C., Bliss, J. G., Harrison, G. I., Powell, K. T., and Williams, G. B. (1991) Microdrop technology: a general method for separating cells by function and composition. *Methods* **2,** 234–247.
47. Weaver, J. C. (1994) Molecular basis for cell membrane electroporation. *Ann. NY Acad. Sci.* **720,** 141–152.
48. Okino, M. and Mohri, H. (1987) Effects of a high-voltage electrical impulse and an anticancer drug on *in vivo* growing tumors. *Jpn. J. Cancer Res.* **78,** 1319–1321.
49. Mir, L. M., Orlowski, S., Belehradek, J., Jr., and Paoletti, C. (1991) In vivo potentiation of the bleomycin cytotoxicity by local electric pulses. *Eur. J. Cancer* **27,** 68–72.
50. Dev, S. B. and Hofmann, G. A. (1994) Electrochemotherapy—a novel method of cancer treatment. *Cancer Treatment Rev.* **20,** 105–115.
51. Prausnitz, M. R., Bose, V. G., Langer, R. S., and Weaver, J. C. (1992) Transdermal drug delivery by electroporation. Abstract, Proc. Intern. Symp. Control. Rel. Bioact. Mater. 19, Controlled Release Society, July 26–29, Orlando, FL, pp. 232,233.
52. Prausnitz, M. R., Bose, V. G., Langer, R., and Weaver, J. C. (1993) Electroporation of mammalian skin: a mechanism to enhance transdermal drug delivery. *Proc. Natl. Acad. Sci. USA* **90,** 10,504–10,508.
53. Titomirov, A. V., Sukharev, S., and Kistoanova, E. (1991) In vivo electroporation and stable transformation of skin cells of newborn mice by plasmid DNA. *Biochim. Biophys. Acta* **1088,** 131–134.
54. Sukharev, S. I., Titomirov, A V., and Klenchin, V. A. (1994) Electrically-induced DNA transfer into cells. Electrotransfection in vivo, in *Gene Therapeutics* (Wolff, J. A., ed.), Birkhäuser, Boston, pp. 210–232.
55. Gaylor, D. C., Prakah-Asante, K., and Lee, R. C. (1988) Significance of cell size and tissue structure in electrical Trauma. *J. Theor. Biol.* **133,** 223–237.
56. Bhatt, D. L., Gaylor, D. C., and Lee, R. C. (1990) Rhabdomyolysis due to pulsed electric fields. *Plast. Reconstr. Surg.* **86,** 1–11.
57. Hughes, K. and Crawford, N. (1989) Reversible electropermeabilisation of human and rat blood platelets: evaluation of morphological and functional integrity "in vitro" and "in vivo." *Biochim. Biophys. Acta* **981,** 277–287.
58. Mouneimne, Y., Tosi, P.-F., Barhoumi, R., and Nicolau, C. (1991) *Biochim. Biophys. Acta* **1066,** 83–89.

59. Zeira, M., Tosi, P.-F., Mouneimne, Y., Lazarte, J., Sneed, L., Volsky, D. J., and Nicolau, C. (1991) *Proc. Natl. Acad. Sci. USA* **88**, 4409–4413.
60. Belehradek, M., Domenge, C., Orlowski, S., Belehradek, J., Jr., and Mir, L. M. (1993) *Cancer* **72**, 3694–3700.
61. Riviele, J. E., Monterio-Riviere, N. A., Rogers, R. A., Bommannan, D., Tamada, J. A., and Potts, R. O. Pulsatile Transdermal Delivery of LHRH Using Electroporation: Drug Delivery and Skin Toxicology (submitted).
62. Potts, R. O. and Francoeur, M. L. (1990) Lipid biophysics of water loss through the skin. *Proc. Natl. Acad. Sci. USA* **87**, 3871–3873.
63. Bach, D. and Miller, I. R. (1980) Glyceryl monooleate black lipid membranes obtained from squalene solutions. *Biophys. J.* **29**, 183–188.
64. Sugar, I. P. (1981) The effects of external fields on the structure of lipid bilayers. *J. Physiol. Paris* **77**, 1035–1042.

CHAPTER 2

Effects of Pulse Length and Strength on Electroporation Efficiency

Sek Wen Hui

1. Introduction

Electroporation is now a standard method of transfection and cell loading. There is a variety of commercial electroporation equipment, and many published and manufacturer-supplied protocols. Many of these protocols are results of trial and error. These empirical protocols are valuable guides for successful applications of electroporation.

Because experimental conditions vary case by case, new and modified protocols are constantly needed to optimize the transfection yield. Developing new protocols for new cases by trial and error in each laboratory is wasteful in terms of time and energy. This chapter is intended to present a guide, based on known theories of electroporation, to help users modify existing protocols and develop new protocols for new applications. Succeeding in doing so would take some guess work out of experimental trials in tailoring protocols for individual case needs.

2. Theoretical Guide
2.1. General Considerations

Although the detailed molecular mechanism of electroporation is still not completely understood, recent fundamental studies give us a general concept of the events happening during the electroporation process. We know that membrane permeation is the result of the electric breakdown of the lipid bilayer when the induced transmembrane potential exceeds the breakdown potential of the bilayer. Pores in membranes are formed and maintained by the electric pulse field. In contrast to lipid bilayers that reseal immediately after the breakdown, the permeated state of cell membranes may last tens of minutes after the termination of the pulse field. During this time, molecules may enter the cell through a number of pathways, including electrophoresis (of charged molecules, such as DNA), electro-osmotic and colloid-osmotic flow, as well as diffusion. Therefore, several physical factors may affect the electro-transfection efficiency:

1. The transmembrane potential created by the imposing pulse electric field.
2. The extent of membrane permeation (number and size of pores or affected areas).
3. The duration of the permeated state.
4. The mode and duration of molecular flow.
5. The global and local (surface) concentrations of DNA.
6. The form of DNA.
7. The tolerance of cells to membrane permeation.
8. The heterogeneity of the cell population.

We now examine these factors more quantitatively.

2.2. Creating Electropores

A membrane pore is created when the energy stored in the membrane capacitor exceeds the energy required to keep the membrane intact against pore expansion. The energy E_p of forming a pore of a given radius r in a membrane is determined by the balance between the line tension γ of the pore edge and the surface tension Γ of the membrane.

$$E_p = 2\pi r \gamma - \pi r^2 \Gamma \qquad (1)$$

This pore energy reaches a maximum at a critical value of pore radius $r_c = \gamma/\Gamma$ (1,2). The line tension of the pore edge depends on the molecular packing of the membrane. Pores with radii $<r_c$ tend to reseal, whereas those with radii $>r_c$ tend to expand if the membrane is under tension.

When subject to an imposed electric field E, which causes the charging of the equivalent membrane capacitor, the energy stored in this membrane capacitor over a precursor area of the pore is:

$$E_e = \pi r^2 \varepsilon_0 (\varepsilon_w - \varepsilon_m) V^2 / 2d \qquad (2)$$

where ε denotes the dielectric constant and the subscripts 0, w, and m refer to free space, water, and membrane, respectively. V is the transmembrane voltage (membrane potential) imposed by the pulse field across the membrane of thickness d. For a given membrane, an electropore of radius r will form if the electric energy E_e given in Eq. (2) is greater than the energy E_p required to form a pore of such size. This energy defines the breakdown voltage V_b, over which the membrane will break down and pores of diameter r will be formed. If $r < r_c$, the electric breakdown is reversible. Otherwise the pore will expand once it is formed. A macroscopic relationship has been described by Zhelev and Needham *(2)*.

In considering the simple case of a spherical cell, the electric conductance of the cell interior is much higher than that of the cell membrane. The imposed membrane potential, or transmembrane voltage V experienced by the cell is:

$$V = 1.5 \, a \, E \cos \theta \, [1 - \exp(-t/\tau)] \qquad (3)$$

where E is the pulse electric field, a is the radius of the cell, and θ is the angle between the field direction and the radial vector of the surface point where membrane potential is considered. The charging or relaxation time τ of the membrane is determined by the internal and external conductivities of the cell *(3)*. The highest V is, of course, at the poles along or against the field direction, where θ is 0 and π respectively. Since the breakdown voltage, V_b, of most biomembranes is about 1 V, for a cell 10 μm in diameter, a 0.7 kV/cm pulse is sufficient to produce a breakdown potential at the poles. As the imposed field strength E increases, the area where membrane breakdown potential is experienced extends further from the pole, i.e., the wider the breakdown area. The percentage of breakdown area is given by $(1 - E_b/E)$ where E_b is the pulse field strength needed to produce the membrane breakdown voltage V_b *(4)*.

2.3. DNA Transport

Apart from the extent of membrane permeation, there are several other factors that control the intake of exogenous DNA by the cell. It is believed that the majority of exogenous DNA transported into cells in electroporation is through electrophoresis *(5,6)*. Even if cells remain permeated long after the pulse, as determined by dye penetration, adding DNA immediately after the pulse usually results in a much lower transfection efficiency compared to adding DNA before the pulse. Shielding the charge of DNA by cations also reduces the transfection efficiency. It has been reported that more DNA enters the cells during a second pulse of lower field strength, once the cell is permeated by the first pulse *(7,8)*. Uptake of DNA adsorbed on cell surfaces has also been suggested *(9)*. For a given pore-forming pulse voltage, the transfection efficiency depends more on the total length of a pulse than on the time span when cells remain permeable. Thus, the diffusion of DNA into cells through electropores is not an important contribution in transfection efficiency. Other physical factors, such as the form and concentration of DNA, are also important *(10)*. Because the physical forms of DNA affect their electrophoresis mobility, among other effects, these factors are important with regard to the pulse length and strength.

If the majority of DNA enters the cell during the pulse, the transfection efficiency should be proportional to the time integral of the applied field multiplied by the extent of membrane permeation. For simple rectangular pulses or exponential decay pulses, the time integral is the pulse length (or decay time constant) T, which is usually much greater than the membrane relaxation time τ. Thus, we expect the transfection efficiency to be proportional to:

$$\int E(1 - E_b/E)dt = (E - E_b)T \qquad (4)$$

If multiple rectangular pulses are used, T represents the sum of all pulse periods. If an exponentially decaying pulse from a capacitor-type pulse generator is used, T represents the decay time constant of the pulse. The relationship should hold as long as E is not too much greater than E_b, such that the electropores so created are reversible, the permeated areas stay in the vicinity of cell poles, and the viability of the cell population is not significantly compromised *(4,11)*.

It should be pointed out that many electroporation protocols give electric parameters in terms of voltage and capacitor value (in µF, for instance). These quantities are meaningful only if the sample resistance and interelectrode distance are given as well. For a uniform sample in a cuvet, the pulse field strength E is the applied voltage divided by the interelectrode distance. The pulse decay time $T = RC$ is the capacitance C used in the pulse generator multiplied by the total resistance R of the sample and any circuit elements. The formulae relating voltage and capacitance to the relevant electric parameters E and T are sample- and instrument-dependent. Therefore, voltage and capacitance should not be cited as universal electric parameters. Because sample resistance varies case by case, users should convert the capacitance setting to decay time constant for reporting purposes.

2.4. Sample Homogeneity and Cell Viability

From experience, we know that not every cell in a given sample is transfected, even if the above conditions are satisfied. In fact, only a small percentage of cells are transfected in each run. Within a given sample, cells vary in size and shape, as well as in the dielectric and conductive properties of cellular components, such that the critical field and membrane relaxation time differ from cell to cell. In addition to this physical variation are the more important biological variations in cell cycle, age, and gene expression controls. Therefore, these theoretical considerations should be treated only as a guide. The actual response will depend on the homogeneity of the cell population.

The above analysis applies only to reversible breakdown of cell membranes by the electric pulses, such that membranes reseal in time, and cells recover from the traumatic event of electroporation. In cases where the applied field is high enough to trigger irreversible membrane breakdown, i.e., $E \gg E_b$ such that $r > r_c$, electropores do not reseal, and the cell viability is low. As a consequence, cell viability imposes an upper limit on the transfection efficiency.

3. Experimental Evidence

Several experiments using fluorescent molecules as tracers have shown that macromolecule leakage or intake during electroporation is indeed proportional to the quantity $(E - E_b)T$ *(4,12,13)*. Figure 1 shows the uptake of fluorescein isothiocyanate-labeled dextran by mouse C3H/10T$_{1/2}$ cells as a function of ET. Because cells take up dextran spontaneously, the x-axis intercept does not correspond to the value of E_bT. Even though the experimental data represent a wide range of E and T combinations, the linear relationship is obvious, regardless of voltage and time ranges, as long as the reversible breakdown limit is not exceeded *(12)*. A recent measurement of bovine serum albumin uptake by erythrocyte ghosts shows the same trend *(13)*.

Short-term transfection efficiency is sometimes expressed in terms of the percentage of cells transfected or the number of transfected cells per

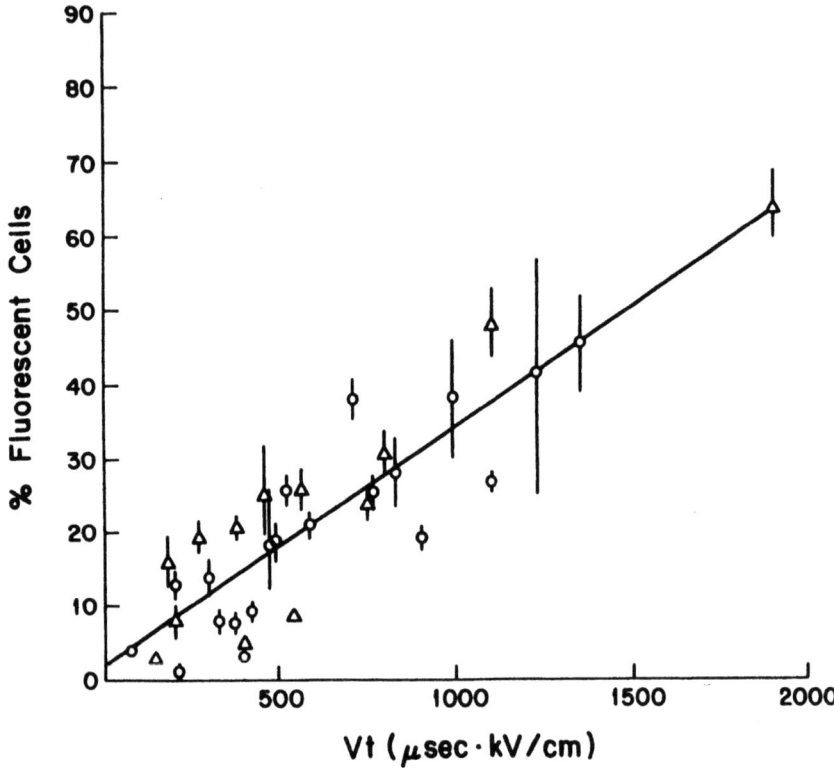

Fig. 1. Macromolecular uptake is proportional to ET. The percentage of fluorescent C3H/10T 1/2 cells after electroporation is plotted against the product of pulse length T and field strength E, for both rectangular (O) and exponential decay (△) pulses, using FITC dextran of mol wt 41 kDa. The straight line represents linear regression of all points *(12)* (courtesy of Eaton Publishing Co.).

given amount of DNA. Both short- and long-term transfection efficiencies are often quoted as the number of clones from an initial population of cells. Whichever transfection efficiency values are measured, these values depend on the percentage of permeated, viable cells, as well as the number of copies of plasmid DNA delivered into those cells.

If the transfection efficiency is proportional to the percentage of permeated cells that receives DNA, then the transfection efficiency would also be a linear function of $(E - E_b)T$. For a given pulse duration, T, the transfection efficiency is expected to be proportional to E. Similarly, for a given E, the transfection efficiency is expected to be proportional to T.

The effects of these electric parameters on pBR322 transfection of *E. coli* JM105 were reported by Xie and Tsong *(14)*. Figure 2 shows such relations plotted as log[transfection efficiency*(TE)*] against E or log*[T]*. The first plot (Fig. 2A) is not expected to be linear if DNA enters the cell mainly by electrophoresis rather than by diffusion *(4)*, but if log*[TE]* is plotted against log*[E]*, a more linear relationship is found. A linear relationship is also apparent in plots of log*[TE]* against log*[T]* plot (Fig. 2B). Furthermore, when cell viability is affected by irreversible membrane breakdown, the linear relationship yields to the viability limit.

The transfection efficiency of HeLa cells by pRSVgpt plasmid DNA *(15)* was measured as a function of *ET* (*T* is given as the exponential decay half time $\tau_{1/2}$ of pulses generated by a capacitor-type generator). Figure 3 shows that the approximately linear relationship is obeyed by both the short (0.275–0.310 ms) and the long (2.2–4.4 ms) pulse groups. Apparently, even at the highest voltage applied, the cell viability limit had not been reached. The *x*-axis intercept of the least-square-fit line gives $E_bT = 0.5$ kV ms/cm. This value implies that, for a 0.75-ms pulse, the threshold applied field strength to cause reversible breakdown in some cells is 0.7 kV/cm. This value is approximately equal to that given by Eq. (3), and agrees with most threshold field strength values for electroporation and electrofusion *(16)*.

Human lymphoid cells were transfected by pCP4-fucosidase plasmids. Cells were subjected to three consecutive exponentially decay pulses while in Baker and Knight *(17)* medium. The cells were incubated at 37°C in the same medium for 30 min after the pulses, before transferring back to the normal culture medium. The fucosidase activity was assayed after 48 h. The transfection efficiency given by the enzyme activity of transfected cells is shown in Fig. 4. Data points were taken within the range of field strengths of 0.4–4.0 kV/cm, and durations of 0.14–3.4 ms. Apparently, the viability limit is reached at 1.5 kV ms/cm. Below this limit, the transfection efficiency is approximately linear with *ET*. The threshold breakdown condition is again about $E_bT = 0.5$ kV ms/cm. A single point (solid square) obtained using 4 kV/cm field strength pulses is exceptionally low in transfection efficiency, perhaps owing to too much cell death caused by irreversible membrane permeabilization.

The effect of pulse strength and duration on the transfection of CHO cells was investigated by Wolf et al. *(6)*. The transfection efficiency of CHO cells in suspension, by pSV2CAT or pBR322-βgal plasmids, was

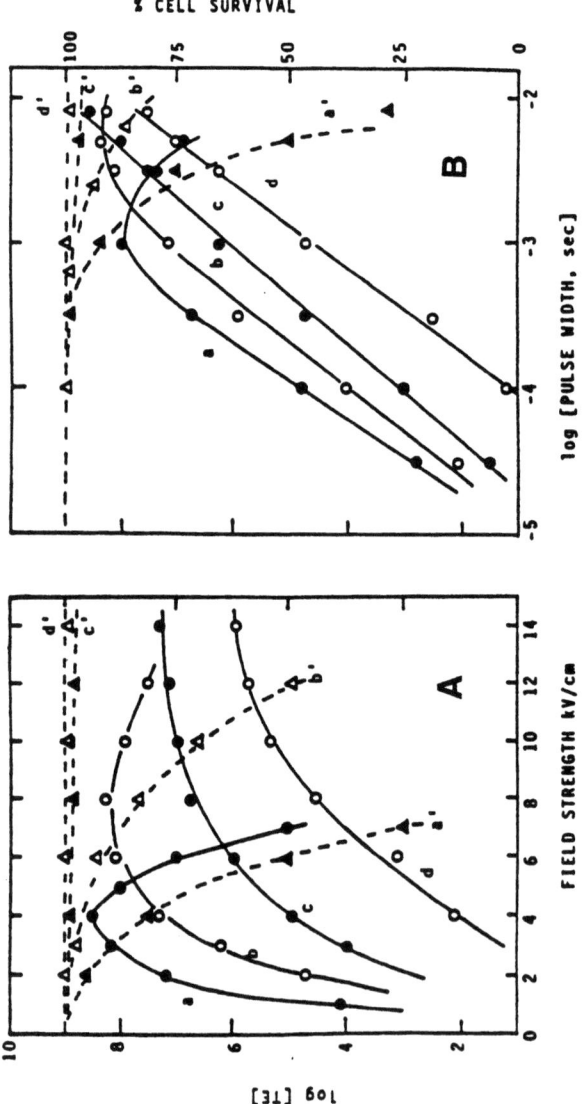

Fig. 2. Effects of electric parameters on pBR322 transfection of *E. coli* JM105. (A) Dependence of log[TE] on the pulse field strength. The pulse durations were 5 ms, 1 ms, 200 μs, and 40 μs for curves *a*, *b*, *c*, and *d*, respectively. The corresponding curves (- - -) for percent cell survival are given in curves *a'*, *b'*, *c'*, and *d'*. The left ordinate indicates logarithm of the transfection efficiency (TE) and the right ordinate the percent cell survival. (B) Dependence of log[TE] on pulse width. The field strengths were 6, 4, 3, and 2 kV/cm, respectively, for curves *a*, *b*, *c*, and *d*. The corresponding curves (- - -) for percent cell survival are given in curves *a'*, *b'*, *c'*, and *d'*. Experimental conditions are identical to (A) *(14)* (courtesy of the Rockefeller University Press and the Biophysical Society).

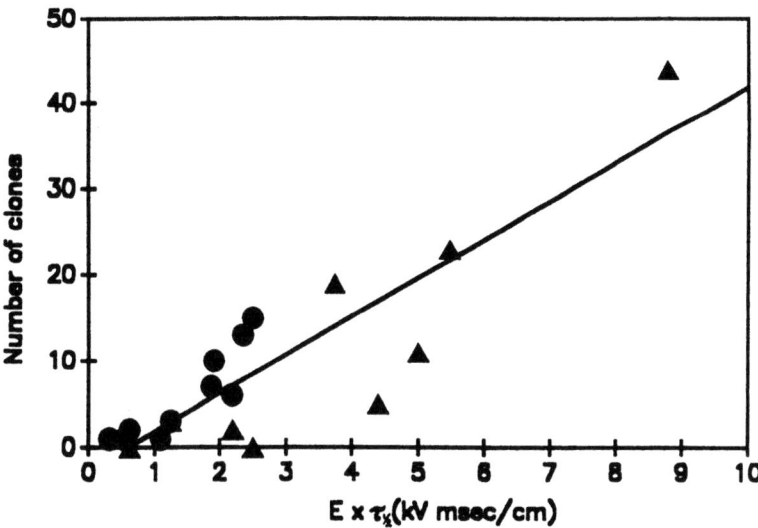

Fig. 3. Effect of electric parameters ET on pRSVgpt transfection of HeLa cells. The number of successfully transfected HeLa cells with pRSVgpt plasmid by electroporation is plotted as a function of the product of peak field strength E and decay half-time $\tau_{1/2}$ of an applied exponential pulse. Data points for different pulse lengths (● $\tau_{1/2}$ = 0.275–0.310 ms, ▲ $\tau_{1/2}$ = 2.2–4.4 ms) are plotted together *(15)* (courtesy of the Eaton Publishing Co.).

found to be a linear function of either E or T, when the other parameter was held constant. The minimum field strength E_b for detectable permeability as well as transfection was found to be 0.6 kV/cm. Interestingly, the transfection efficiency decreases with increasing delay between repeating pulses, indicating that DNA is collected on the cell surface and driven through the electropores by electrophoretic force, in agreement with previous experiments using two consecutive pulses to increase transfection yield *(8)*.

4. Conclusion

The existing theory for reversible electric breakdown of cell membranes and the transport of DNA across the plasma membranes through electropores adequately describes a linear relation between transfection efficiency and $(E - E_b)T$. E_b is determined by the electric energy derived from the applied pulses and the energy cost to form an electropore in the

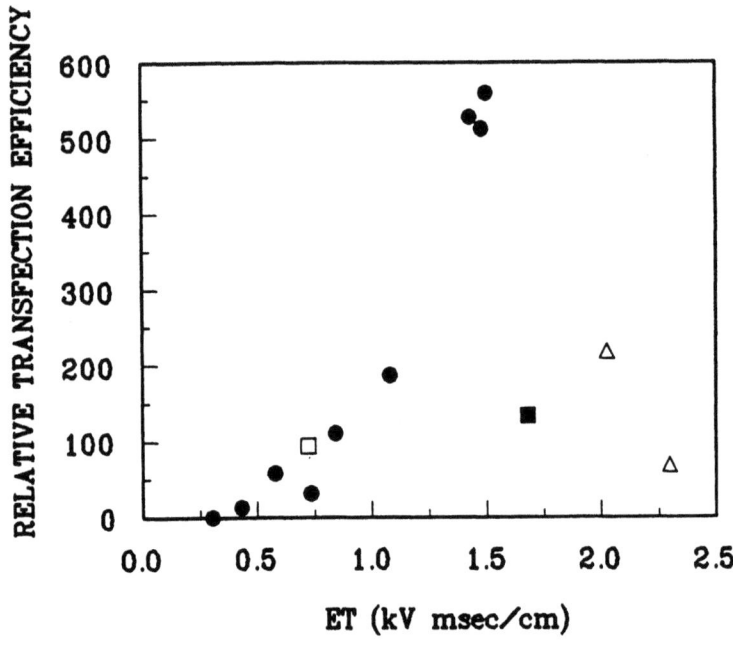

Fig. 4. The relative transfection efficiency of human JTL lymphoid cells transfected by pCEP4-FUC. The activity of α-fusosidase is plotted as a function of the product of the pulse field strength E and the equivalent integral pulse time T. The samples were subjected to three exponentially decaying pulses of given decay half-times: (□) 0.2 ms, (●) 0.24–0.28 ms, (△) 0.34–0.35 ms, at E = 0.5–2 kV/cm, and (■) 0.14 ms at E = 4 kV/cm.

membrane of cells of a given size. Although the viability limit varies from cell line to cell line, within this limit, the linear relationship between transfection efficiency and $(E - E_b)T$ seems to hold over a wide range of E and T combinations, and is applicable to most cells. Therefore, this relationship serves as a guideline for selection of electric parameters in various applications.

Since the transfection efficiency is proportional to ET, there is a choice of optimizing either E or T. Excessive field strength leads to permeation of a wider area, or forming larger pores that may exceed the reseal limit $r_c = \gamma/T$. As long as pores remain reversible, the required T value may be satisfied by applying longer and multiple pulses. Cell membranes do not reseal as rapidly as lipid bilayers, so that multiple pulses may not form additional pores in the membranes. There are reported advantages of multiple pulses over a single pulse, and AC bursts over rectangular pulses *(18,19)*. A bipolar oscillating field was reported to be more effective than a unipolar oscillating field, since both poles of the cells are permeated *(20)*. Clearly there is merit in using lower applied field strength, and increasing the duration and number of pulses to achieve the required *ET* value.

Protocols for electroporation continue to develop as cases demand. The present knowledge of electroporation mechanisms is capable of guiding rational design of new protocols. Instead of describing a cumulation of successful protocols, this chapter gives the basics to develop one's own protocols. It is hoped that the above analysis will take some guesswork out of applying electrotransfection in new situations.

Acknowledgments

The results from my laboratory cited in this chapter are the effort of R. T. Kubiniec, D. A. Stenger, H. Liang, and N. G. Stoicheva. The work was supported by a grant GM 30969 from the National Institutes of Health.

References

1. Abidor, I. G., Arakelyan, V. B., Chernomordik, L. V., Chizmadzhev, Y. A., Pastushenko, V. F., and Tarsevich, M. R. (1979) Electric breakdown of bilayer membranes. I. The main experimental facts and their qualitative discussion. *Bioelectrochem. Bioenerg.* **6,** 37–52.
2. Zhelev, D. V. and Needham, D. (1993) Tension-stabilized pores in gaint vesicles: determination of pore size and pore line tension. *Biochim. Biophys. Acta* **1147,** 89–104.
3. Stenger, D. A., Kaler, K. V. I. S., and Hui, S. W. (1991) Dipole interaction in electrofusion. Contributions of membrane potential and effective dipole interaction pressures. *Biophys. J.* **59,** 1074–1084.
4. Schwister, K. and Deuticke, B. (1985) Formation and properties of aqueous leaks induced in human erythrocytes by electrical breakdown. *Biochim. Biophys. Acta* **816,** 332–348.
5. Klenchin, V. A., Sukharev, S. I., Serov, S. M., Chernomordik, L. V., and Chidmadzhev, Y. A. (1991) Electrically induced DNA uptake by cells is a fast process involving DNA electrophoresis. *Biophys. J.* **60,** 804–811.
6. Wolf, H., Rols, M. P., Boldt, E., Neumann, E., and Teissie, J. (1994) Control by pulse parameters of electric field-mediated gene transfer in mammalian cells. *Biophys. J.* **66,** 524–531.
7. Andreason, G. L. and Evans, G. A. (1989) Optimization of electroporation for transfection of mammalian cells. *Anal. Biochem.* **180,** 269–275.

8. Sukharev, S. I., Klenchin, V. A., Serov, S. M., Chernomordik, L. V., and Chizmadzhev, Y. A. (1992) Electroporation and electrophoretic DNA transfer into cells. The effect of DNA interaction with electropores. *Biophys. J.* **63,** 1320–1327.
9. Xie, T. D., Sun, L., and Tsong, T. Y. (1990) Study of mechanisms of electric field-induced DNA transfection. I. DNA entry by surface binding and diffusion through membrane pores. *Biophys. J.* **58,** 13–19.
10. Nickoloff, J. A. and Reynolds, R. J. (1992) Electroporation-mediated gene transfer efficiency is reduced by linear plasmid carrier DNAs. *Anal. Biochem.* **205(2),** 237–243.
11. Rols, M. P. and Teissie, J. (1990) Electropermeabilization of mammalian cells. Quantitative analysis of the phenomenon. *Biophys. J.* **58(5),** 1089–1098.
12. Liang, H., Purucker, W. J., Stenger, D. A., Kubiniec, R. T., and Hui, S. W. (1988) Uptake of fluoroscence-labeled dextrans by 10T 1/2 fibroblasts following permeation by rectangular and exponential-decay electric field pulses. *BioTechniques* **6,** 550–558.
13. Prausnitz, M. R., Milano, C. D., Gimm, J. A., Langer, R., and Weaver, J. C. (1994) Quantitative study of molecular transport due to electroporation: uptake of bovine serum albumin by erythrocyte ghosts. *Biophys. J.* **66,** 1522–1530.
14. Xie, T. D. and Tsong, T. Y. (1992) Study of mechanisms of electric field-induced DNA transfection III. Electric parameters and other conditions for effective transfection. *Biophys. J.* **63,** 28–34.
15. Kubiniec, R. T., Liang, H., and Hui, S. W. (1990) Effects of pulse length and pulse strength on transfection by electroporation. *BioTechniques* **8,** 1–3.
16. Chang, D. C., Chassy, B. M., Saunders, J. A., and Sowers, A. E. (1992) *Guide for Electroporation and Electrofusion.* Academic, San Diego.
17. Baker, P. F. and Knight, D. E. (1983) High-voltage techniques for gaining access to the interior of cells: application to the study of exocytosis and membrane turnover. *Methods Enzymol.* **98,** 28–37.
18. Xie, T. D. and Tsong, T. Y. (1990) Study of mechanisms of electric field-induced DNA transfection. II. Transfection by low amplitude low frequency alternating electric fields. *Biophys. J.* **58,** 897–903.
19. Chang, D. C., Gao, P. Q., and Maxwell, B. L. (1991) High efficiency gene transfection by electroporation using a radio-frequency electric field. *Biochim. Biophys. Acta* **1992,** 153–160.
20. Tekle, E., Austumian, R. D., and Chock, P. B. (1991) Electroporation by using bipolar oscillating electric field: an improved method for DNA transfection of NIH 3T3 cells. *Proc. Natl. Acad. Sci. USA* **88,** 4230–4234.

CHAPTER 3

Instrumentation

Gunter A. Hofmann

1. Introduction

The techniques of electroporation and electrofusion require that cells be subjected to brief pulses of electric fields of the appropriate amplitude, duration, and wave form. In this chapter, the term electro cell manipulation (ECM) shall describe both techniques. ECM is a quite universal technique that can be applied to eggs, sperm, platelets, mammalian cells, plant protoplasts, plant pollen, liposomes, bacteria, fungi, and yeast—generally to any vesicle surrounded by a membrane. The term "cells" will be used representatively for any of the vesicles to be manipulated unless specific requirements dictate otherwise.

Electroporation is characterized by the presence of one membrane in proximity to molecules that are to be released or incorporated. One or several pulses of the appropriate field strength, pulse length, and wave shape will initiate this process.

Electrofusion is characterized by two membranes in close contact that can be joined by the application of a pulsed electric field. The close contact can be achieved by mechanical means (centrifuge), chemical means (PEG), biochemical means (avidin-biotin [1]), or by electrical means (dielectrophoresis [2]). Only the electric method is discussed as it relates to ECM instrumentation.

The intent of this chapter is to provide the researcher with a basic understanding of the hardware components and electrical parameters of ECM systems to allow intelligent, economical choices about the best instrumentation for a specific application and to understand its limitations. Commercial instruments have been available for more than 10 years; the commercial ECM technology has matured and become more costeffective. Rarely is it economical to build one's own instrument. Although articles occasionally appear on how to build an instrument for a few hundred dollars, the plans are generally of poor design, and the cost estimates often do not take into account the researcher's time for electronic development. Furthermore, today's commercial instruments often incorporate measuring circuits for important parameters, which are difficult to develop. The difficulty is that in one housing, there are voltages of many kilovolts and currents of hundreds of Amperes (A) flowing next to low signal/control voltages, typically between 5 and 20 V. Sophisticated design is needed to prevent crosstalk or electromagnetic interference between these different circuits. Thus, it is usually not costeffective to build an instrument unless specific parameters are needed that are not available commercially. Another very important issue is safety. Voltages and currents generated in efficient ECM generators are large enough to induce cardiac arrest. Generators need to be constructed to be safe and foolproof against accidental wrong settings. They must also deliver the pulse to the chamber in such a way that the operator will not, under any circumstances, come in contact with parts carrying high voltage.

A database of over 2500 publications in the field of electroporation and electrofusion is maintained and updated continuously by BTX (San Diego, CA) as a service to the research community. Any researcher may inquire about the BTX Electronic Genetics® Database and request a database search.

2. Components of an ECM System and Important Parameters

Generally, ECM systems consist of a generator providing the electric signals and a chamber in which the cells are subject to the electric fields created by the voltage pulse from the generator. A third optional component is a monitoring system, either built into the generator or connected in line between the generator and the chamber, which measures the electrical parameters as the pulse passes through the system. Each component is discussed in Sections 4.–6. In this section, we discuss the relationship between the electrical parameters, which the ECM system provides, and the parameters that the cells experience.

Instrumentation

The biophysical process of electropermeabilization is caused by the electrical environment of the cell in a medium. The main parameter, which describes this environment, is the electric field strength E, measured in V/cm. Though the presence of the cell itself modifies the field in close proximity, knowledge of the average field strength at the location of the cells is sufficient for the purpose of ECM experiments. The electric field is generally created by the application of a potential difference (voltage) between metallic electrodes immersed in the medium containing the cells. For the simple electrode geometry of parallel plates located at a distance d (cm), the electric field is calculated from the applied voltage V as:

$$E = V/d \text{ (V/cm)} \tag{1}$$

Practical values of E used in ECM range from a few hundred V/cm for mammalian cells to many kV/cm for bacteria.

The electric field in the medium gives rise to currents depending on the medium specific resistivity r, which is measured in $\Omega \cdot$ cm. The specific resistivity ranges from a low of about 100 $\Omega \cdot$ cm for saline solutions to many k$\Omega \cdot$ cm for nonionic solutions, such as mannitol. The resulting current density j is:

$$j = E/r \text{ (A/cm}^2\text{)} \tag{2}$$

The current produced results in heating of the medium. Saline solutions with a low value of r experience severe heating effects as compared to nonionic solutions for the same electric field and pulse length.

The temperature rise ΔT (°C) can influence the permeabilization mechanism, or lead to excessive heating and evaporation of the medium. It can be calculated for different pulse wave shapes:

Square pulse: $\Delta T = E^2 t/4.2\, r$, where t is the pulse length in s.
Exponential pulse: $\Delta T = E^2 \tau/8.4\, r$, where τ is the 1/e time constant (s) (*see* Section 4.2.1.).

Having defined the parameters at the location of the cells, we can relate them to the electrical parameters at the chamber electrodes:

Electrode voltage: $V = E \cdot d$ (V) (plane parallel electrodes)
Chamber current: $J = j \cdot F$ (A), where F is the electrode area, cm^2
Chamber resistance: $R = f(r)\, \Omega$, where $f(r)$ is a function of the chamber geometry.
For plane parallel electrodes, $R = r\, d/F$.

The voltage at the chamber electrodes is not necessarily equal to the voltage that is indicated or even measured in the generator. This relationship is discussed in Section 4.1. The range of the electric field for optimum yield is quite narrow. Deviation of 5–10% from the optimum value can lead to a drop of an order of magnitude in yield. The pulse length is a less sensitive parameter. It is desirable to assure that the optimum electric field is established in the chamber.

3. Volume Requirements

Small volumes of 100 µL to a few milliliters can be treated in a batch mode: fill the chamber, electroporate, or fuse, and empty the chamber. Larger volumes (many milliliters to 1 L) require chambers that might not be available and a high output power level, which generators typically cannot deliver. A good solution to this problem is the use of a flow-through system in which the generator periodically pulses in synchronism with a pump, so that every volume element of cell/transformant mixture is exposed to the desired electric fields and number of pulses as it passes through the chamber. This method requires flowthrough chambers and generators that can pulse automatically, either at a fixed or adjustable repetition rate. Such generators and chambers are available (*see* Tables 1 and 6). For fusion, a continuous flow is not desirable, because fused cells need to be undisturbed for a period of time to round off and complete the fusion process. In this case, a pulsating (stop and go) flowthrough system would be appropriate.

4. Generators

The relationship between the electrical parameters in a generator and the parameters actually delivered to the chambers is important because substantial differences can exist. Following this discussion, different types of generators are described. Table 1 presents a survey of commercially available generator types.

4.1. Actual Voltage Delivered to the Chamber

The momentary power the generators are required to deliver to chambers can far exceed the electrical power available from laboratory outlets. To overcome this limitation, electrical energy is stored in capacitors by charging them slowly at low power to a preset voltage and then discharging them at high power level into the chamber. The voltage V_0 to which the capacitors will be charged can be set and is typically indicated

Instrumentation

Table 1
Survey of Electroporation and Electrofusion Generators

		Electroporation				
		Exponential discharge wave form				
Manufacturer	No PS	With PS one pulse length	With PS multiple pulse lengths	Square wave	Electro cell fusion	Stand-alone monitor
IBI (New Haven, CT)			X			
Invitrogen (San Diego, CA)	X					
Bio-Rad (Richmond, CA)		X	X			
BRL (Grand Island, NY)		X	X			
BTX	X	X^a	X^a	X^a	X	X

Abbreviation: PS, power supply.
^{a}R, optional version available with repetitive pulsing for flowthrough applications.

at the front panel of the generators. The actual voltage delivered to the chamber can be substantially lower than what is normally assumed to be the generator output voltage. This effect is caused by the internal resistance of the generator (typically around 1 Ω), which absorbs part of the charging voltage during discharge and is more pronounced in larger chambers (several milliliters) and low resistivity medium. Some generators are also designed with a relatively high internal resistance, which is undesirable, to protect the output switch against high currents. These generators can exhibit a drastic drop in actual voltage delivered to the chamber under certain circumstances. If the internal resistance R_i and the chamber resistance R_c are known, the actual voltage V on the chamber can be calculated as:

$$V = V_0 \cdot R_c/(R_c + R_i) \tag{3}$$

4.2. Generators for Electroporation

The two types of generators commonly encountered differ by the wave shape of their output: exponential decay wave form or square pulses. Though both can in principle be used for electroporation, it appears that bacteria are transformed more efficiently by exponential wave forms (with some exceptions [3]), whereas some mammalian cell types (4) and plant protoplasts (5) show generally superior transformation results with square waves.

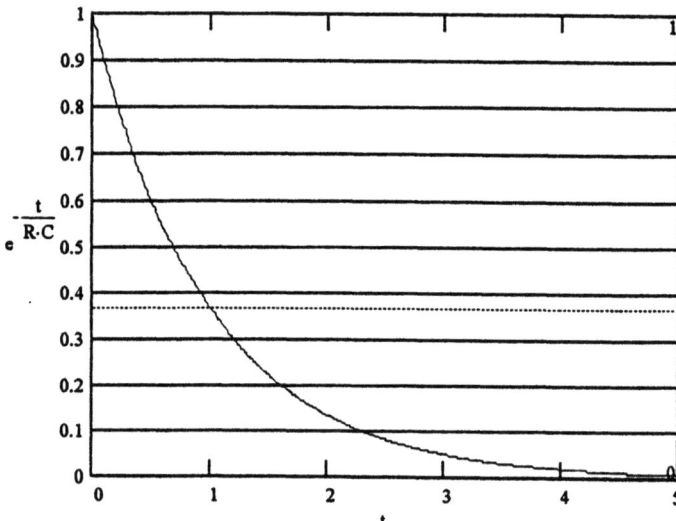

Fig. 1. Exponential decay wave form, representative of the complete discharge of a capacitor into a resistor.

4.2.1. Exponential Wave Form Generators

The voltage of a capacitor C (capacity measured in Farad or, more conveniently, in microfarad) discharging into a resistor R (Ω) follows an exponential decay law (Fig. 1):

$$V = V_o \cdot exp\,(-t/RC) \qquad (4)$$

The pulse length of such a discharge wave form is commonly characterized by the "$1/e$ time constant." This is the time required for the initial voltage to decay to $1/e \approx 1/3$ of the initial value ($e = 2.718\ldots$ is the basis of natural logarithms). This time constant can be conveniently calculated from the product of R and C, where C is the storage capacitor in the generator and R is the total resistance into which the capacitor discharges, which can have several components. Figure 2 shows a general circuit diagram of an exponential decay generator.

The power supply slowly charges the capacitor to the desired voltage and does not play a role during the discharge. The internal resistance R_i of the capacitor is on the order of 0.5–1 Ω for electrolytic capacitors and, in normal operation, is much smaller than any other resistance in the circuit and can therefore be neglected. The resistor R_L is installed in some instruments to limit the current in the circuit, especially in case of an arc in the chamber, which would result in high currents because the chamber

Instrumentation

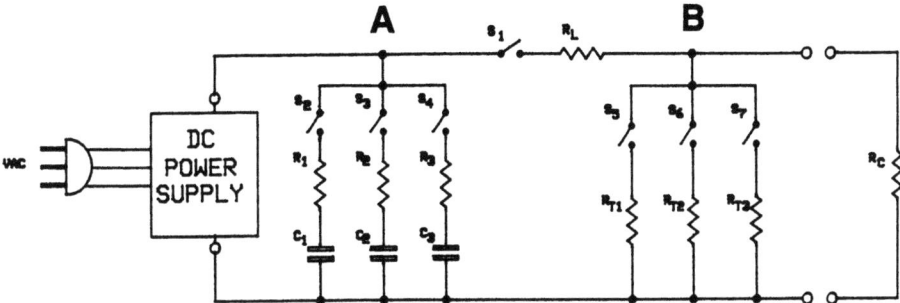

Fig. 2. General circuit diagram of an exponential decay wave form generator. $C_{1,2,3}$ are the energy storage capacitors, which have an internal resistance $R_{1,2,3}$. They can be added to the circuit by switches $S_{2,3,4}$ in order to vary the total capacitance. Closing switch S_1 allows the charged capacitors to discharge to the output and into the chamber, represented by the resistor R_c. R_L is a discharge current-limiting resistor, which is needed in some designs. R_{T1}, R_{T2}, and R_{T3} are timing resistors, which can be added to the circuit by switches $S_{5,6,7}$. If the output voltage is measured at (**A**), instead of (**B**), incorrect readings of the actual voltage on the chamber will result.

resistance drops to very low values during an arc. The size of this resistor is determined by the maximum current capability of the switch. As a result of the presence of R_L, the voltage at the chamber is reduced by the voltage drop across R_L, which can be substantial. Furthermore, some instruments measure the peak discharge voltage at the point A instead of directly across the chamber at point B, resulting in incorrect readings. Use of instruments that do not have a built-in current-limiting resistor provides advantages. One needs to be aware that without a current-limiting resistor, arcs appear more violent because of higher current flow. However, if the instrument and chamber stand are designed correctly, this should be of no consequence. It should be noted here that arcing in the chamber occurs mostly at high field strengths (above 10 kV/cm) and is a statistical effect.

The resistance R_t is a timing resistor that, typically, can be selected to adjust the pulse length. Maintaining a low value, relative to the chamber resistance R_c, serves the function of determining pulse length. Often, the size of the capacitance can be changed by connecting one or more capacitors in parallel. Since the time constant is determined by the product of resistance and capacitance, either variable can be used to adjust it. Keeping the resistance as low as possible, well below the chamber resis-

Table 2
Comparison of Electroporation Generators
with Built-in Power Supply and Multiple Pulse Length

Manufacturer of electroporation system	IBI	Bio-Rad	BRL	BTX
Model number	geneZAPPER™ 450/2500	Gene Pulser®	Cell-Porator™	ECM® 600
# of Unit components required	2	3	2	1
Voltage range	50–2500	50–2500	0–2500	50–2500
# of Pulse length settings	34	8	8	126
Maximum field strength (kV/cm)	12.5	25	16.6	40
Maximum current, A	130	120	40	>1000
Monitoring	Partial	Partial	Partial	Full
Actual voltage	No	No	Yes	Yes
Actual pulse length	Yes	Yes	No	Yes
Safety design	Operator shock proof	Safety interlock	Safety interlock	Operator shock, arc, and short circuit proof
Data base	No	No	No	Yes
Warranty	1 yr	1 yr	1 yr	2 yr

tance, is generally desirable. Sometimes the chamber resistance is too low for the timing resistors to be effective. In this case, the chamber resistance itself will determine the pulse length, which then can be adjusted only by varying the capacitance.

For the characterization of the pulse into the chamber, only two parameters need to be known: the peak voltage and the $1/e$ pulse length. It is convenient to use a generator with a built-in measuring circuit that measures the pulse parameters at the output of the instrument (Point B in Fig. 2). Table 2 shows a comparison of the main features of commercially available exponential discharge generators with built-in power supply, multiple pulse length capability, and at least some monitoring.

If only a limited number of applications are planned, such as *E. coli* transformation, a generator with a fixed pulse length will be sufficient. This simplifies the generator design and reduces costs. To reduce costs

Table 3
Exponential Decay Generator Options and Costs

Fixed time constant t	Fixed t	Variable t
No power supply (PS)	With PS	With PS and monitoring
~$1000	$1000–2000	$4000–5000

even further, it is also possible to use an external electrophoresis power supply and eliminate a built-in supply. Table 3 shows generators available with increasing flexibility and cost options. The maximum voltage is typically 2500 V. The time constant for fixed pulse length is typically 5 ms.

4.2.2. Square-Wave Generators

Square-wave pulses appear to have advantages for certain applications, such as transfection of mammalian cell lines and plant protoplasts, though no generalization can be made. Each cell line needs to be individually investigated to determine whether use of square-wave pulses would be advantageous. In general, square waves do not appear to result in higher transformation yields for bacteria, although there are some protocols that give good results (6; Xing Xin, Texas Heart Institute, personal communication). Square waves are used almost exclusively for in vivo applications of electroporation, such as electrochemotherapy, where drugs are electroporated into tumor cells. These generators are more difficult to build because the square-wave pulse is produced by a partial discharge of a large capacitor, which requires the interruption of high currents against high voltages. In the past, their costs were higher than exponential discharge generators, and the range of parameters was more limited. However, recent advances in solid-state switching technology have lowered costs. A square-wave generator is now available that can deliver up to 3000 V into a 20-Ω load at costs comparable to exponential discharge units (see Table 1).

4.3. Generators for Electrofusion

If nonelectrical means of cell–cell contact are used, any electroporation generator can also be used for electrofusion. If it is desirable to induce cell–cell contact by dielectrophoresis, the generator needs to produce an alternating wave form (ac) over a longer period of time, typically seconds, before the fusion pulse is applied.

The optimal frequency appears to be around 1 MHz (7). Above and below this frequency, the viability of mammalian cells, at least, appears

Table 4
Mouse Egg Fusion with Different Wave Forms and Chambers

AC wave form and chamber type	# of Eggs	% Fusion	% Developed	Stability of development
Nonsinusoidal, wire chamber	20	100	75	Stable
Sinusoidal, rectangular bar chamber	24	87.5	52.4	Less stable

to suffer. Nonionic fusion media are desirable to reduce the generation of heat and turbulence. A pure sinusoidal wave form is not necessary and, possibly, not even advantageous. It is, however, important that there is no net dc component in the wave form. Higher harmonics in the wave form appear to produce better fusion results. Table 4 compares results obtained with different wave forms and chambers (8).

Commercial fusion generators are available (Table 1) that allow the sequential application of ac wave forms and fusion pulses, which are generally of the square-wave type.

4.4. Generators with Other Wave Forms

Researchers have experimented with wave forms other than exponential and square. It is apparent that for some applications, special wave forms have certain advantages. Bursts of radio frequency electric fields (a few 100 kHz) appear to be more benign to cells and might be advantageous when fusing cells of widely different sizes (9,10). However, such generators are not presently available commercially, and are difficult and expensive to build with high-power levels.

5. Chambers

There are many choices in chambers for ECM. In general, chambers need to create the required field strength from the voltage delivered to the electrodes by the generator; they need to contain the appropriate volume, and need to be sterilized or sterilizable, easily filled, emptied, and if reused, easily cleaned. Table 5 gives the main trade-off parameters in the selection of chambers. The following describes only the more frequently used chambers in the field.

Table 5
Chamber Trade-off Parameters

Small volume	Large volume
Disposable	Reusable
Homogeneous field	Inhomogeneous field
Visualization of cells	Cells obscured
Batch process	Flowthrough
Aluminum electrodes	Stainless steel or noble material
Presterilized	Sterilizable
Small gap	Large gap

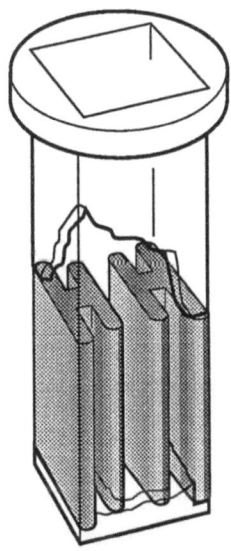

Fig. 3. Disposable electroporation cuvet with molded-in aluminum electrodes.

5.1. Small-Volume Chambers

Disposable, presterilized cuvets with molded-in aluminum electrodes (Fig. 3) are most frequently used for electroporation. They are available with different gap sizes, typically 1 mm (for bacteria), and 2 and 4 mm (for mammalian cells and plant protoplasts). The electric field in these cuvets is quite homogeneous. Some workers clean and reuse cuvets to reduce costs.

Reusable, parallel plate electrode assemblies (Fig. 4) that fit into spectrophotometer cuvets are available. They are also available in different gap sizes.

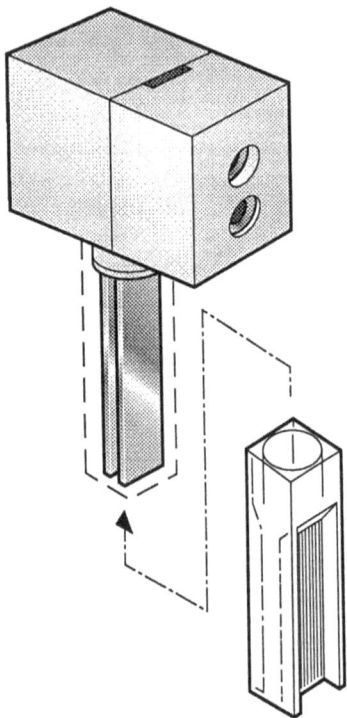

Fig. 4. Reusable parallel plate electrode assembly to fit into spectrophotometer cuvets.

Note that chemical cleaning or even autoclaving might not remove cell debris, transformant, and medium breakdown products that might have been deposited onto the electrodes during a pulse. Only a good mechanical cleaning will remove the debris.

Electrodes on microslides are used to visualize the fusion process under a microscope. Parallel wires (Fig. 5), separated by 1 mm or less, produce divergent fields that favor dielectrophoretic pearl chain formations of cells. For small gaps (<1 mm), a meander-type electrode configuration (Fig. 6) allows visualization of the fusion process. Electrodes with square bars (Fig. 7), which provide a more homogeneous electric field, can also be mounted on microslides for visualization of embryo manipulation.

5.2. Large-Volume Chambers

Intermediate-size chambers with a volume of a few milliliters can be built with parallel bars (Fig. 8). The electrodes can be flat to create

Instrumentation

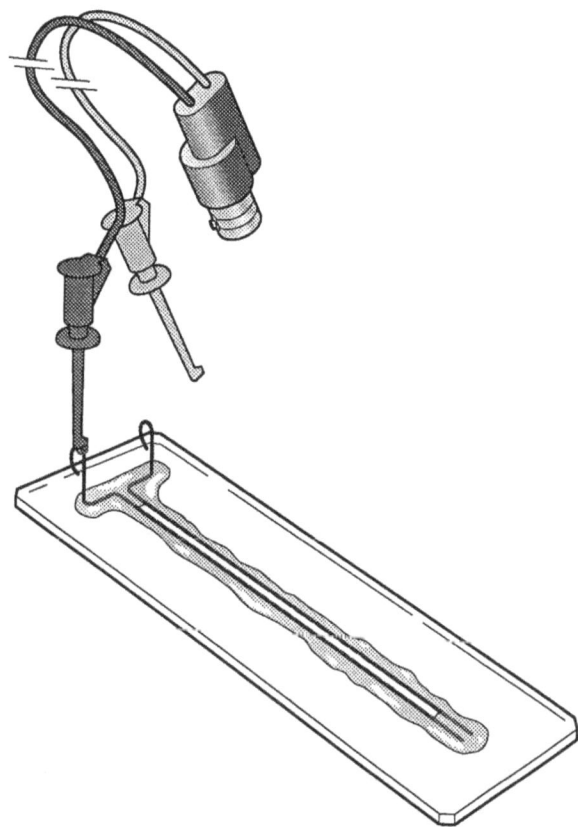

Fig. 5. Parallel wire electrodes mounted on a microslide for visual observation of the electrofusion process. These electrodes create an inhomogeneous electric field, which is preferable for dielectrophoresis.

homogeneous fields or they can have grooves to create divergent fields for fusion. A convenient implementation of a large-volume chamber with a volume up to 50 mL is an array of parallel plate electrodes fitted into a plastic Petri dish (Fig. 9). The gap between the electrodes can be 2 mm for mammalian cells or 10 mm for embryo and fish egg electroporation. Generally, such large volumes need a high resistivity medium because the chamber resistance with saline solution, such as PBS, would be very low. Partial filling of the chamber will reduce the resistance proportionally. As an example, 10 mL of PBS in a 10-cm diameter Petri dish with 2-mm spaced electrodes resulted in a resistance of 0.4 Ω. Some generators can generate sufficient voltage to transform mammalian cells even

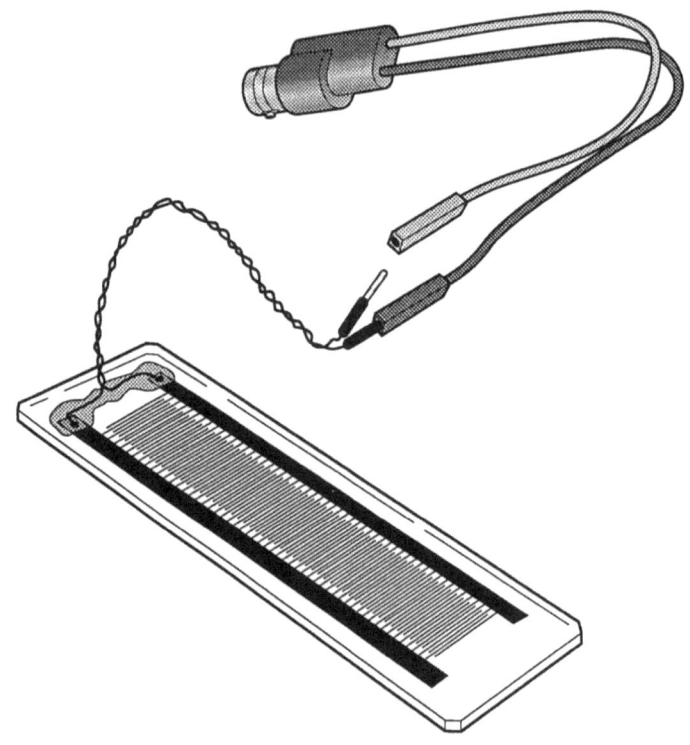

Fig. 6. Meander-type chamber for visual observation of fusion.

with PBS. The parallel plate electrode configuration in a Petri dish is also very useful for electroporation of adherent cells, if the electrodes are situated so they touch the Petri dish bottom. Instead of parallel plates, an array of concentric electrodes can also be used to create a large-volume ECM chamber in a Petri dish *(11)*.

5.3. Small-Volume Flowthrough Chambers for the ECM of Large Volumes

If it is required to transform large volumes (above 50 mL), it is economical to pulse the generator repetitively in synchrony with a pump that pushes the medium with the cells and transformants through a relatively small chamber. The repetition rate and pumping speed can be arranged so that every volume element receives one or, if desired, multiple pulses. Care needs to be taken in the design of the flowthrough chamber to minimize dead volume. Repetitive pulse generators are com-

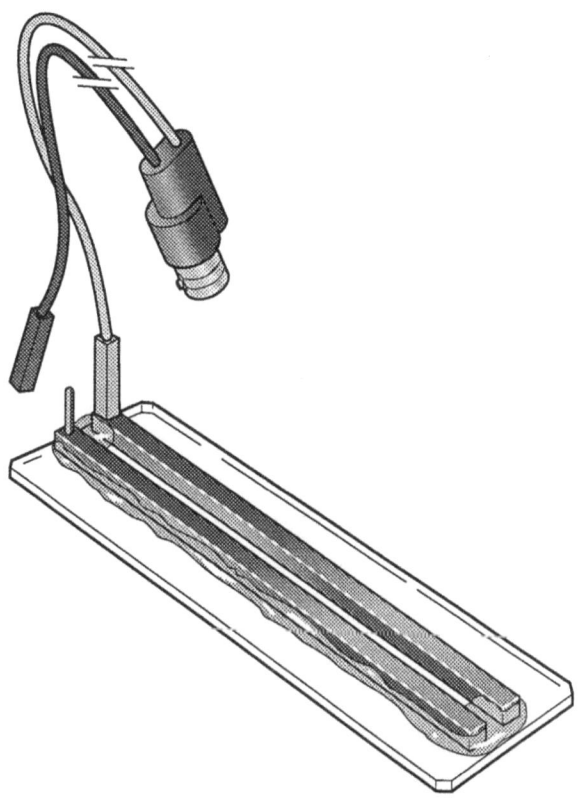

Fig. 7. Rectangular electrodes mounted on a microslide for visual observation, generating homogeneous electric fields.

mercially available for exponential decay, as well as for square-wave form output.

5.4. Chamber Material

Despite an oxide layer present on the surface, aluminum (Al) electrodes appear to give satisfactory results in disposable chambers. The commercially available presterilized cuvets use embedded Al electrodes. Stainless steel (SS) is used more often for reusable chambers. SS can be mechanically cleaned more easily. Gold plating is an option for SS as well as Al, but it appears that the increase in yield does not justify the additional costs. Comparative electrofusion experiments of embryos in either SS or gold-plated chambers did not show a substantial difference in fusion yield. Over 100 fusion experiments were performed using gold-plated electrodes, with a

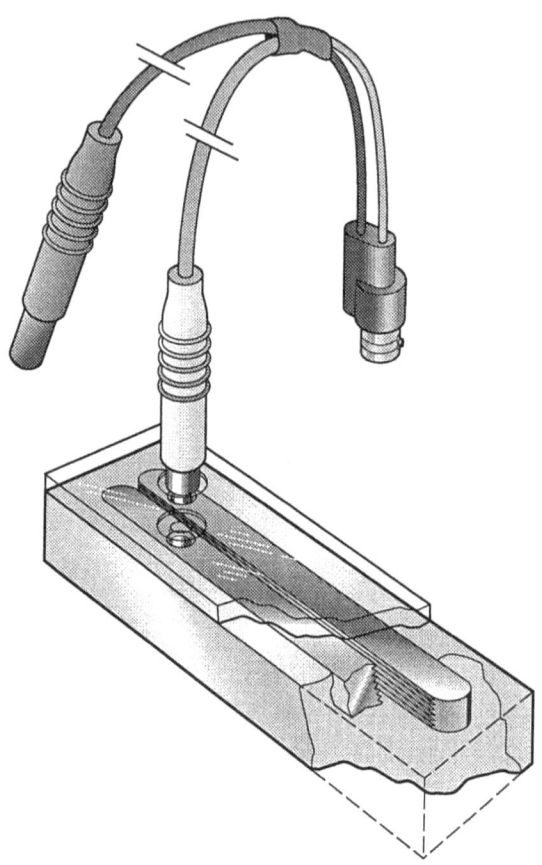

Fig. 8. Intermediate-volume chamber with parallel bar electrodes. The electrodes can be flat to create homogeneous fields or have grooves to create inhomogeneous fields.

fusion yield of >90%; with care, similar results can be achieved with SS electrodes (James M. Robl, U. of Massachusetts, personal communication). A comparison of plant protoplast fusion yields using a large-volume parallel plate chamber made of SS or a gold-plated concentric ring chamber (both for Petri dishes) showed a consistently higher yield for the gold-plated chamber *(10)*. Table 6 shows a survey of commercially available chambers.

6. Measuring ECM Parameters

By following an established protocol, it is generally not necessary to measure the ECM parameters, especially if the same types of generator

Instrumentation

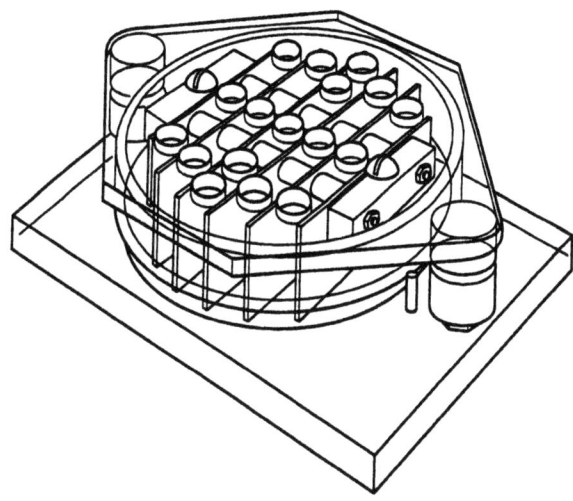

Fig. 9. Petri dish electrodes for large-volume electroporation of mammalian cells in suspension or adherence, and electroporation of fish eggs.

and chamber are used (different generators might give different output voltages for the same charging voltage setting). However, when pursuing new applications with an instrument that does not have built-in monitoring, it is desirable to measure the voltage and pulse actually delivered to the chamber to allow accurate reporting and reproducible performance. A commercial instrument is available to monitor ECM parameters specifically, display them, and print them out (Table 1). A measuring system can be assembled consisting of a digital oscilloscope (bandwidth should be 100 MHz), and a high-voltage probe attenuating the voltage signal 1:1000, with a voltage range up to 3 kV. Commercial generators and chambers are typically constructed so that at no place in the circuit is the high-voltage potential easily accessible. Therefore, adapters need to be placed in line between the generator and the chamber so that a voltage probe can be connected. Before any measurements are performed, the grounding situation must be understood and verified with the manufacturer. Sometimes neither of the two outputs of the generator is at the ground potential of the oscilloscope (which is normally tied to the power line ground), depending on the design of the discharge circuit. It is still possible to perform measurements in this case by disconnecting the oscilloscope from the power line earth/ground, either by inserting an iso-

Table 6
Comparison of Commercially Available Chambers

Chamber types	Manufacturers				
	IBI	Invitrogen	Bio-Rad	BRL	BTX
Cuvets: disposable	3	3	3	3	3
Cuvets: reusable, homogeneous field	3				2
Cuvets: reusable, divergent field	1				1
Microslides homogeneous field					2
Microslides divergent field					2
Meander					1
Flat electrode divergent field					1
Petri dish electrodes					2
96-Well plate electrode					1
Flowthrough					2
Sandwich					2
In vivo electrodes					3

*The numbers indicate available variations, typically in the gap size.

lation transformer between the line and the oscilloscope or by disconnecting the oscilloscope ground lead to the power line with an insulating plug. These plugs can be recognized by regular three prongs on one side and only two receptacles on the other side with the earth/ground wire separate, which should not be connected. During the pulse, the chassis of the oscilloscope will attain a potential difference to the laboratory ground and should not be touched. These kinds of measurements obviously are hazardous, and should be performed with extreme care and only by trained personnel. If it is desirable to measure both the voltage output and the current, a convenient, contactless way is to route one lead to the chamber through a current transformer. Several manufacturers provide these elements (e.g., Pearson Electronics Inc. [Palo Alto, CA], Current Transformer Model Nr. 411). The most important specifications to verify the usefulness of a current transformer are the peak current capability (e.g., 5000 A for the 411) and the limit of the product of current × pulse length (e.g., 0.2 A · s for the 411) to avoid saturation of the current transformer before the pulse has passed completely. Measuring current and voltage allows one to determine the chamber resistance as a function of time from Ohm's law, $R = V/I$. Through the geometry of the chamber, the specific resistivity of the cell/medium suspension as a function of time can then be determined, which might be of interest for biophysical investigations of the ECM process, because lysis of cells results in an increase of the medium conductivity.

Acknowledgments

I want to thank my colleagues for helpful comments, and especially Linda Hull for editing the manuscript.

References

1. Tsong, T. Y. and Tomita, M. (1993) Selective B lymphocyte-myeloma cell fusion. *Methods in Enzymol.* **220,** 238–246.
2. Pohl, H. A. (1978) *Dielectrophoresis.* Cambridge University Press, London.
3. Meilhoc, E., Masson, J.-M., and Teissie, J. (1990) High efficiency transformation of intact yeast cells by electric field pulses. *Biotechnology* **8(3),** 223–227.
4. Takahashi, M., Furukawa, T., Saito, H., Aoki, A., Koike, T., Moriyama, Y., Shibata, A. (1991) Gene transfer into human leukemia cell lines by electroporation: experiments with exponentially decaying and square wave pulse. *Leukemia Res.* **15(6),** 507–513.
5. Saunders, J., Rhodes, S. C., and Kaper, J. (1989) Effects of electroporation profiles on the incorporation of viral RNA into tobacco protoplasts. *Biotechniques* **7(10),** 1124–1131.
6. Xie, T. and Tsong, T. (1992) Study of mechanisms of electric field-induced DNA transfection III, Electric parameters and other conditions for effective transfection. *Biophys. J.* **63,** 28–34.
7. Hofmann, G. H. (1989) Cells in electric fields—physical and practical electronic aspects of electro cell fusion and electroporation, in *Electroporation and Electrofusion in Cell Biology* (Neumann, E., Sowers, A., and Jordan, C., eds.), Plenum, New York, pp. 389–407.
8. Nagata, K. and Imai, H. (1992) The difference of electro fusion rate for the pronuclear transplantation of mouse eggs between three different electro generators. The 7th Eastern Japan Animal Nuclear Transplantation Research Conference.
9. Chang, D. C. (1989) Cell fusion and cell poration by pulsed radio-frequency electric fields, in *Electroporation and Electrofusion in Cell Biology* (Neumann, E., Sowers, A., and Jordan, C., eds.), Plenum, New York, pp. 215–227.
10. Tekle, E., Astumian, R. D., and Chock, P. B. (1991) Electroporation by using bipolar oscillating electric field: an improved method for DNA transfection of NIH 3T3 cells. *PNAS* **88,** 4230–4234.
11. Motumura, T., Akihama, T., Hidaka, T., and Omura, M. (1993) Conditions of protoplast isolation and electrical fusion among citrus and its wild relatives, in *Techniques on Gene Diagnosis and Breeding in Fruit Trees* (Hayashi, T., et al., eds.), FTRS, Japan, pp. 153–164.

PART II

ELECTROPORATION PROTOCOLS

CHAPTER 4

The Introduction of Proteins into Mammalian Cells by Electroporation

William F. Morgan and Joseph P. Day

1. Introduction

Proteins can be introduced into cells in several ways, including permeabilization by Sendai virus, trypsinization, osmotic shock, microinjection, electroporation, or after transfection of cells with expression vectors containing the gene(s) of interest. Unlike most of these other methods, electroporation is a simple, relatively inexpensive, and efficient method for rapidly introducing DNA, small molecules, and proteins into bacterial or mammalian cells *(1–10)*. Electroporation has the advantage over other techniques in that many cells can be treated simultaneously and one or more compounds or macromolecules, such as proteins, can be introduced into cells at the same time (*see* Note 1).

In this chapter, we describe our experience electroporating restriction endonucleases into cells *(4,5,10,11)*. The purpose of these studies was to determine the cellular responses to induced DNA damage, in particular, how DNA double-strand breaks can lead to chromosomal rearrangements. We used type II restriction endonucleases, which recognize and cleave specific DNA sequences, to produce DNA double-strand breaks with defined end structures. We describe electroporation conditions that optimize uptake of a restriction enzyme into the cell nucleus, as measured by chromosome aberration formation *(5,10–15)*. These methods are optimal for Chinese hamster ovary cells, but the same techniques have been used to electropermeabilize human *(1)* and mouse *(16)* cells with various degrees of success. We used the rearrangement of metaphase chromosomes as an end point of restriction endonuclease cleavage; therefore, we also describe our method for chromosomal analysis.

It is our experience that nuclear targeting requires electroporation conditions different from those required for cell permeabilization. For example, under the electroporation conditions in which β-galactosidase could enter a cell, no restriction enzyme-induced chromosome aberrations could be observed. This suggests that the cell membrane had been electropermeabilized, but that the enzyme was not reaching the nucleus. Possible reasons for this have been discussed previously *(9)*.

Although electroporation can be used effectively to introduce biologically active molecules into the nucleus of a cell, an inherent problem in our studies of the cellular consequences of restriction enzyme-induced DNA damage has been variability between experiments performed under identical conditions. For example, electroporation of the same amount of enzyme in duplicate experiments can result in significant differences in chromosome aberration yields *(5,15)*. We do not know the precise reason for the observed variation, but presumably it has to do with the amount of enzyme entering the nucleus. Although the concentration of enzyme in the extracellular medium can be controlled, the amount that actually enters the cell nucleus cannot be easily measured.

We have observed that electroporation results in cell-cycle delays *(5)*. The length of the delay is cell-type-dependent and is largely owing to a prolonged S phase, usually resulting in about a twofold increase in the length of a cell cycle *(5)*.

Recent studies of the basis of electroporation *(8,17,18)* may be helpful in determining optimal electroporation conditions for the introduction of a particular protein (*see* Note 2). The creation of an actual pore by electroporation has not been demonstrated, and evidence suggests that pores do not form *(17,18)*. Nevertheless, there are at least two distinct processes by which molecules translocate across the electropermeabilized membrane. Small molecules, regardless of charge, pass relatively freely through the permeabilized membrane for a time much longer than the duration of the electric field pulse. This time corresponds to the resealing of the membrane and is a temperature-dependent process. Electropermeabilization itself is not temperature-dependent and can be performed with similar efficiency from 37 to near 0°C *(19)*. Chinese hamster ovary cells maintained on ice remain permeable for hours without affecting cell viability, thus allowing small molecules to traverse the membrane freely (*see* Note 3). As the temperature rises, cell membranes reseal more quickly *(20)*. Similarly, DNA can enter cells for many minutes after electroporation, but probably not through the same mechanism by which small molecules enter cells.

In our experience, the electroporation conditions described here for the introduction of restriction enzymes into cells do not result in successful transfection of DNA into the same cells. Generally, DNA is introduced more efficiently after a long pulse of fixed voltage, or after repeated short pulses, near the minimum voltage that causes permeabilization *(21)*. Higher voltages result in cell death. It appears that there is a multistep process of DNA translocation in which DNA is first reversibly taken up by the membrane *(3)*. The low voltage appears to slow the off-rate from this intermediate state. This result also explains the apparent superiority of "exponential" vs "square-wave" pulsing *(2)*. The exponential pulse consists of a high-voltage initial spike followed by a much longer decay at lower voltage.

It is not clear whether proteins pass freely through the permeabilized membrane like small molecules, are translocated through an intermediate state like DNA, or have a unique mode of translocation possibly depending on the particular structural features of each protein molecule. Signal sequences guide the passage of a protein from the interior of the cell to the extracellular medium, but the cellular machinery and processes involved in excretion do not necessarily work in reverse during or after electroporation. On a cautionary note, the chaperones that help proteins fold, pass through membranes, and refold could renature proteins introduced by electroporation. Chaperones may be invaluable in helping to maintain the maximum activity of any protein after electroporation, but could cause problems if a denatured protein control is introduced.

2. Materials

1. Phosphate-buffered sucrose (PBS): 7 mM KH_2PO_4, pH 7.4, 1 mM $MgCl_2$, 272 mM sucrose. Store at 4°C; replace after 2 mo. Use commercially available stock solutions of $MgCl_2$ (*see* Note 4).

2. HEPES-buffered saline (HeBS): 21 mM HEPES, pH 7.05, 137 mM NaCl, 5 mM KCl, 0.7 mM Na$_2$HPO$_4$, 6 mM glucose. Store at 4°C; replace after 2 mo (see Note 5).
3. Culture medium: Chinese hamster ovary cells are cultured as monolayers in McCoy's 5A medium supplemented with 10% fetal bovine serum, 2 mM L-glutamine, penicillin (50 U/mL), and streptomycin (50 µg/mL). Human cells are cultured in RPMI-1640 medium with 10% fetal bovine serum, 2 mM L-glutamine, penicillin (50 U/mL), and streptomycin (50 µg/mL). Cells are cultured at 37°C in an atmosphere of 5% CO$_2$ in air.
4. Saline A: 8% NaCl, 0.4% KCl in H$_2$O.
5. Trypsin/EDTA: Stock solutions of 0.5 g/L trypsin, 0.2 g/L EDTA, 1.0 g/L glucose, and 0.58 g/L NaHCO$_3$ in saline A are stored at –20°C.

3. Methods

3.1. Electroporation with the Bio-Rad (Richmond, CA) Gene Pulser

1. Cells are cultured as described in Section 2.
2. Treat the required number of flasks containing monolayers of exponentially growing cells with 5 mL of trypsin/EDTA until the cells can be brought into suspension. If working with suspension cultures, proceed to step 3.
3. Remove the trypsin/EDTA by pouring off the supernatant after centrifugation at approx 300g.
4. Wash the cell pellet, either in serum-free medium or in PBS, at ambient temperature or at 37°C. This wash step is important because trypsin and EDTA can inhibit or destroy restriction enzymes or other proteins.
5. Recentrifuge the cells and resuspend them in PBS at ambient temperature to 1–10 × 10^6 cells/mL (see Note 6).
6. Put 800 µL of the cell suspension into a 0.4-cm electroporation cuvet, and add restriction enzyme. Homogenize by gently inverting the cuvet ten times.
7. Electroporate the cells as soon as possible with 0.3 kV and 125-µF capacitance. Observe the time constant, which should be between 12 and 22 ms under these conditions.
8. Remove the electroporated cells gently from the electroporation cuvet by using a clean, sterile Pasteur pipet. To avoid cell clumping, do not draw bubbles up into the cell suspension.
9. Add the suspension to a flask containing prewarmed medium (37°C).
10. Replace the medium with fresh medium after 1.5–2 h to eliminate any unattached cells and the PBS (11,12).

Introduction of Proteins into Cells

3.2. Electroporation with the Bethesda Research Laboratories (BRL, Gaithersburg, MD) Cell Porator

1. Treat monolayers of exponentially growing cells with trypsin/EDTA until the cells can be brought into suspension.
2. Remove the trypsin/EDTA by pouring off the supernatant after centrifugation at approx 300g.
3. Wash the cell pellet, either in serum-free medium or in HeBS, at 4°C.
4. Recentrifuge the cells and resuspend them in HeBS at 4°C to 2×10^6 cells/mL.
5. Put 800 µL of the cell suspension in a disposable electroporation cuvet, and add restriction enzyme. Homogenize by gently inverting the cuvet ten times.
6. Electroporate the cells under the following conditions: field strength 650 V/cm (indicated voltage of 260 divided by 0.4-cm electrode gap), capacitance 1600 µF, and the electroporator set at low resistance.
7. Carefully remove the cell suspension from the cuvet, and mix with 10 mL of medium in a tissue-culture dish.
8. Incubate cells at their optimal growth temperature for 1 h, and then disperse cell clumps by pipeting up and down.
9. Incubate cells for 1 h, and then replace the medium with fresh medium *(5,10)*.

3.3. Metaphase Cell Chromosome Aberration Assay

1. At various times after electroporation, add colcemid to exponentially growing cells at a final concentration of $2 \times 10^{-7}M$ (*see* Note 7).
2. Collect metaphase cells after 1–3 h by gently shaking the flask.
3. Centrifuge the metaphase cells at approx 300g, and resuspend them in hypotonic (0.075M) KCl prewarmed to 37°C.
4. Incubate the cells at 37°C for 5–15 min, then recentrifuge, and resuspend them in methanol for 2 min.
5. Centrifuge the cells, and resuspend them in methanol-acetic acid (3:1) for 5 min.
6. Centrifuge the cells, and resuspend them in a small volume of methanol-acetic acid.
7. Using a pipet, drop the metaphase cells onto clean glass microscope slides and allow them to air-dry.
8. Stain the slides with 4% Giemsa for 2–5 min, dry, and permanently mount them with a coverslip for microscopic analysis of cytogenetic abnormalities.

Table 1
Chromosomal Aberration Yields, Plating Efficiency, and Cell Survival in CHO Cells After Exposure to *Pvu*II (Adapted from Yates et al. [11])

PvuII, U	Aberrant cells, %[a]	Plating efficiency[b]	Cell survival, %[c]
0[d]	8.8	0.850	100.0
10	37.2	0.280	33.0
25	60.9	0.110	12.5
50	69.1	0.045	5.0
100	81.9	0.030	3.5
250	89.4	0.020	2.5
500	84.2	0.025	3.0
750	86.1	0.030	3.5
1000	84.7	0.020	2.5

[a] Number of cells showing ≥ 1 chromosomal rearrangement. Each point represents the mean of two independent experiments. One hundred metaphase spreads were analyzed for each experiment.

[b] Mean number of colonies per plate (three plates/point) divided by 200, the number of cells plated.

[c] Surviving fraction expressed as a percentage of control (electroporation alone). Surviving fraction of the control with this CHO cell line is consistently between 0.7 and 0.9.

[d] Electroporation of the storage buffer alone (i.e., the *Pvu*II shipping buffer without *Pvu*II).

4. Notes

1. Electroporation can be used to introduce a variety of proteins into cells, including monoclonal antibodies (MAb) *(22,23)*, functional enzymes *(24)*, other endonucleases in addition to restriction enzymes *(25)*, and a host of other proteins and protein products *(26–28)*.
2. Both the Bio-Rad Gene Pulser and the BRL Cell Porator are effective in the introduction of enzymes into the cell, but the most effective conditions for the BRL machine also result in relatively high levels of cell death *(10)*. Consequently, we prefer to use the Bio-Rad Gene Pulser, because we achieve efficient uptake of the restriction enzyme with very little electroporation-induced cell death (Table 1).
3. In our experience, incubating cells on ice for 5 min immediately following electroporation is important when the BRL electroporation apparatus is used, but incubation on ice after electroporation has no noticeable effect when the Bio-Rad Gene Pulser is used.
4. For optimal efficiency of restriction enzyme uptake after electroporation with the Bio-Rad Gene Pulser, we use PBS as the electroporation medium.

When preparing this solution, it is best to use a premade $MgCl_2$ stock solution (e.g., $1M$ concentration; Sigma, St. Louis, MO) for reasons we do not understand.
5. For optimal efficiency of restriction enzyme uptake after electroporation with the BRL Cell Porator, we use HeBS as the electroporation medium.
6. We generally use 2×10^6 cells/mL, although this technique works efficiently with up to 10^7 cells/mL.
7. The metaphase cell chromosome aberration assay is included as one method available for confirming restriction enzyme action in the cell nucleus. It is not necessary to include this step if electroporation of the protein of interest does not result in DNA damage.

Acknowledgments

We thank J. Phillips and J. King for helpful discussions and M. McKenney for editing the manuscript. This work was supported by the Office of Health and Environmental Research, US Department of Energy, contract DE-AC03-76-SF01012, by the National Institutes of Health National Research Service Award 5-T32-ES07016 from the National Institute of Environmental Health Sciences (J. P. D.), and by the Radiology Research Foundation, University of California, San Francisco.

References

1. Abella Columna, E., Giaccia, A. J., Evans, J. W., Yates, B. L., and Morgan, W. F. (1993) Analysis of restriction enzyme-induced chromosomal aberrations by fluorescence in situ hybridization. *Environ. Mol. Mutagen.* **22,** 26–33.
2. Andreason, G. L. and Evans, G. A. (1988) Introduction and expression of DNA molecules in eukaryotic cells by electroporation (published erratum appears in *BioTechniques* **6,** 853). *BioTechniques* **6,** 650–660.
3. Eynard, N., Sixou, S., Duran, N., and Teissie, J. (1992) Fast kinetics studies of Escherichia coli electrotransformation. *Eur. J. Biochem.* **209,** 431–436.
4. Morgan, W. F. and Winegar, R. A. (1990) The use of restriction endonucleases to study the mechanisms of chromosome damage, in *Chromosomal Aberrations: Basic and Applied Aspects* (Obe, G. and Natarajan, A. T., eds.), Springer-Verlag, Berlin, pp. 70–78.
5. Morgan, W. F., Ager, D., Chung, H. W., Ortiz, T., Phillips, J. W., and Winegar, R. A. (1990) The cytogenetic effects of restriction endonucleases following their introduction into cells by electroporation, in *Mutation and the Environment, Part B: Metabolism, Testing Methods, and Chromosomes* (Mendelsohn, M. L. and Albertini, R. J., eds.), Wiley-Liss, New York, pp. 355–361.
6. Shigekawa, K. and Dower, W. J. (1988) Electroporation of eukaryotes and prokaryotes: a general approach to the introduction of macromolecules into cells. *BioTechniques* **6,** 742–751.

7. Smith, M., Jessee, J., Landers, T., and Jordan, J. (1990) High efficiency bacterial electroporation: 1×10^{10} E. coli transformants/µg. Focus **12,** 38–40.
8. Teissie, J. and Rols, M. P. (1993) An experimental evaluation of the critical potential difference inducing cell membrane electropermeabilization. Biophys. J. **65,** 409–413.
9. Winegar, R. A. and Lutze, L. H. (1990) Introduction of biologically active proteins into viable cells by electroporation. Focus **12,** 34–37.
10. Winegar, R. A., Phillips, J. W., Youngblom, J. H., and Morgan, W. F. (1989) Cell electroporation is a highly efficient method for introducing restriction endonucleases into cells. Mutat. Res. **225,** 49–53.
11. Yates, B. L., Valcarcel, E. R., and Morgan, W. F. (1992) Restriction enzyme-induced DNA double-strand breaks as a model system for cellular responses to DNA damage. Int. J. Radiat. Oncol. Biol. Phys. **23,** 993–998.
12. Morgan, W. F., Yates, B. L., Rufer, J. T., Abella Columna, E., Valcarcel, E. R., and Phillips, J. W. (1991) Chromosomal aberration induction in CHO cells by combined exposure to restriction enzymes and X-rays. Int. J. Radiat. Biol. **60,** 627–634.
13. Ager, D. D., Phillips, J. W., Abella Columna, E., Winegar, R. A., and Morgan, W. F. (1991) Analysis of restriction enzyme-induced DNA double-strand breaks in Chinese hamster ovary cells by pulsed-field gel electrophoresis: implications for chromosome damage. Radiat. Res. **128,** 150–156.
14. Morgan, W. F., Phillips, J. W., Chung, H.-W., Ager, D. D., and Winegar, R. A. (1990) The use of restriction endonucleases to mimic the cytogenetic damage induced by ionizing radiations, in *Ionizing Radiation Damage to DNA: Molecular Aspects* (Wallace, S. S. and Painter, R. B., eds.), Wiley-Liss, New York, pp. 191–199.
15. Yates, B. L. and Morgan, W. F. (1993) Nonhomologous DNA end rejoining in chromosomal aberration formation. Mutat. Res. **285,** 53–60.
16. Chang, C., Biedermann, K. A., Mezzina, M., and Brown, J. M. (1993) Characterization of the DNA double strand break repair defect in scid mice. Cancer Res. **53,** 1244–1248.
17. Genco, I., Gliozzi, A., Relini, A., Robello, M., and Scalas, E. (1993) Electroporation in symmetric and asymmetric membranes. Biochim. Biophys. Acta **1149,** 10–18.
18. Hibino, M., Itoh, H., and Kinosita, K., Jr. (1993) Time courses of cell electroporation as revealed by submicrosecond imaging of transmembrane potential. Biophys. J. **64,** 1789–1800.
19. Lopez, A., Rols, M. P., and Teissie, J. (1988) 31P NMR analysis of membrane phospholipid organization in viable, reversibly electropermeabilized Chinese hamster ovary cells. Biochemistry **27,** 1222–1228.
20. Rols, M. P., Dahhou, F., Mishra, K. P., and Teissie, J. (1990) Control of electric field induced cell membrane permeabilization by membrane order. Biochemistry **29,** 2960–2966.
21. Rols, M. P. and Teissie, J. (1990) Electropermeabilization of mammalian cells. Quantitative analysis of the phenomenon. Biophys. J. **58,** 1089–1098.
22. Chakrabarti, R., Wylie, D. E., and Schuster, S. M. (1989) Transfer of monoclonal antibodies into mammalian cells by electroporation. J. Biol. Chem. **264,** 15,494–15,500.

23. Berglund, D. L. and Starkey, J. R. (1989) Isolation of viable tumor cells following introduction of labelled antibody to an intracellular oncogene product using electroporation. *J. Immunol. Methods* **125,** 79–87.
24. Dagher, S. F., Conrad, S. E., Werner, E. A., and Patterson, R. J. (1992) Phenotypic conversion of TK-deficient cells following electroporation of functional TK enzyme. *Exp. Cell Res.* **198,** 36–42.
25. Tsongalis, G. J., Lambert, W. C., and Lambert, M. W. (1990) Correction of the ultraviolet light induced DNA-repair defect in xeroderma pigmentosum cells by electroporation of a normal human endonuclease. *Mutat. Res.* **244,** 257–263.
26. Lambert, H., Pankov, R., Gauthier, J., and Hancock, R. (1990) Electroporation-mediated uptake of proteins into mammalian cells. *Biochem. Cell Biol.* **68,** 729–734.
27. Graziadei, L., Burfeind, P., and Bar-Sagi, D. (1991) Introduction of unlabeled proteins into living cells by electroporation and isolation of viable protein-loaded cells using dextran-fluorescein isothiocyanate as a marker for protein uptake. *Anal. Biochem.* **194,** 198–203.
28. Negrutskii, B. S. and Deutscher, M. P. (1991) Channeling of aminoacyl-tRNA for protein synthesis in vivo. *Proc. Natl. Acad. Sci. USA* **88,** 4991–4995.

CHAPTER 5

Electroporation of Antigen-Presenting Cells for T-Cell Recognition and Cytotoxic T-Lymphocyte Priming

Weisan Chen and James McCluskey

1. Introduction

Specific cytotoxic T-lymphocytes (CTL) are important in the protection against viral and parasitic infections *(1,2)*, as well as monitoring and even eradicating tumor cells *(3–5)*. CTL normally recognize endogenously processed peptide antigens (Ag) complexed to MHC class I molecules *(6,7)*. Immunization with virus-infected antigen-presenting cells (APC) or Ag-expressing transfectants, and even APC osmotically loaded with antigens can elicit $CD8^+$ cytotoxic T-cells *(8–10)*. However, APC pulsed with native protein antigens usually do not elicit CTL, unless specialized APC are used *(11,12)*. Generally, noninfectious forms of viral antigens or other extracellular proteins ("exogenous Ag") do not efficiently enter the class I processing pathway, whereas proteins synthesized *de novo* ("endogenous Ag") have ready access.

Antigen presentation is one of the key events regulating the cellular immune system, thereby shaping the specific T-cell repertoire. Because almost all proteins begin their existence in the cytosol, the class I processing and presentation pathway is potentially able to monitor all newly synthesized proteins, including viral proteins *(8,13,14)*, tumor antigens, and autoantigens *(3,15)*. Peptide fragments of protein antigens are generated through proteolysis by proteasomes in the cytoplasm, and then imported into the endoplasmic reticulum via the peptide transporters (TAPs), where they facilitate the proper assembly and transportation of class I molecules to the cell surface *(7,16,17)*.

The mechanism by which specialized APC deliver soluble antigen into the class I pathway is not clear *(11,12,18,19)*, and these cells are not well defined. Macrophages may be one of the members of this group *(20)*. Thus, in order to sensitize conventional target cells for recognition by class I restricted CTL or to prime a naive CTL, intact antigen must be introduced into the cytoplasm either biosynthetically following viral infection *(21)* or gene transfection *(7)*, or as exogenous soluble antigen delivered to the cytoplasm by osmotic loading *(10)*, liposome encapsulation *(22–25)*, packaging in immunostimulating complexes (ISCOMS) *(26)*, or linkage to lipophilic carrier molecules *(27,28)*. In general, these methods of antigen delivery are cumbersome and require specialized techniques or customized reagents for sensitizing target cells.

However, target cells can be efficiently and easily sensitized for the class I restricted recognition of soluble antigens by electroporation. Electroporation delivers soluble Ag into live cells *(29,30)* by disrupting the lipid bilayer of the cell membrane *(31,32)*. This antigen-delivery method can also be used to prime CTL responses to soluble Ag in naive mice either in vivo or in vitro *(30,33)*. The advantages of this system are its reproducibility, economic use of Ag, which can be reused, and overall simplicity.

We have modified this approach to generating CTL using the chicken ovalbumin (OVA) system in which CTL are easily generated in H-2^b mice. The major peptide antigen recognized by OVA-specific CTL in these mice is OVA$_{257-264}$. Therefore, this peptide can be used as a positive control for sensitizing APC for recognition by OVA-specific CTL in H-2^b mice. Conditions for loading APC with soluble OVA are described here; however, the principles are similar for other soluble Ag.

2. Materials

1. Antigen and antigenic peptide: Dissolve OVA (mol wt 42.8 kDa, grade VI, Sigma, St. Louis, MO) in PBS containing 10 mM MgCl$_2$ immediately before use (*see* Note 1). The peptide octamer OVA$_{257-264}$ can be assembled on an Applied Biosystems 431A peptide synthesizer, after which it is deprotected, cleaved from PAM resin, and then HPLC-purified. Peptides are dissolved in PBS at ~500 µM as stock solutions and kept at –20°C.

2. Cell lines and APC: Theoretically, all cells with the appropriate MHC class I molecule expressed on their surface can be used as APC. In practice, different cell types have different capacitance and voltage requirements for optimal electroporation. Cell types that have been successfully loaded with Ag by electroporation include mouse spleen cells, L-cells, the mouse thymoma cell line EL-4 *(10)*, tumor cell line RMA *(34)*, human cell lines, such as Jurkat-K^b *(35)* and EBV-transformed lymphoid cell lines (unpublished), and K^b-transfected muscle cell line (K^b-5, unpublished). Mouse spleen cells and L-cell transfectants *(36)* have been studied in most detail.
3. Cell-culture medium (DME-10): DMEM with 10% fetal calf serum (FCS, ICN), 0.5 m*M* 2-mercaptoethanol (Sigma), 2 µ*M* glutamine (ICN), and 50 µg/mL of gentamicin. For CTL priming in vitro and CTL restimulation in vitro, RPMI-1640 containing the above additives (RP-10) plus 5–10 U/mL of IL-2 should be used.
4. HAT: 100 µ*M* Hypoxanthine (ICN), 0.4 µ*M* aminopterin (ICN), and 16 µ*M* thymidine (ICN).
5. G418: 0.3–0.5 mg/mL of final active concentration (Gibco, Grand Island, NY) depending on the drug sensitivity of the transfected cell line.
6. Red blood cell lysis buffer: 0.83% ammonium chloride in 17 m*M* Tris-HCl, pH 7.2 *(37)*.
7. Electroporation loading buffer: PBS containing 10 m*M* $MgCl_2$ (*see also* Note 5).
8. ^{3}H-Thymidine (Amersham Life Science, Little Chalfont, UK): 1 mCi/mL, 5 Ci/mmol.
9. Sodium [^{51}Cr] chromate (Dupont, Steverage, UK): 1 mCi/mL, 564.71 mCi/mg.

3. Methods

3.1. Antigen Loading of APC by Electroporation (Modeled on the OVA System in H-2^b Mice)

1. Remove spleens from C57BL/6J mice and disrupt the spleens gently on a metal mesh in RPMI medium. Pass the suspension through a 21-gage needle two to three times to achieve a single-cell suspension. Pellet the cells and resuspend in red blood cell lysis buffer (2–5 mL/spleen) for 2–5 min at 37°C. Pellet the cells through 2 mL of FCS, and then wash once in medium free of FCS. X-irradiate the cells with 30 Gy; alternatively, grow L-cells in DME-10, harvest in exponential phase by trypsin, wash once, and then irradiate with 60–100 Gy.
2. Resuspend the APC in 0.4 mL of electroporation loading buffer containing soluble Ag. Concentrations of OVA as low as 0.5 mg/mL (~11 µ*M*) are used to load APC for priming OVA-specific CTL (*see* Notes 2 and 3).

3. Preincubate the APC-antigen mixture at 4°C for 10 min in a test tube and then transfer into a precooled electroporation cuvet with a 0.4-cm electrode gap immediately before electroporation (*see* Note 5). The volume of the APC–Ag mixture should be 0.4–0.8 mL.
4. Set the electroporator capacitance to 250 µF and voltage to 0.45 kV. Preliminary killing curves are useful in establishing the settings, especially with different APC (*see* Note 4).
5. Electroporate the cells according to the instructions of the manufacturer of the electroporation device (*see* Notes 4 and 5). Always check the actual capacitance, voltage delivered, and the time constant after the electroporation.
6. Place the electroporated cells on ice for a further 10 min, and wash once with 10 mL of medium. At this stage, the cells are ready to be used as APC.

3.2. CTL Priming In Vivo

1. Prepare fresh spleen cells and X-irradiate them at 30 Gy as in Section 3.1. Incubate 5×10^7 cells in 0.4 mL of loading buffer containing 2–5 mg/mL of OVA (or 45–112 µM) for 10 min on ice.
2. Load the cells by electroporation as described in steps 3–6 in Section 3.1.
3. After washing the cells, reinject them into a syngeneic C57BL/6J mouse by iv injection at the tail base.
4. Ten days later, harvest spleen cells from the primed mouse and restimulate 5×10^7 cells in an upright T-25 flask by coculturing with the same number of antigen-loaded, X-irradiated splenic APC in 10 mL of RP-10 medium for 5 d (*see* Note 6).
5. Assay these responder cells for antigen-specific killing by using the ^{51}Cr-release assay (*see* Section 3.5.).

3.3. CTL Priming In Vitro

1. Follow steps 1–6 as described in Section 3.1. to prepare antigen-loaded APC.
2. Coculture 5×10^7 fresh spleen cells in 10 mL of RP-10 with the same number of antigen-loaded, X-irradiated splenic APC in an upright T-25 flask for 5–7 d.
3. Recover all the live responder cells by Ficoll-Hypaque gradient centrifugation and restimulate these cells with Ag-loaded splenic APC in 24-well plates. Thus, 10^5 responder cells are cocultured with 2.5×10^6 antigen-loaded and X-irradiated splenic cells as stimulators, and the same number of unloaded splenic cells as feeders in 2 mL of RP-10 containing 10 U/mL IL-2 for 5 d (*see* Note 6). Following restimulation, the responder cells are ready to assay for antigen-specific killing by using the ^{51}Cr-release assay (*see* Section 3.5.).

3.4. IL-2 Assay for Antigen Presentation Using T-Hybridoma Reporter Cells

1. Set up APC and electroporated APC at different cell densities (10^4–10^6/well) in flat-bottom 96-well plates (Costar, Cambridge MA). APC alone without loading antigen should be assayed as a negative control, and as a positive control, use APC pulsed with 1 μM of OVA$_{257-264}$ peptide (or equivalent antigenic peptide) washed prior to adding T-hybridoma cells.
2. Add 10^5cells/well of OVA$_{257-264}$ specific hybridoma cells, such as GA4.2, to the APC, and coculture for 24 h at 37°C in a CO_2 incubator.
3. Transfer 50 µL of supernatants into a new flat-bottom 96-well plate, and freeze for at least 30 min at –70°C to kill cells carried over during harvesting.
4. T-hybridoma activation is measured by IL-2 production assaying proliferation of the IL-2-dependent cell line, CTLL (see Note 7). Briefly, wash CTLL cells three times with medium, and then aliquot 5000 CTLL cells in 150 µL DME-10 into each well containing 50 µL of supernatants. After 18 h, add 0.5–1 µCi ^{3}H-thymidine (0.5–1 µL) to the culture and 6 h later, harvest the cells onto a nylon wool membrane. Dry the membrane and measure the incorporation of ^{3}H-thymidine by β-scintillation.

3.5. ^{51}Cr-Release Assay for CTL Primed by Electroporated APC

1. Recover effector CTL cells by Ficoll-Hypaque gradient centrifugation, and add to round-bottom 96-well plates in 100 µL of RP-10 medium at different effector/target ratios (E:T), normally including 50:1, 17:1, and 6:1.
2. Label target cells (10^6) with 100 µCi (100–150 µL) of sodium [^{51}Cr]chromate in an equal volume of conditioned (after overnight culture) DME-10 medium at 37°C for 50–60 min (see Note 8). After three washes in medium, aliquot 10^4-labeled target cells in 100 µL vol to the wells containing expected responder T-cells according to E:T ratios (see Note 9). The cell mixture is gently centrifuged at ~175g for 1 min to allow immediate effector-target contact.
3. Centrifuge the microtiter plate at ~175g for 5 min, and collect the supernatants. Measure the radioactivity in the supernatant by a γ-counter, and determine the percent specific lysis as following:

% Specific lysis = [(CTL-induced release – spontaneous release)/ (maximum release by detergent – spontaneous release)] × 100

4. Notes

1. Soluble antigen (such as OVA) should be freshly made to avoid proteolysis and peptide contamination. To exclude Ag activity owing to peptide contamination, fix some APC with 1% paraformaldehyde in Hank's Balanced Salt Solution (HBSS; 38) directly after loading. These cells should not present antigenic peptides, such as $OVA_{257-264}$, because the processing and presentation for class I take at least 30–60 min after electroporation (29, unpublished data). Any T-cell activation by fixed APC should reflect peptide contamination of the antigen source.
2. The APC number and buffer volume for electroporation are important. The electroporation conditions described above are based on 0.4-mL vol and defined APC number. Alteration of electroporation volume will change the conductance and resistance of the system leading to modified discharging by the internal capacitance. Similarly, if the cell density is too high, it will substantially change the conductance of the loading buffer and lower the Ag-loading efficiency. If large numbers of APC are needed, it is easiest to carry out multiple Ag- loading electroporations and combine these cells, rather than attempting to scale up the procedure.
3. Using a range of Ag concentrations is recommended, and in general, higher concentrations are more efficient. The optimal cell densities for Ag loading are $0.5–5 \times 10^7/0.4$ mL for spleen cells, and $0.5–5 \times 10^6/0.4$ mL for L-cells. The range of OVA concentrations giving linear T-cell responses following electroporation of these APC was 0.1–5 mg/mL (2.2–112.5 μM).
4. Different cell types may have different Ag-loading properties. We recommend trying different voltage and capacitance settings for new cell types. After the electroporation, culture the cells for 4–6 h in the incubator, count the dead cells by trypan blue exclusion, and determine the voltage/capacitance settings that cause about 30–50% cell death for antigen-presentation experiments.
5. We use the Bio-Rad (Hercules, CA) Gene Pulser and Bio-Rad Gene-Pulser cuvets (cat. no. 165-2088) in our electroporation experiments. Optimal electroporation buffer was PBS containing 10 mM $MgCl_2$ (loading buffer, pH ~7.2). However, other buffers, including HEPES-buffered culture medium and DNA transfection buffers, are also suitable.
6. In vivo primed CTL normally show antigen-specific killing after one restimulation *in vitro*, but in vitro primed CTL generally require several restimulations to expand the responder cell population and reduce nonspecific killing activity.
7. CTLL cell culture is notoriously capricious. It is important that all cell-culture reagents be fresh and contain 2-mercaptoethanol (0.05 mM) and

glutamine. It is necessary to screen FCS batches for their suitability in supporting CTLL and T-cell growth. CTLL should always be kept at a reasonable density (10^5/mL) and split every 2 d, preferably into a new culture dish. A concentration of between 40 and 80 U/mL of IL-2 is recommended for supporting the growth of CTLL.

8. For ^{51}Cr-loading, it is important to keep the isotope loading mixture at neutral pH and to avoid alkaline pH. It is therefore advisable to label the targets with an equal volume of preconditioned medium (a 12–24 h culture supernatant) in a CO_2 incubator. Also the target cells should be growing in exponential phase and be very healthy. Otherwise, the spontaneous release of ^{51}Cr can exceed 25% rather than the average 10–15%.

9. In studies of OVA recognition, targets are normally EL-4 cells or the OVA gene transfected EL-4 cell, EG7. $OVA_{257-264}$-pulsed EL-4 can serve as a positive control target. For other antigen systems, histocompatible T-lymphoblasts or cell lines are suitable target APC. The incubation is carried out at 37°C in a CO_2 incubator for 3.5–4 h for T-lymphoblasts and most cell lines. However, some cell lines, including L-cells, sometimes require a 6 h incubation.

References

1. Zinkernagel, R. M. and Doherty, P. D. (1979) MHC-restricted cytotoxic T cells, studies on the biologicial role of polymorphic major transplantation antigens determining T cell restriction specificity, function, and responsiveness. *Adv. Immunol.* **27**, 52–180.
2. Tarleton, R. L., Koller, B. H., Latour, A., and Postan, M. (1992) Susceptibility of β2-microglobulin-deficient mice to *Trypanosoma cruzi* infection. *Nature* **356**, 338–340.
3. Schendel, D. J., Gansbacher, B., Oberneder, R., Kriegmair, M., Hofstetter, A., Riethmuller, G., and Segurado, O. G. (1993) Tumor-specific lysis of human renal cell carcinomas by tumor-infiltrating lymphocytes I. HLA-A2-restricted recognition of autologous and allogeneic tumor lines. *J. Immunol.* **151**, 4209–4220.
4. Sahasrabudhe D. M., Burstyn, D., Dusel, J. C., Hibner, B. L., Collins, J. L., and Zauderer, M. (1993) Shared T cell-defined antigens on independently derived tumors. *J. Immunol.* **151**, 6302–6310.
5. Storkus, W. J., Zeh, H., III, Maeurer, M. J., Salter, R. D., and Lotze, M. T. (1993) Identification of human melanoma peptides recognized by class I restricted tumor infiltrating T lymphocytes. *J. Immunol.* **151**, 3719–3727.
6. Townsend, A. R. M., Rothbard, J., Gotch, F. M., Bahadur, G., Wraith, D., and McMichael, A. J. (1986) The epitopes of influenza nucleoprotein recognized by cytotoxic T lymphocytes can be defined with short synthetic peptides. *Cell* **44**, 959–968.
7. Townsend, A. and Bodmer, H. (1989) Antigen recognition by class I-restricted T lymphocytes. *Ann. Rev. Immunol.* **7**, 601–624.

8. Bennink, J. R., Yewdell, J. W., Smith, G. L., Moller, C., and Moss, B. (1984) Recombinant vaccinia virus primes and stimulates influenza haemagglutinin-specific cytotoxic T cells. *Nature* **311**, 578–579.
9. Townsend, A. R. M., McMichael, A. J., Carter, N. P., Huddleston, J. A., and Brownlee, G. G. (1984) Cytotoxic T cell recognition of the HLA influenza nucleoprotein and haemagglutinin expressed in transfected mouse L cells. *Cell* **39**, 13–25.
10. Moore, M. W., Carbone, F. R., and Bevan, M. J. (1988) Introduction of soluble protein into the class I pathway of antigen processing and presentation. *Cell* **54**, 777–785.
11. Rock, K. L., Gamble, S., and Rothstein, L. (1990) Presentation of exogenous antigen with class I major histocompatibility complex molecules. *Science* **249**, 918–921.
12. Debrick, J. E, Campbell, P. A., and Staerz, U. D. (1991) Macrophages as accessory cells for class I MHC-restricted immune responses. *J. Immunol.* **147**, 2846–2851.
13. Townsend, A. R. M., Gotch, F. M., and Davey, J. (1985) Cytotoxic T cells recognize fragments of the influenza nucleoprotein. *Cell* **42**, 457–476.
14. Gotch, F., McMichael, A., Smith, G., and Moss, B. (1987) Identification of viral molecules recognized by influenza-specific human cytotoxic T lymphocytes. *J. Exp. Med.* **165**, 408–416.
15. Lehmann, P. V., Forsthuber, T., Miller, A., and Sercarz, E. E. (1992) Spreading of T-cell autoimmunity to cryptic determinants of an autoantigen. *Nature* **358**, 155–157.
16. Parham, P. (1990) Transporters of delight. *Nature* **348**, 674,675.
17. Yewdell, J. W., Esquivel, F., Arnold, D., Spies, T., Eisenlohr, L. C., and Bennink, J. R. (1993) Presentation of numerous viral peptides to mouse major histocompatibility complex (MHC) class I-restricted T lymphocytes is mediated by the human MHC-encoded transporter or by a hybrid mouse-human transporter. *J. Exp. Med.* **177**, 1785–1790.
18. Staerz, U. D., Karasuyama, H., and Garner, A. M. (1987) Cytotoxic T lymphocytes against a soluble protein. *Nature* **329**, 449–451.
19. Grant, E. P. and Rock, K. L. (1992) MHC class I-restricted presentation of exogenous antigen by thymic antigen-presenting cells *in vitro* and *in vivo*. *J. Immunol.* **148**, 13–18.
20. Rock, K. L., Rothstein, L., Gamble, S., and Fleischacker, C. (1993) Characterization of antigen-presenting cells that present exogenous antigens in association with class I MHC molecules. *J. Immunol.* **150**, 438–446.
21. McMichael, A. J., Askonas, B. A., and Frelinger, J. A. (1991) Antigen processing and presentation by class I MHC, in *Antigen Processing and Recognition* (McCluskey, J., ed.), CRC, Boca Raton, FL, pp. 85–108.
22. Noguchi, Y., Noguchi, T., Sato, T., Yokoo, Y., Itoh, S., Yoshida, M., Yoshiki, T., Akiyoshi, K., Sunamoto, J., Nakayama, E., and Shiku, H. (1991) Priming for *in vitro* and *in vivo* anti-human T lymphotropic virus type I cellular immunity by virus-related protein reconstituted into liposomes. *J. Immunol.* **146**, 3599–3603.
23. Reddy, R., Zhou, F., Huang, L., Carbone, F. R., Bevan, M. J., and Rouse, B. T. (1991) pH sensitive liposomes provide an efficient means of sensitizing target cells to class I restricted CTL recognition of a soluble protein. *J. Immmmol. Methods* **141**, 157–163.

24. Harding, C. V., Collins, D. S., Kanagawa, O., and Unanue, E. R. (1991) Liposome-encapsulated antigens engender lysosomal processing for class II MHC presentation and cytosolic processing for class I presentation. *J. Immunol.* **147**, 2860–2863.
25. Nair, S., Zhou, F., Reddy, R., Huang, L., and Rouse, B. T. (1992) Soluble proteins delivered to dendritic cells via pH-sensitive liposomes induce primary cytotoxic T lymphocyte responses *in vitro*. *J. Exp. Med.* **175**, 609–612.
26. Takahashi, H., Takeshita, T., Morein, B., Putney, S., Germain, R. N., and Berzofsky, J. A. (1990) Induction of $CD8^+$ cytotoxic T cells by immunization with purified HIV-1 envelope protein in ISCOMS. *Nature* **344**, 873–875.
27. Deres, K., Schild, H., Wiesmuler, K.-H., Jung, G., and Rammensee, H.-G. (1989) *In vivo* priming of virus-specific cytotoxic T lymphocytes with synthetic lipopeptide vaccine. *Nature* **32**, 561–564.
28. Schild, H., Rotzschke, O., Kalbacher, H., and Rammensee, H.-G. (1990) Limit of T cell tolerance to self proteins by peptide presentation. *Science* **247**, 1587–1589.
29. Harding, C. V. (1992) Electroporation of exogenous antigen into the cytosol for antigen processing and class I major histocompatibility complex (MHC) presentation: weak base amines and hypothermia (18°C) inhibit the class I MHC processing pathway. *Eur. J. Immunol.* **22**, 1865–1869.
30. Chen, W., Carbone, F. R., and McCluskey, J. (1993) Electroporation and commercial liposomes efficiently deliver soluble protein into MHC class I presentation pathway: priming *in vitro* and *in vivo* for class I-restricted recognition of soluble antigen. *J. Immunol. Methods* **160**, 49–57.
31. Mir, L. M., Banoun, H., and Paoletti, C. (1988) Introduction of definite amounts of nonpermeant molecules into living cells after electropermeabilization, direct access to cytosol. *Exp. Cell. Res.* **175**, 15–25.
32. Lambert, H., Pankov, R., Gauthier, J., and Hancock, R. (1990) Electroporation-mediated uptake of proteins into mammalian cells. *Biochem. Cell. Biol.* **68**, 729–734.
33. Chen, W., Khilko, S., Fecondo, J., Margulies, D. H., and McCluskey, J. (1994) Determinant selection of MHC class I-restricted antigenic peptides is explained by class I-peptide affinity and is strongly influenced by non-dominant anchor residues. *J. Exp. Med.* **180**, 1471–1483.
34. Karre, K., Ljunggren, H., Piontek, G., and Kiessling, R. (1986) Selective rejection of H-2-deficient lymphoma variants suggests alternative immune defence strategy. *Nature* **319**, 675–678.
35. Hosken, N. A. and Bevan, M. J. (1990) Defective presentation of endogenous antigen by a cell line expressing class I molecules. *Science* **248**, 367–370.
36. Schulze, D. H., Pease, L. R., Obata, Y., Nathenson, S. G., Reyes, A. A., Ikuta, S., and Wallace, R. B. (1983) Identification of the cloned gene for the murine transplantation antigen H-$2K^b$ by hybridization with synthetic oligonucleotides. *Mol. Cell. Biol.* **3**, 750–755.
37. Coligan, J. E., Kruisbeek, A. M., Margulies, D. H., Shevach, E. M., and Strober, W. (1992) *Current Protocols In Immunology*. Wiley Interscience, New York.
38. Rock, K. L., Rothstein, L. E., Gamble, S. R., Gramm, C., and Benacerraf, B. (1992) Chemical cross-linking of class I molecules on cells creates receptive peptide binding sites. *J. Immunol.* **148**, 1451–1457.

CHAPTER 6

Electroporation of Antibodies into Mammalian Cells

Paul L. Campbell, James McCluskey, Jing Ping Yeo, and Ban-Hock Toh

1. Introduction

The introduction into mammalian cells of antibodies with specificities for endogenous cellular factors permits the functional assessment of such factors in the context of living cells. Antibodies have been successfully introduced into several cell types by various methods, including fine-needle microinjection *(1)*, osmotic lysis of pinocytotic vesicles *(2–4)*, liposome-mediated delivery *(5)*, and fusion of red cell ghosts loaded with protein *(5–8)*. Each of these techniques has its associated drawbacks: Microinjection is very time-consuming and is inappropriate for transfer of antibodies into large numbers of cells; osmotic lysis of pinocytotic vesicles results in massive cell damage and thus requires very large numbers of cells and long recovery periods *(4)*. Red cell ghosts or liposomes require either the use of targeting molecules likely to modify the nature of the target cell membrane *(5)* or the use of potentially hazardous virus to stimulate fusion *(6,7)*.

More recently, electroporation has been used to introduce monoclonal antibodies (MAb) into various cell types for the purpose of modulating the activities of intracellular molecules. As well as avoiding many of the drawbacks inherent in the alternative procedures mentioned above, electroporation has the added advantage that it permits the rapid entry of antibody into large numbers of cells with few nonspecific effects and thus allows the study of short-term metabolic effects. For example, using this methodology, specific in vivo inhibition of asparagine synthetase has been demonstrated in three different cell lines with minimal effects on cell viability and on nonspecific indices of cellular dysfunction *(9)*. Electroporation of MAb to known epitopes of the human insulin receptor has been used to study the C-terminal domain function of this protein in mouse embryo fibroblast transfectants expressing this human gene *(10)*. The same group has also successfully employed this technique to investigate the C-terminal domain function of the Glut 1 glucose transporter in the same cells *(11)*. Very recently, autoantibodies recognizing a cell-cycle-dependent antigen were purified and used to electroporate cultured cell lines. This had the dramatic effect of arresting the division of the growing cells in mitosis *(12)*. Use of this technique to inhibit lipoxygenase activity in lentil protoplasts *(13)* demonstrates that the method is well suited to applications in plant as well as animal cells, although this will not be discussed further in this chapter.

As discussed elsewhere in this book, the use of electroporation for transient or stable DNA-mediated gene transfer into mammalian cells often requires the use of high field strengths or long pulse durations, which result in significant pulse-induced mortality, often ranging from 50–75%. In general, transfection efficiency in these procedures increases with cell death, and optimization of the procedure requires a trade-off between these variables. However, in the generation of stable or episomal gene transfectants, the use of stringent selection procedures after electroporation means that preservation of cell viability is not the overriding consideration. Furthermore, when used for transient transfection in the absence of selection, experiments can often be conducted over very short periods of time using large numbers of cells and amounts of DNA, coexpressed controls can be used to monitor transfection efficiency, and very sensitive readout systems are available for monitoring the expression of transfected genes *(14)*.

In contrast, the electroporation of antibodies into cells requires both that the introduction of antibody into cells should take place at high efficiency and that a high proportion of cells survive the procedure in a physiologically unperturbed state so that the effects of the introduced antibody on specific cellular functions can be measured. Furthermore, the conditions used should be such that the electroporated antibody is incorporated in its native conformation, thus retaining its activity. Therefore, electroporation conditions must be optimized for each cell line to meet these criteria. Optimization of these parameters has been most comprehensively reported by Chakrabarti et al. *(9)*, and this report will form the basis of the methods described below.

2. Materials
2.1. Stock Solutions

1. Phosphate-buffered saline (PBS): 0.23 g NaH_2PO_4 (anhydrous; Sigma, St. Louis, MO), 1.15 g Na_2HPO_4 (anhydrous; Sigma), 9.00 g NaCl (Sigma), add H_2O to 900 mL; adjust to pH 7.4 with $1M$ NaOH; add H_2O to 1 L. Sterilize by autoclaving or by filtration though a 0.2-µm Millipore membrane filter.
2. 2X Trypan blue: 0.4% trypan blue (Gibco BRL, Gaithersburg, MD) w/v in PBS. Store in dark bottle and filter after prolonged storage (every few months).
3. Propidium iodide (PI): 10 mg/mL propidium iodide as an aqueous solution (Sigma, St. Louis, MO). Store at 4°C in the dark. **Caution:** Propidium iodide is a suspected carcinogen and should be handled with extreme care.
4. Antifade solution: [1,4-diazabicyclo(2,2,2)octane] (DABCO) (Sigma). **Caution:** DABCO is toxic; wear gloves when handling.

2.2. Antibodies

Antibodies should be purified by physical or affinity-based methods; if necessary, exchange buffer with PBS by either dialysis or gel filtration. Concentration to around ten times working concentration should be carried out by standard methods, or alternatively, the purified protein can be lyophilized. The physical integrity of purified antibodies should be checked by SDS-PAGE under nonreducing and reducing conditions. The affinity of the purified antibody for its antigen and any inhibitory effect on the function of the target antigen should first be established in vitro where possible, since these parameters are among those that are likely to determine the ability of the antibody to modulate the activity of its target in vivo (*see* Note 1). Details of these techniques can be found in standard methodological textbooks of immunology (e.g., *15*). Prior to electroporation, antibodies so purified should be sterilized by filtration through a Millipore 0.2-µm membrane filter. To maximize the incorporation of antibody and the likelihood of interfering with the function of the target antigen, working concentrations of antibody should be made as high as is practically achievable (*see* Note 1). Obviously, this will vary according to the particular antibody preparation used. As a guide, working concentrations of antibodies successfully employed by various groups are given, together with references, in Table 1.

Should the source of antibody be patient serum, prior to further handling, this should first be heat-inactivated by heating to 56°C for 30 min to destroy complement components and to disable infectious organisms. Debris should be removed by centrifugation for 30 min in an airfuge or by ultracentrifugation.

Caution: Human body fluids should always be regarded as potentially infectious, and Universal Precautions should be followed at all times when handling such material.

2.3. Cell Lines

Cell lines that have been used in experiments involving the electroporation of antibodies, together with electroporation conditions and references, are given in Table 1. Medium requirements differ between cell lines, and so the formulation that is routinely used for the maintenance of any given cell line should be used; this is unlikely to influence the efficiency of electroporation. However, cells should be harvested during the exponential growth phase, since the efficiency of electroporation decreases significantly when cells reach stationary phase *(9)*.

2.4. Solutions Used for Electroporation

Various solutions used by different groups for electroporation of antibodies are listed in Table 1. In comparisons among HEPES-buffered saline (HeBS), Dulbecco's Modified Eagle's Medium (DMEM), and opti-MEM (Gibco; a modified Eagle's medium supplemented with HEPES buffer, insulin, transferrin, and β-mercaptoethanol), opti-MEM supplemented with 5% FCS (fetal calf serum) gave the best results *(9)*, and is the solution used in the method described below.

3. Methods

3.1. Optimization of Electroporation Conditions

The two main electroporation parameters that need to be optimized are the electric field strength and the duration of the electric pulse. The former is a function of the applied voltage and the distance between the electrodes, whereas the pulse duration is generally an exponential decay characteristic with a time constant given by the product of the capaci-

Table 1
Conditions Used for Electroporation of Antibodies into Several Cell Lines

Cell lines	Solution	Field strength, V/cm	Capacitance, μF	Time constant, ms	Antibody, mg/mL	Reference
HeLa	opti-MEM	750	50–60	0.9	2	9
HT-5	opti-MEM	750	50–60	0.9	2	9
L5178Y D10/R	opti-MEM	750	50–60	0.7	2	9
NHIR	PBS	750	NR	0.93	0.25	10
NHIR	PBS	750	NR	0.93	0.3–0.4	11
L-cells	PBS + MgCl$_2$ (10 mM)	563	250	NR	0.5	12

NR; not reported.

tance and the buffer resistance. Thus, the field strength can be altered by varying the voltage, whereas control over the pulse duration can be achieved by varying the capacitance. For any given combination of field strength and pulse duration, the extent of macromolecular entry and the degree of mortality will vary between cell lines. These parameters must therefore be optimized for each cell line used.

1. Harvest cells in exponential growth phase by centrifugation at 1000g for 5 min. Adherent cell lines should be harvested at no greater than 50% confluency. Wash cells once in opti-MEM containing 5% FCS.
2. Resuspend cells at a density of 10^7 cells/mL in opti-MEM containing 5% FCS. Set aside half the cell suspension without antibody and process in parallel as a control for antibody toxicity. To the remainder, add antibody as a sterile stock in PBS in a volume <10% of total volume. The working concentration of antibody should be made as high as is practicable (see Section 2.2. and Note 1). Add an equivalent volume of PBS to the cell suspension without antibody. Distribute cell suspensions to a range of electroporation cuvets in volumes specified by the electroporator manufacturer. For subsequent determination of antibody incorporation, a similar volume of the cell suspension containing antibody, but not electroporated should be processed in parallel with the electroporated samples. This will allow correction for antibody association with cell-membrane components, both nonspecifically and via Fc receptors where present.
3. Electroporate aliquots at a range of voltages and capacitances. Those used must be empirically determined for the particular apparatus and cell line. As a general guide, conditions used by various groups are given in Table 1. For each set of electroporation parameters, two aliquots of cells, one containing and one not containing antibody, should be electroporated. Time constants should be recorded where the apparatus permits as a means of monitoring reproducibility in subsequent experiments.
4. Allow cells to recover in the cuvet for 5–10 min at room temperature. Resuspend cells in the medium formulation that is routinely used for maintenance, and plate at a density normally used for maintenance of the line. Prior to morphological assessment of adherent cells electroporated with antibody, dead cells should be removed by a change of medium after allowing the living cells to adhere over a period of up to 3 h.

3.2. Assessment of Cell Viability

Cell viability can be assessed by trypan blue exclusion or by flow cytometry using propidium iodide (15). The latter is described in Section 3.3.2. The toxicity of the introduced antibody can be determined by comparison of cells electroporated in its presence with cells electroporated under identical conditions in the absence of antibody. Various methods are available to assess the degree of nonspecific toxicity of the procedure, including the determination of DNA, RNA, and protein synthesis postelectroporation by the incorporation of [^{3}H]thymidine, [^{3}H]uridine, and [^{35}S]methionine *(9,15)*.

1. Pellet 5–10 × 10^5 cells by centrifugation at 1000g for 5 min and discard supernatant.
2. Resuspend the pellet in 0.5 mL of PBS. Mix well.
3. Transfer an aliquot of 10–20 µL cells to a 1.5-mL tube, and add an equal volume of trypan blue solution.
4. Apply a drop of the trypan blue/cell suspension to a hemocytometer, and count the stained (nonviable) and unstained (viable) cells separately. Cells should be counted within 3–5 min of addition of trypan blue, since longer incubations will lead to further cell death and an underestimation of cell viability. The counting error (CE%) for each subset of cells is given by

$$CE\% = (1/n^{1/2}) \times 100\% \tag{1}$$

   where n is the number of cells counted. It follows that for a 5% counting error, ~400 cells should be counted.
5. To obtain the total number of viable cells per milliliter of aliquot, multiply the total number of viable cells by 2 (the dilution factor for the addition of trypan blue), and then by the correction factor for the volume of the hemocytometer used. To obtain the total number of cells per milliliter of aliquot, sum the total number of viable and nonviable cells per milliliter and multiply by 2 and then by the correction factor for the volume of the hemocytometer used.
6. Calculate the percentage of viable cells as follows:

$$\text{Viable cells } (\%) = [(\text{total number of viable cells/mL of aliquot}) / (\text{total number of cells/mL of aliquot})] \times 100 \tag{2}$$

Although this is a simple technique for determining viability and can be performed in 5–10 min, the sensitivity of the technique for detecting nonviability is low, since it depends on loss of membrane integrity, which may be a late event in cell death. Thus, it is possible that many cells may be present that have compromised viability, but have not lost membrane integrity at the time of staining. This, together with ongoing cell death for several hours after electroporation, makes it advisable to check viability at several time-points after electroporation (e.g., 0, 6, 24, and 48 h postelectroporation) *(see* Note 2).

3.3. Assessment of Antibody Incorporation

Incorporation of antibody can be simply, but nonquantitatively detected by direct immunofluorescence or flow cytometry, but this necessitates the use of an antibody that is conjugated to an appropriate fluor (e.g., fluorescein isothiocyanate; FITC) during the electroporation procedure. Commercially available antibody conjugates of the same class and preferably from the same species as the test antibody can be used for this purpose, in which case similar absolute concentrations as the intended test antibody should be used. Alternatively, the test antibody can be conjugated to an appropriate fluor by established methods *(15)*. Fluorescence of electroporated cells should be compared to the control aliquot, which was incubated with antibody, but not subjected to electroporation. Similarly, this control should be used to set background fluorescence values for flow cytometry. Quantitative determinations of antibody incorporation can be made by enzyme-linked immunosorbent assay (ELISA) of cell lysates *(9,15)* or by scintillation counting of lysates from cells electroporated in the presence of ^{125}I-labeled antibodies of known specific activity *(10)* (*see* Note 2).

3.3.1. Determination of Antibody Incorporation by Direct Immunofluorescence Microscopy

1. Cytocentrifuge cells onto a microscope slide. Apply a drop of antifade solution and a coverslip.
2. View cells through a fluorescent microscope with appropriate filters for the fluor used.
3. Count fluorescent cells and total cells within a typical field. Counting errors should be estimated according to the considerations mentioned in Section 3.2.
4. Calculate the percentage of fluorescent cells as a fraction of the total cells.

3.3.2. Determination of Antibody Incorporation and Cell Viability by Flow Cytometry

The procedures and settings for flow cytometric analysis of electroporated cells will depend on the instrument, fluorescent antibody, and cell line used, and is beyond the scope of this chapter. Background fluorescence should be set using cells that were incubated with antibody, but not electroporated. For propidium iodide (PI) staining, compensation for overlap from the antibody-conjugated fluor should be employed. Otherwise, viable cells will be excluded from the analysis. These considerations require the input of an experienced flow cytometrist.

1. Harvest an aliquot of approx 10^6 cells by centrifugation at $1000g$. Remove and discard supernatant.
2. Wash pellet once with cold (4°C) PBS.
3. Resuspend pellet in 0.5 mL of cold PBS.
4. Immediately prior to flow cytometric analysis, add 20 µL of PI solution.
5. Load sample on instrument and acquire data.
6. Calculate percentage of cells that are negative for PI staining (viable cells).
7. Gate on this population, and measure incorporation of fluorescent antibody.

4. Notes

1. Detection of a specific effect of the electroporated antibody on some cellular function is expected to depend on many variables. These include the concentration and affinity of the introduced antibody for its target in the context of the intracellular milieu; whether binding of the antibody is inhibitory to the function of the target molecule; the absolute intracellular concentration of the target molecule and whether its cellular distribution renders it accessible to the introduced antibody; the complexing of the target molecule with other cellular factors, such that the epitope(s) recognized by introduced antibodies are unavailable for binding, or are subject to competition by such factors; whether the activity of the factor targeted is rate-limiting for the process being measured or whether kinetic processes (e.g., rapid synthesis of new target molecules) preclude an effect being detected. Of all of these variables, only the concentration and affinity of the antibody for its target are open to manipulation, and in this context it would be preferable to maximize these parameters to the greatest extent possible. If no effect on a measured process is observed despite optimization of these parameters, choice of a different means of measuring the process should be considered.
2. For the further experimental use of this technique, select values for voltage and capacitance that give postelectroporation viability of 80–90%, but that provide clearly detectable incorporation of antibody.

References

1. Arnheiter, H. and Haller, O. (1988) Antiviral state against influenza virus neutralised by microinjection of antibodies to interferon-induced Mx proteins. *EMBO J.* **7,** 1315–1320.
2. Okada, C. Y. and Rechsteiner, M. (1982) Introduction of macromolecules into cultured mammalian cells by osmotic lysis of pinocytotic vesicles. *Cell* **29,** 33–41.

3. Morgan, D. O. and Roth, R. A. (1987) Acute insulin action requires insulin receptor kinase activity: introduction of an inhibitory monoclonal antibody into cells blocks the rapid effects of insulin. *Proc. Natl. Acad. Sci. USA* **84,** 41–45.
4. Chakrabarti, R., Pfeiffer, N. E., Wylie, D. E., and Schuster, S. M. (1989) Incorporation of monoclonal antibodies into cells by osmotic permeabilization. Effect on cellular metabolism. *J. Biol. Chem.* **264,** 8214–8221.
5. Guyden, J., Godfrey, W., Doe, B., Ousley, F., and Wofsy, L. (1984) Immunospecific vesicle targeting facilitates fusion with selected cell populations. *Ciba Found. Symp.* **103,** 239–253.
6. Hosoi, T., Ozawa, K., Tsao, C. J., Urabe, A., Uchida, T., and Takaku, F. (1985) Introduction of macromolecules into hemopoietic stem cells with an erythrocyte-ghost-mediated system. *Biochem. Biophys. Res. Commun.* **128,** 193–198.
7. Ozawa, K., Hosoi, T., Tsao, C. J., Urabe, A., Uchida, T., and Takaku, F. (1985) Microinjection of macromolecules into leukemic cells by cell fusion technique: search for intracellular growth-suppressive factors. *Biochem. Biophys. Res. Commun.* **130,** 257–263.
8. Ohara, J., Sugi, M., Fujimoto, M., and Watanabe, T. (1982) Microinjection of macromolecules into normal murine lymphocytes by means of cell fusion. II. Enhancement and suppression of mitogenic responses by microinjection of monoclonal anti-cyclic AMP into B-lymphocytes. *J. Immunol.* **129,** 1227–1232.
9. Chakrabarti, R., Wylie, D. E., and Schuster, S. M. (1989) Transfer of monoclonal antibodies into mammalian cells by electroporation. *J. Biol. Chem.* **264,** 15,494–15,500.
10. Baron, M., Gautier, N., Kaliman, P., Dolais-Kitabgi, J., and Van Obberghen, E. (1991) The carboxyl-terminal domain of the insulin receptor: its potential role in growth-promoting effects. *Biochemistry* **30,** 9365–9370.
11. Tanti, J.-F., Gautier, N., Cormont, M., Baron, V., Van Obberghen, E., and Le Marchand-Brustel, Y. (1992) Potential involvement of the carboxy-terminus of the Glut 1 transporter in glucose transport. *Endocrinology* **131,** 2319–2324.
12. Yeo, J. P., Alderuccio, F., and Toh, B. H. (1994) A new chromosomal protein essential for mitotic spindle assembly. *Nature* **367,** 288–291.
13. Maccarrone, M., Veldink, G. A., and Vliegenthart, J. F. G. (1992) Inhibition of lipoxygenase activity in lentil protoplasts by monoclonal antibodies introduced into the cells via electroporation. *Eur. J. Biochem.* **205,** 995–1001.
14. Campbell, P. L., Kulozik, A. E., Woodham, J. P., and Jones, R. W. (1990) Induction by HMBA and DMSO of genes introduced into mouse erythroleukemia and other cell lines by transient transfection. *Genes and Dev.* **4,** 1252–1266.
15. Coligan, J. E., Kruisbeek, A. M., Margulies, D. H., Shevach, E. M., and Strober, W. (eds.) (1992) *Current Protocols in Immunology.* Wiley, New York.

CHAPTER 7

Electroporation of Adherent Cells *In Situ* for the Introduction of Nonpermeant Molecules

Leda H. Raptis, Kevin L. Firth, Heather L. Brownell, Andrea Todd, W. Craig Simon, Brian M. Bennett, Leslie W. MacKenzie, and Maria Zannis-Hadjopoulos

1. Introduction

Electroporation has been used for the introduction of DNA *(1–4)*, proteins *(5–7)*, and various nonpermeant drugs and metabolites into cultured mammalian cells *(8,9)*, as described in this and other volumes *(10,11)*. Most electroporation techniques for adherent cells involve the delivery of the electrical pulse while the cells are in suspension *(5,12)*. However, the detachment of these cells from their substratum by trypsin or EDTA can cause metabolic alterations that may lead to cell death or increase the cells' sensitivity to additional damaging agents *(13)*. The efficient incorporation of proteins without cell damage is an especially crucial requirement, since for most proteins of interest and contrary to DNA, no convenient large-scale method exists for the selection of viable from damaged cells or cells where no protein introduction took place after electroporation. Therefore, for studies using adherent cells, it is imperative to deliver the pulse while the cells are attached to their solid substratum. This has been achieved using a variety of methods *(14,15)*.

In this chapter, we describe a technique where cells are grown and electroporated on a glass surface coated with electrically conductive, optically transparent indium-tin oxide. This coating allows the direct visualization of the electroporated cells and offers the possibility of ready examination owing to their extended morphology. Moreover, it is very durable, inert, nontoxic to the cells, and promotes excellent cell adhesion and growth. An added advantage is the fact that, unlike many plastics, indium-tin oxide does not exhibit any spontaneous fluorescence, making the examination of the electroporated cells under a fluorescent microscope possible.

The procedure described below is applicable to a wide variety of nonpermeant molecules. In a modified version, it can be used for the study of intercellular, junctional communication. Cells are grown on a glass slide, half of which is conductive. An electric pulse is applied in the presence of the fluorescent dye, Lucifer yellow, causing its penetration into the cells growing on this conductive surface. After removing the excess unincorporated dye by washing, the movement of Lucifer yellow can be microscopically observed over time from the electroporated cells to the neighboring ones growing on the nonconductive area by using fluorescence illumination. In this way, the transfer of dye through gap junctions can be precisely quantitated in a large number of cells simultaneously, without any detectable disturbance to cellular metabolism.

2. Materials

2.1. Solutions

1. Hank's Balanced Salt Solution (HBSS, Gibco, Gaithersburg, MD).
2. Phosphate-buffered saline (PBS): 137 mM NaCl, 2.7 mM KCl, 9.5 mM sodium phosphate, pH 7.4.
3. HEPES buffered saline: 10 mM HEPES, pH 7.0, 140 mM NaCl.
4. Lucifer Yellow CH (Lithium salt, Sigma, St. Louis, MO), 2 mg/mL in PBS or HEPES-saline. Stable at 4°C for at least a month.
5. Trypan blue (Sigma): 0.5% in PBS or HEPES-saline. Stable for at least a month at room temperature. It tends to precipitate on longer storage.
6. Basilen Blue E-3G (inhibitor of glutathione-S transferases, Sigma) solution: 1 mM in HBSS. Stable at 4°C for at least a month.
7. Secondary antibodies: fluorescein isothiocyanate (FITC)-conjugated goat antimouse IgG and fluorescein isothiocyanate (FITC)-conjugated rabbit antichicken IgG (Sigma).

2.2. Antibodies

In principle, the introduction of a wide variety of antibodies should be possible. To test the technique, we have introduced chicken IgG, or a monoclonal antibody (MAb) against cruciform DNA structures (2D3 [16,17]) into a variety of cells.

1. 2D3: Since this MAb (IgG_1) is not stable after ammonium sulfate precipitation, it was purified and concentrated as follows (see Note 1): Hybridoma cells secreting 2D3 were grown in serum-free culture using HB102 media and supplement (Du Pont-New England Nuclear, Boston, MA), and passed through an Amicon XM-100 ultrafiltration filter (mol wt cutoff: 100,000). The antibody was washed extensively with PBS, effectively removing the antibiotics from the culture supernatant, and concentrated approx 24-fold by ultrafiltration (final concentration, 120–240 µg/mL). The antibody was stored in lyophilized form.
2. Chicken IgG (Sigma): 2 mg/mL, extensively dialyzed against PBS or HEPES-saline.

2.3. Equipment

1. Electroporation apparatus: The system for electroporation *in situ* includes the following elements:
   a. Pulse generator.
   b. Glass slides coated with conductive and transparent material.
   c. A negative electrode above the cells and a positive contact bar.
   d. Means to contain the cells and the material to be introduced and to define the space between the cells and the negative electrode.

   Various methods of making each of the above components can be used for electroporation *in situ*. A number of commercially available pulse generators can be successfully employed. However, the features that are important for this application are the ability to select a stable voltage level in 0.1-V increments and the availability of relatively small capacitance values (1–50 µF). A suitable instrument, the Epizap model EZ-10 pulse generator (Ask Science Products Inc., Kingston, Ont., Canada), made specifically for electroporation *in situ*, has a power supply internal to the unit. The voltage to which the capacitor is charged can be controlled to within 0.1 V of the target setting and can be adjusted between 0 and 200 V. Cells are grown on glass coated with electrically conductive, optically transparent indium-tin oxide at the time of pulse delivery (Donnelly Corp., Boulder, CO; see Note 2).
2. Phase-contrast and fluorescence microscope, upright or inverted.

2.4. Cell Lines

The technique can be applied to many types of adherent cells, able to attach to and grow on glass. We have tested a number of lines, such as the Fisher rat fibroblast F111 and its polyoma virus-transformed derivatives *(18)*, mouse fibroblast NIH 3T3 *(19)*, mouse Balb/c 3T3, mouse NIH 3T6, mouse C3H10T½ fibroblast derivatives expressing a *ras*-antimessage (e.g., Rev3 *[20]*), and the porcine kidney epithelial line LLC-PK$_1$ (American Type Culture Collection, Rockville, MD). Cells are grown as described in Note 3.

3. Methods
3.1. Electropermeation Technique

Two different assemblies have been used for *in situ* electroporation, each with its own advantages (*see* Note 4). In either setup, the electroporation device consists of a circuit for charging and discharging a capacitor, and an assembly for delivering the pulse to the cells. The cells are growing on a glass slide [1] coated with electrically conductive, optically transparent indium-tin oxide [2], which permits growth of adherent cells (Figs. 1 and 2). A digital storage oscilloscope, such as the Tektronix 2212, can be used if necessary, to record the pulse shape as the capacitor is discharged through the electroporation assembly.

In the first setup (Fig. 1), cells are grown in an area defined by a "dam" of silicone [3] (*see* Note 5). The pulse is delivered through a stainless-steel negative electrode [4] with an insulating spacer [6] at the edge distal to the positive contact bar [5]. The negative electrode and the positive contact bar are identical blocks of stainless steel with the bottom faces polished to a mirror-like finish.

In the second configuration, a piece of electrically insulating adhesive tape (*see* Note 5) is attached to the slide after having been cut to form a window through which the conductive surface is exposed so that cells can grow on it. This frame (Fig. 2 [14]) serves the purposes of defining the area for cell growth, supporting the negative electrode, and holding the electroporation fluid in place. The pulse is delivered through a stainless-steel negative electrode that is slightly larger than the cell growth area and is placed on top of the cells [4], and an identical stainless-steel or aluminum block is used as a positive contact bar [5].

Electroporation of Adherent Cells

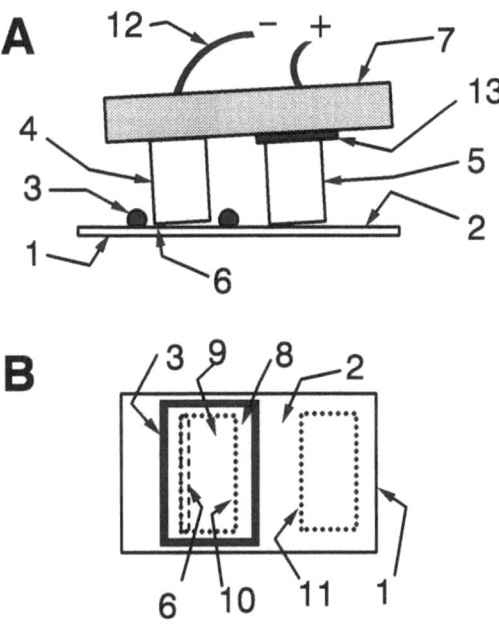

Fig. 1. Electroporation electrode assembly-I. (**A**) Side view: A coating of indium-tin oxide [2] on the upper surface of the glass slide [1] makes a conductive path between the positive electrode [5] and the base of the negative electrode [4], through the cells and the electroporation medium. A spacer [6] insulates the negative electrode from direct contact with the conductive surface. When an electric pulse is applied, current passes from the positive contact bar [5] to the conductive layer, through the cells growing in area [8] and [9] through the liquid medium (not shown) to the negative electrode [4]. A shim [13] at the top of the positive contact bar serves to create the angle, whereas a Plexiglas™ carrier [7] serves to keep the electrode assembly together. A ring of silicone [3] acts as a dam to hold the medium around the negative electrode. [12] shows connecting wires to the pulse generator. (**B**) Top view: The outline of the negative [10] and the positive [9] electrodes relative to the silicone dam [3] are indicated. Note that the cells grow over the entire area enclosed by the dam, underneath [9] as well as outside [8] the electrode.

A complete circuit is formed by placing the electrode set on top of the slide as shown in Figs. 1 and 2. This set is linked to a circuit such that during a pulse, current flows through the positive contact bar [5] along

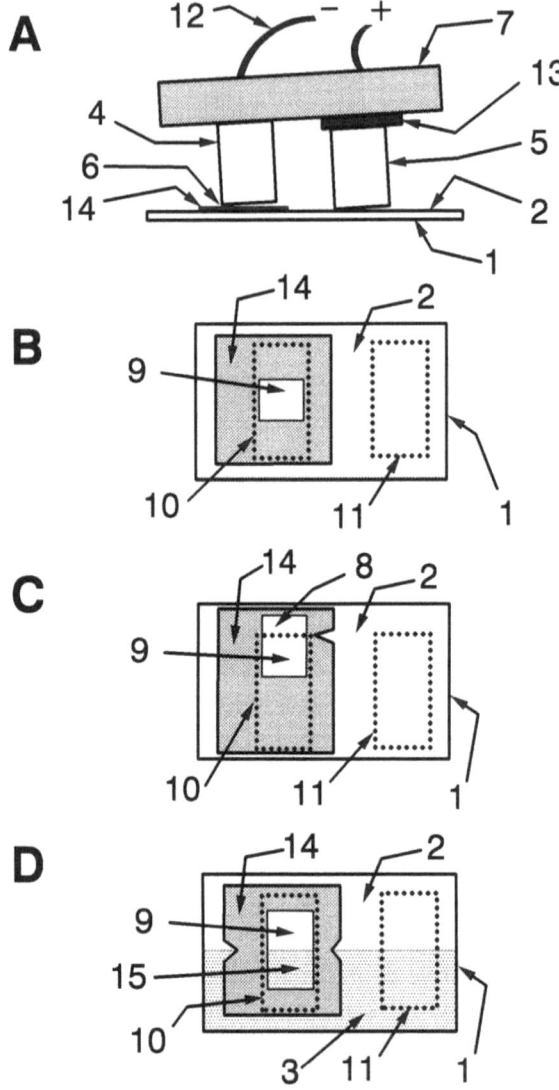

Fig. 2. Electroporation electrode assembly-II. (**A**) Side view: A coating of indium-tin oxide [2] on the upper surface of the glass slide [1] makes a conductive path between the positive contact bar [5] and the base of the negative electrode [4], through the cells and the electroporation medium. An insulating window frame [14] defines an area of cell growth and supports the negative electrode. When an electric pulse is applied, current passes from the positive contact bar to the conductive layer, through the electroporation medium and

the conductive coating [2] to the cells growing in area [9] and the electroporation solution, to the bottom surface of the negative electrode [4] and back to the pulse source. As shown in Fig. 2B, the electrically insulating frame [14] creates a gap between the conductive coating [2] and the negative electrode so that current can only flow through the electroporation fluid and the cells growing in the window [9].

In either of the above assemblies it is important to obtain a uniform electric field strength over the area below the negative electrode, in spite of the fact that the conductive coating exhibits a significant amount of electrical resistance. To achieve this, the negative electrode is inclined relative to the glass surface, rising in the direction of the positive contact bar. This angle is set by inserting a shim [13] between the positive contact bar and the electrode carrier [7]. Note that the positive contact bar only contacts the slide along an edge when the electrode set is correctly placed. In this configuration, current traveling a path from the positive contact bar [5] to the nearest edge of the negative electrode [4] transverses the smallest distance through the conductive coating and the longest distance through the electroporation fluid. On the other hand, current following the path to the most distant part of the negative electrode encounters more resistance through the conductive coating and less across the smaller electroporation fluid gap (*see* Note 6).

---

the cells growing in the window, to the negative electrode. A shim [13] at the top of the positive contact bar serves to create the angle, whereas a Plexiglas™ carrier [7] serves to keep the electrode assembly together. Wire leads [12] connect the electrode set to the pulse generator. (**B**) Top view: The outline of the negative electrode [10] and the outline of the positive contact bar [11] are shown to indicate their positions on the conductive glass slide [1] in relation to the window where the cells are grown [9]. (**C**) Top view, alternate placement of frame and electrode: The negative electrode is placed [10] such that only part of the cells growing in the window are covered by it [9], and are therefore electroporated, whereas cells in area [8] are not subjected to a pulse. For reproducibility, a notch on the plastic frame helps position the electrode. (**D**) Electroporation electrode assembly for the study of intercellular communication—top view. The etched [3] and conductive parts [2] of the slide are shown relative to the outlines of the electrodes [10] and [11] and the insulating tape window where the cells are grown on conductive [9] and nonconductive [15] areas. The notches in the window frame aid in lining up the frame on the glass such that the conductive area in the window is 8×8 mm.

1. Make sure the glass is clean and free of fingerprints (*see* Note 7). Place the plastic frame (or a silicone dam) on the conductive side of the slide, and press firmly (*see* Note 5). You can test which side is conductive using an ohmmeter.
2. Plate the cells: Uniform spreading of the cells is very important, since the optimal voltage depends in part on the degree of cell contact with the conductive surface. A few days before the experiment, place the glass slide inside a 6- or 10-cm Petri dish (*see* Note 8), and sterilize by adding 70% ethanol for 20 min. Remove the ethanol by rinsing with PBS. Add a sufficient amount of medium to cover the slide (approx 9 mL for a 6-cm [diameter] dish), or add 1 mL of medium to the square enclosed by the dam [3], if the setup described in Fig. 1 is used. Plate the cells in the window (Fig. 2B [9] or 2C [8] and [9]) or the space inside the dam (Fig. 1B [8] and [9]). Since in either case the growth area is rectangular rather than round, the cells cannot be rocked back and forth to achieve a uniform spreading. Therefore, this must be accomplished by adding the trypsinized cell suspension through the medium to the edges of the area first, and then the center. Do not disturb the dishes for 10 min, so that the cells can attach to the conductive surface, and then place in a CO_2 incubator.
3. Prior to pulse application, remove the growth medium, and gently wash the cells twice with the electroporation buffer (PBS or HEPES-saline; *see* Note 9).
4. Carefully wipe the plastic frame with a folded Kleenex to create a dry area on which a meniscus can form.
5. Add the same electroporation buffer, supplemented with the molecules to be introduced to the cells, in sufficient quantity as to form a meniscus rising from the inner edges of the dam (Fig. 1 [8] and [9]), or the window defined by the insulating tape (Fig. 2 [8] and [9]).
6. Carefully place the electrode on top of the cells and hold it in place. This is easier if the electrode leads are flexible. Make sure there is a sufficient amount of electroporation buffer under the positive contact bar to ensure electrical contact (*see* Note 10).
7. Deliver a single pulse of 40–120 V from a capacitor of a size of 1–5 μF for a conductive cell growth area of 8 × 8 mm (*see* Note 11).
8. Incubate the assembly at room temperature for 1 min to permit pore closure (*see* Note 12).
9. Remove the electrode set.
10. The solution can be carefully aspirated and reused a number of times (*see* Note 13).
11. To remove the unincorporated material and facilitate pore closure, wash the cells with the same growth medium containing serum *(21)*.

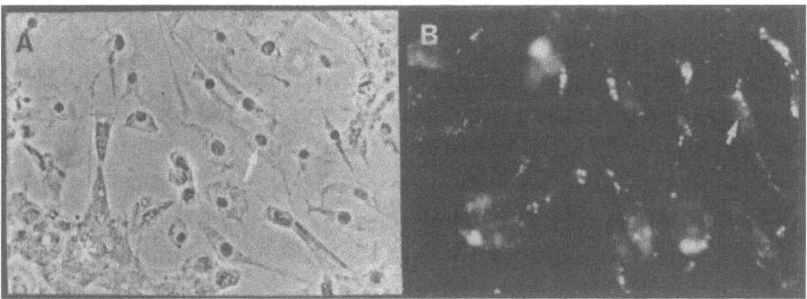

Fig. 3. (**A**) Phase-contrast micrograph of rat F111 cells that have been killed by the pulse. Note the dark, prominent nuclei (white arrow) and the flat, nonrefractile appearance. Such cells do not retain any electroporated material (*see* Fig. 6C). White star denotes an area of dense cell growth. Note that these cells are not killed by the pulse. (**B**) Fluorescence micrograph of rat F111 cells incubated with 2 mg/mL Lucifer yellow for 1 h at 37°C. Note the brightly fluorescing pinocytotic vacuoles in the cytoplasm (arrow).

12. If Lucifer yellow or other fluorescent dye was included in the electroporation buffer, place a coverslip on the window and observe the cells under phase-contrast and fluorescence microscopy using a standard or inverted microscope (*see* Note 4). With the setup shown in Fig. 1, the dam makes observation using a standard microscope under high power difficult. However, the cells can be observed under high power using an inverted microscope.
13. Cell viability can be assessed visually (Fig. 3), by calculating the cloning efficiency of the cells by trypsinizing and plating, or by the addition of trypan blue 2 h after electroporation, at which time the pores should have completely healed (*see* Note 14).
14. Test for the presence of the introduced molecules inside the cells using an appropriate assay (*see* Note 15).

4. Notes

1. The purity of the material to be electroporated is of paramount importance. Substances like detergents, preservatives, or antibiotics from the culture media could kill the cells into which they are electroporated.
2. The glass is purchased as 14-in. square sheets and cut to size. Since the resistance can substantially vary from sheet to sheet, care should be taken that each series of experiments is performed using slides of closely matched conductivities, identified by engraving a number on the underside.

3. Fisher rat fibroblast F111 and its polyoma virus transformed derivatives *(18)*, mouse fibroblast NIH 3T3 *(19)*, and mouse NIH 3T6 can be grown in plastic Petri dishes in Dulbecco's Modification of Eagle's Medium (DMEM) supplemented with 5% calf serum in a humidified 5% CO_2 incubator. Mouse C3H10T½ fibroblast derivatives expressing a *ras*-antimessage (e.g., Rev3 *[20]*) are grown in DMEM supplemented with 10% fetal calf serum. The porcine kidney epithelial line LLC-PK_1 (American Type Culture Collection) grows in a 1:1 mixture of Ham's F12 and DMEM with 10% calf serum and 5 µg/mL insulin.
4. In many instances, it may be desirable to observe nonelectroporated cells side by side with the electroporated ones. This can be achieved with the electrode configuration in Fig. 1 by designing the area of cell growth enclosed by the dam as to be slightly larger than the electrode (Fig. 1B [8] and [9]). In this setup, however, observation of the cells under high power necessitates the use of an inverted microscope (e.g., Olympus MT-2). For the cells to be visible with a 20× objective, the glass must be <1 mm thick, which makes it more fragile and more difficult to manipulate and wash. Another disadvantage is that the spacer (Fig. 1 [6]) crushes some of the cells, and it can do extensive damage if the electrode is inadvertently moved sideways during electroporation.

    In the configuration outlined in Fig. 2, a coverslip can be placed on top of the frame [14] after the pulse and the cells observed with a standard fluorescent microscope, or an inverted one. To have nonelectroporated cells as a control, the electrode can be placed so that some of the cells are not covered, as in Fig. 2C [8]. However, contrary to the arrangement in Fig. 1, since the optimal voltage and capacitance depend on the conductive area covered by the electrode rather than the total area of the window, the exact position of the electrode, determining the area receiving the pulse, will affect the optimal voltage (*see* Note 11). A notch on the frame can help position the electrode precisely (Fig. 2C). On the other hand, since the electrode rests on the frame in this configuration, the cells are less likely to be damaged if the electrode is inadvertently moved during electroporation.

    In either setup, cells underneath the edge of the electrode are electroporated to a slightly higher degree than the cells growing on the rest of the conductive surface. This may be because of the lack of current flow through the neighboring region causing the current to pass from the edge of the electrode through a larger area of conductive glass, and hence they experience slightly less resistance. This must be taken into account when the optimal voltage is determined (*see* Note 11).
5. The dam can be made using silicones that specify suitability for aquarium use, such as Dow Corning general-purpose exterior sealant. Fresh silicone

was found to be very toxic to most cells. Therefore, at least 48 h of cure time must be allowed before use. In order to make the dam, place the glass slide over a drawn pattern. The material can be applied directly from a small squeeze tube or a caulking tube. However, much better results are obtained if the silicone is put into a 10-mL syringe without a needle and extruded onto the glass. By turning the glass slide over and looking against the light at an angle, it is possible to determine if there are any gaps in the dam where the silicone meets the glass. The silicone, when cured, makes a permanent rugged dam that can stand up to multiple uses. With care, it is possible to make dams that parallel the perimeter of the electrode within a couple of millimeters. If such a tight fit is desired, then it might be useful to make a small jog in the dam away from the electrode at the large gap end of the electrode. This will provide a path for air to pass beneath the electrode and minimize the suction turbulence that can damage the cells when the electrode is removed from a tightly fitting dam full of fluid.

For applications requiring the dam to be removable, for example, to allow for viewing with a standard, noninverted microscope, a removable sealant can be used. A material marketed as DraftStop under the trade name BullDog Grip (Canadian Adhesives, Brampton, Ontario) proved satisfactory. The material is toxic to many cells when uncured, so it is essential to cure it for 48 h. It forms a good seal and is easily removed.

Frames (Fig. 2 [14]) can be made from PVC electrical tape or adhesive-backed vinyl. They can be cut by hand or made by machine. They are not toxic to the cells, but are not reusable. The authors are using a set of standard sizes of frames that are convenient for different applications. These are: 8×8, 8×12, 8×16, 16×16, and 16×32 mm (longitudinal slide cell window dimension × cross slide cell window dimension).

6. The optimal gap and electrode angle can be determined experimentally by visually assessing cell fluorescence and morphology under phase contrast over the entire area after electroporation of Lucifer yellow at different voltages. The optimal angle depends on the conductivities of the coating and the electroporation medium. The more conductive the glass, the smaller the angle and the amount of solution needed. For glass with a surface resistivity of 15–20 Ω/square and using PBS as electroporation fluid, the optimum angle was found to be approx 0.8°, and the gap between the conductive coating and the distal end of the negative electrode 0.457 mm.

7. The coated slides can be washed and used repeatedly without any obvious effect on cell adhesion or electroporation. The coating is very stable in alkali, but it can be eroded by strong acids, causing changes in its conductivity. If the cells have been fixed on the glass, the slides can be cleaned by incubating in a 1% trypsin solution, followed by autoclaving in 20% Extran 300 deter-

gent (BDH, Toronto, Canada). If the cells have not been fixed, then the slides can be washed in a dishwasher, taking care not to scratch the coating.
8. The size of the electrode depends on the number of cells required for each application. In order to save materials, the smallest number of cells is usually chosen. In addition, to preserve sterility, the slide is placed in a 6- or 10-cm Petri dish. For practical purposes, a slide of 25 × 37 mm can accommodate a growth area of 8 × 16 mm and can fit in a 6-cm Petri dish. A slide of 50 × 50 mm can accommodate a growth area of 16 × 32 mm and can fit in a 10-cm Petri dish.
9. Choice of electroporation buffer: PBS or HEPES-saline have given equally good results. For certain applications, however, such as the introduction of radioactive nucleotides for the labeling of cellular components, it may be desirable to starve cells for phosphate prior to electroporation, in which case a phosphate-free buffer, like HEPES-saline, must be used. In the case of DNA electroporation, it may be desirable to save the solution to use a second time (*see* Note 13). In this case, it may be necessary to purify the DNA preparation from cell debris, since these may be difficult in the presence of phosphate ions, which would tend to precipitate with the DNA.
10. If necessary, the electrode can be sterilized with 70% ethanol before the pulse, and the procedure carried out in a laminar-flow hood, using sterile solutions. Placing the Petri dish on a stand with a mirror (e.g., Cooke Engineering Company), so that its underside is visible, can help position the electrode and avoid air bubbles.
11. Factors affecting the optimal voltage and capacitance: Electrical field strength has been shown to be a critical parameter for cell permeation, as well as viability *(22)*. It is generally easier to select a discrete capacitance value and then precisely control the voltage. The optimal voltage depends on the strain and the metabolic state of the cells, as well as the degree of cell contact with the conductive surface *(6)*. Densely growing, transformed cells or cells in a clump require higher voltages for optimum permeation than sparse, subconfluent cells. For example, Lucifer yellow can be introduced into normal mouse C3H10T½ cells growing on a conductive area of 8 × 8 mm with a pulse of 40–70 V (1-µF capacitance), whereas polyoma virus-transformed C3H10T½ cells require 60–90 V. Similarly, cells that have been detached from their growth surface by vigorous pipeting prior to electroporation require substantially higher voltages (20 V higher for C3H10T½ cells). It is especially striking that cells in mitosis remain intact under conditions where most cells in other phases of the cycle are permeated (Fig. 4). The simplest explanation for the above observations is that cells that are in close contact to the conductive growth surface require a lower voltage, possibly owing to the larger amounts of current passing through an extended cell.

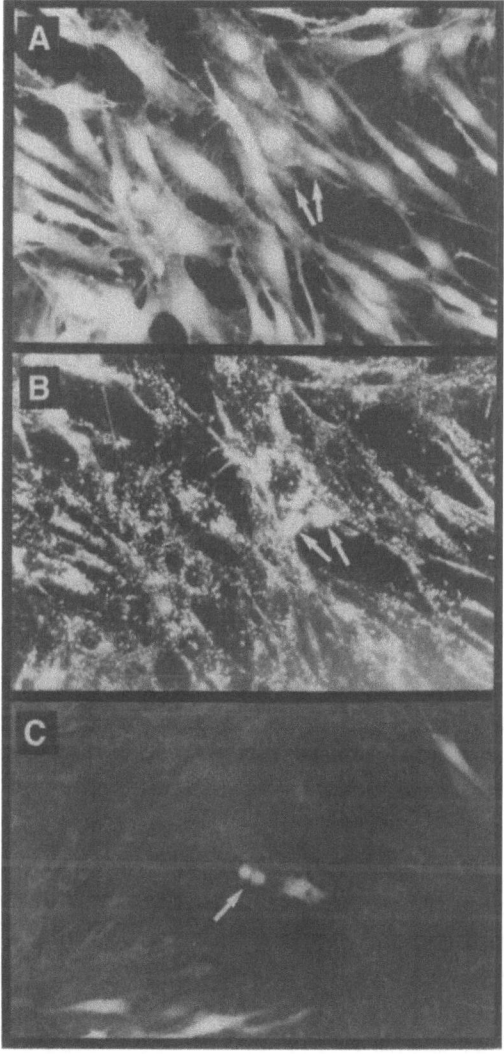

Fig. 4. Electroporation of mitotic F111 cells requires a higher voltage than cells in other phases of the cycle. (**A**) Fluorescence micrograph of F111 cells after introduction of chicken IgG. Cells were fixed and probed with FITC-conjugated antichicken IgG. (**B**) Dark-field micrograph of the same field. Arrows in (A) and (B) point to a cell in late anaphase. This cell exhibits a very faint fluorescence in (A). (**C**) Fluorescence micrograph of F111 cells treated as above with a 98-V pulse. The intensely fluorescent mitotic cell (arrow) stands out against a background of cells, most of which were killed by the pulse. Note that these cells did not retain any electroporated material.

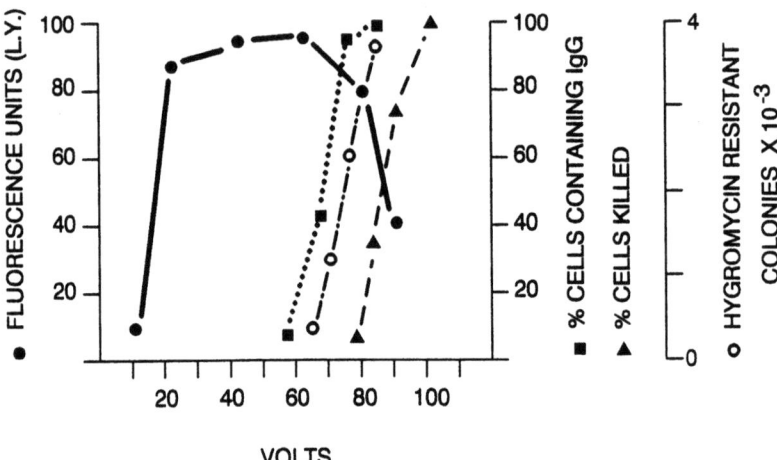

Fig. 5. Effect of field strength on the introduction of different molecules: Pulses of increasing voltage were applied to confluent F111 cells growing on a surface of 8 × 8 mm, from a 0.95-μF capacitor, in the presence of 2 mg/mL Lucifer yellow (●), 2 mg/mL chicken IgG (■), or 100 μg/mL pY3 plasmid DNA (○). Cell killing (▲) was assessed by calculating the plating efficiency of the cells after the pulse. Note that a wider range of voltages (30–65 V) permits efficient introduction of Lucifer yellow with no detectable loss in cell viability than the introduction of IgG or DNA. Points represent averages of four separate experiments. L.Y.: Lucifer yellow.

The limits of voltage tolerance depend on the size and electrical charge of the molecules to be introduced (Fig. 5). For the introduction of a small, uncharged molecule like Lucifer yellow, a wider range of field strengths permits effective permeation with minimal damage to the cells than does the introduction of antibodies or DNA *(6,23–25)*. The results of a typical experiment are shown in Fig. 6. The application of an exponentially decaying pulse of an initial strength of 85 V from a 32-μF capacitor to rat F111 cells growing on a conductive growth area of 17 × 22 mm (gap: 540-μm, angle: 0.8°), resulted in almost 100% of the cells containing the introduced antibodies (Fig. 6). Cells in the area adjacent to that covered by the electrode did not fluoresce, stressing the specificity of the technique. Using exactly the same setup, pulses lower than 65 V did not result in any antibody introduction, whereas pulses of 98 V killed most of the cells, sparing cells in mitosis (Fig. 4). Larger conductive growth areas necessitate higher voltages and/or higher capacitances for optimal permeation. In

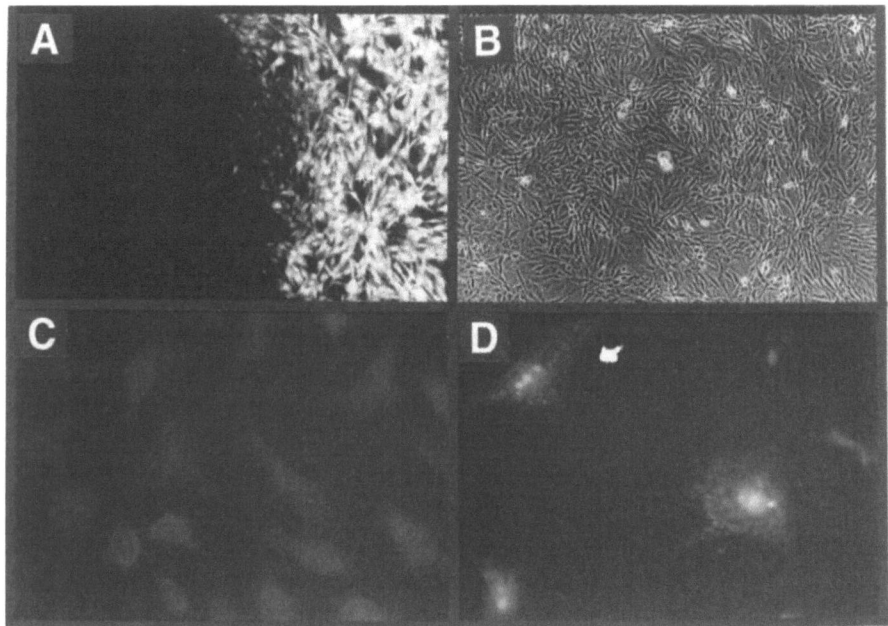

Fig. 6. Electroporation of IgG into F111 cells. (**A**) Low-power fluorescence micrograph of rat F111 cells at the edge of the negative electrode (Fig. 1), electroporated in the presence of chicken IgG. After the pulse, cells were fixed and probed with FITC-conjugated antichicken IgG. Cells at the right side of the picture were covered by the electrode, therefore were submitted to the electrical pulse, whereas cells at the left side were outside this area, hence, they did not receive any current. (**B**) Phase-contrast micrograph of the same field. (**C**) Control cells growing in area [8] (Fig. 1) at the time of pulse delivery were fixed and stained as in (D). Note the very faint, diffuse background fluorescence. (**D**) Fluorescence micrograph of NIH 3T6 cells electroporated with the 2D3 MAb. After the pulse, cells were fixed and probed with FITC-conjugated, antimouse IgG.

the same setup, Lucifer yellow could be introduced into 100% of the cells with pulses as low as 30 V, whereas Basilen blue or nucleotides needed 50–80 V.

Introduction of DNA requires higher voltages for the best yield of stable transfectants than for the introduction of antibodies; a substantial number of cells are killed under these conditions, as evidenced by trypan blue staining (Fig. 5).

Antibodies microinjected into the cytoplasm of adherent cells invariably do not penetrate the nuclei (26). Application of relatively low voltages (70 V) can result in the introduction of antibodies almost exclusively into the cytoplasm of the cells, whereas a slightly higher voltage introduces antibodies into both the cytoplasm and the nucleus. Therefore, careful control of the electric field strength can permit antibody incorporation preferentially into the cytoplasm (Fig. 7A and B). On the other hand, small molecules, such as Lucifer yellow, always penetrate the nuclei, and the fluorescence appears to be concentrated in the nucleus in some lines, possibly because of its volume (Fig. 7C).

The following steps can be used to determine the optimal voltage for the introduction of antibodies into cells growing in an area of 8 × 8 mm: Prepare a series of slides with cells uniformly plated in a 8 × 8 mm window (Fig. 2). Set the apparatus at 1-µF capacitance. Prepare a solution of 2 mg/mL Lucifer yellow, and a solution of 200–2000 µg/mL antibody and 2 mg/mL Lucifer yellow. Use a shim of 0.813 mm and a spacer of two layers of electrical tape (0.457-mm total distance between the electrode and the slide), for a distance of 25 mm between the centers of the electrodes. Electroporate the Lucifer yellow at different voltages to determine the upper limits where a small fraction of the cells (probably the more extended ones) are killed by the pulse, as determined by visual examination under phase-contrast and fluorescence illumination (Figs. 3 and 4C). Depending on the cells, this voltage can vary from 50–110 V. Repeat the electroporation using the antibody solution at different voltages, starting at 10 V below the upper limit and at 2-V increments. The Lucifer yellow offers an easy way to test for cell permeation. Fix the cells with acetone:methanol (1:1) and probe with the fluorescein-labeled secondary antibody.

12. Longer incubation times can result in a considerable amount of pinocytosis (Fig. 3).
13. Since usually only a small fraction of the material penetrates the cells, the solution can be carefully aspirated, filtered, and used again. We have determined through an in vitro binding assay, that the unincorporated 2D3 antibody recovered after electroporation retains cruciform binding activity, and can therefore be filtered and reused.
14. In order to test for cell permeation in initial experiments to determine the optimal voltage, trypan blue can be used, since the demonstration of its presence into the cells does not require a fluorescent microscope. A 0.5% trypan blue solution can be added to the cells immediately after the pulse, and the cells observed under bright-field illumination. Since trypan blue is toxic after introduction into the cells, it should not be included in the electroporation solutions. It is also important that trypan blue does not

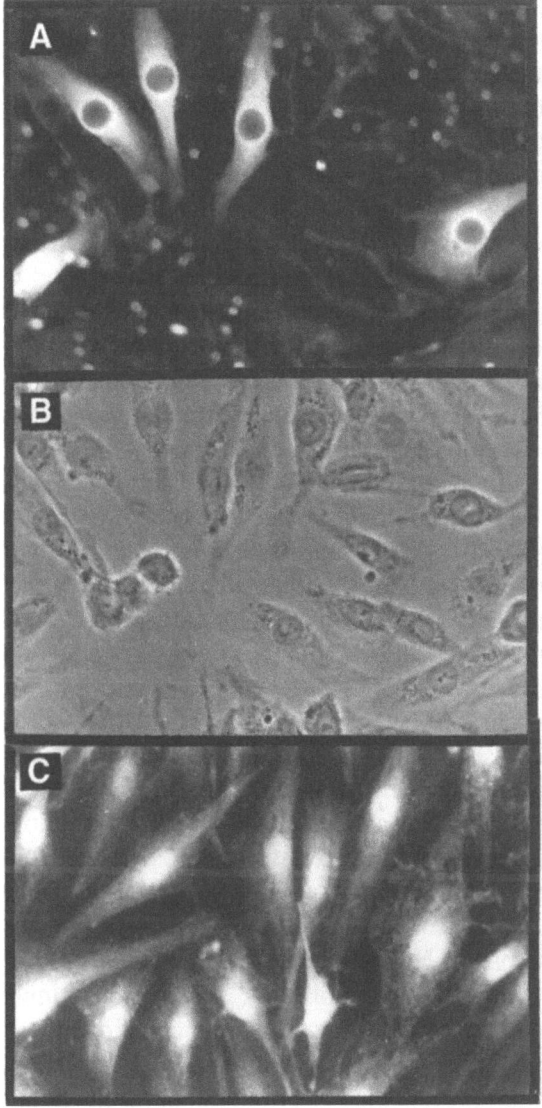

Fig. 7. Lower field strengths can introduce antibodies preferentially into the cytoplasm. (**A**) Fluorescence micrograph of cells electroporated with 2 mg/mL chicken IgG under exactly the same conditions as in Figs. 5 and 6, but with a lower voltage (70 instead of 85 V), fixed and probed with FITC-conjugated antichicken IgG. Note that in most cells, fluorescence is seen only in the cytoplasm. (**B**) Phase-contrast micrograph of the same field. (**C**) Fluorescence micrograph of cells electroporated in the presence of 200 μg/mL Lucifer yellow at 70 V. Note the apparent concentration of the dye into the nuclei.

come into direct contact with the electrode, since washing the electrode for subsequent use in electroporation can be difficult. Lucifer yellow, on the other hand, is not toxic to the cells so that it can be included in all electroporation solutions. Its spectrum is similar to fluorescein, but it is washed off during the fixing and staining of the cells with the FITC-conjugated secondary antibody, so it does not interfere with antibody detection.

15. A variety of nonpermeant molecules can be introduced using the above general method for different applications. As a general rule, the introduction of small, uncharged molecules can be achieved with lower voltages that have minimal effects on cellular metabolism than larger ones (Fig. 5).

For the introduction of antibodies, the cells can be electropermeated as above in the presence of highly purified antibodies, and extensively dialyzed against the electroporation buffer (*see* Note 9). After the pulse, the antibody solution is aspirated and the cells are washed three times with growth medium (*see* Note 12). The cells are fixed with formalin directly on the electroporation glass slides, and the presence of antibodies in the cell can be assessed by immunostaining using a fluorescein-conjugated secondary antibody, or any other suitable detection method (Fig. 6).

For the stable expression of DNA, cells can be electropermeated in the presence of 50–100 µg/mL of DNA (e.g., pY3, containing the gene coding for hygromycin resistance under control of the Moloney Murine Leukemia Virus promotor *[27]*). After the shock, the DNA solution is removed, and the cells washed with medium and placed in the incubator in complete medium. Once the cells have recovered from the pulse, approx 5 h later, they are trypsinized and passaged at a 1:10 dilution into Petri dishes. Selection for hygromycin resistance can start 2–3 d after electroporation and the number of resistant colonies scored 2–3 wk later *(19)*.

Basilen blue can be electroporated into LLC-PK$_1$ cells at a concentration of 1 mM in HBSS. The cells are incubated for 5 min at 37°C and washed with HBSS. One milliliter of serum-containing medium is added to the cells, prior to a further 10-min incubation at 37°C. The delivery of Basilen blue is confirmed by microscopic observation under bright-field illumination. Basilen blue electroporation results in inhibition of glyceryl trinitrate biotransformation, measured by megabore capillary column gas–liquid chromatography *(9)*.

For the introduction of Lucifer yellow, cells can be electroporated in the presence of 2 mg/mL of the dye in PBS or HEPES-saline. If desirable, the amount of Lucifer yellow retained by the cells can be quantitated as follows: Cells are washed three times with PBS and hypotonically lysed by vigorous pipeting in distilled water. The lysate is transferred into quartz cuvets and its fluorescence measured using a Perkin-Elmer model 204A fluorescence spectrophotometer (excitation: 423 nm; emission: 555 nm).

In a modification of the above procedure, Lucifer yellow can be introduced for the study of intercellular, junctional communication. In this case, the conductive coating is removed lengthwise from half the slide, including the area within the insulating window (Fig. 2D [3] and [15]). Cells growing on this nonconductive surface would not receive any pulse. Therefore, they would not be permeated (Fig. 2D [15]). The removal of the conductive coating can be achieved by etching with a solution of $4M$ HCl where zinc powder is freshly added.

Acknowledgments

The authors wish to thank J. F. Whitfield for many helpful discussions, Danwei Wang for expert laboratory assistance, and Stanley Liu for his very enthusiastic technical help. The financial assistance of the Natural Sciences and Engineering Research Council of Canada and the Cancer Research Society Inc. to L. R., as well as grants from the Cancer Research Society Inc. and the Medical Research Council of Canada to M. Z. H., and grants from the Medical Research Council of Canada to B. M. B. and L. W. M., who was also the recipient of an Ontario Ministry of Health career scientist award, are acknowledged. H. B. was the recipient of a studentship from the Medical Research Council of Canada and a Queen's University graduate school travel grant and a Microbix Biosystems Inc. award.

References

1. Satyabhama, S. and Epstein, A. L. (1988) Short-term efficient expression of transfected DNA in human hematopoietic cells by electroporation: definition of parameters and use of chemical stimulators. *DNA* **7**, 203–209.
2. Knutson, J. C. and Yee, D. (1987) Electroporation: parameters affecting transfer of DNA into mammalian cells. *Anal. Biochem.* **164**, 44–52.
3. Potter, H., Weir, L., and Leder, P. (1984) Enhancer-dependent expression of human kappa immunoglobulin genes introduced into mouse pre-B lymphocytes by electroporation. *Proc. Natl. Acad. Sci. USA* **81**, 7161–7165.
4. Neumann, E., Schaefer-Ridder, M., Wang, Y., and Hofschneider, P. H. (1982) Gene transfer into mouse lyoma cells by electroporation in high electric fields. *EMBO J.* **7**, 841–845.
5. Winegar, R. A., Phillips, J. W., Youngblom, J. H., and Morgan, W. F. (1989) Cell electroporation is a highly efficient method for introducing restriction endonucleases into cells. *Mutat. Res.* **225**, 49–53.
6. Raptis, L. and Firth, K. L. (1990) Electroporation of adherent cells in situ. *DNA Cell Biol.* **9**, 615–621.

7. Chakrabarti, R., Wylie, D. E., and Schuster, S. M. (1989) Transfer of monoclonal antibodies into mammalian cells by electroporation. *J. Biol. Chem.* **264**, 15,494–15,500.
8. Mir, L. M., Banoun, H., and Paoletti, C. (1988) Introduction of definite amounts of nonpermeant molecules into living cells after electropermeation: direct access to the cytosol. *Exp. Cell Res.* **175**, 15–25.
9. Simon, W. C., Raptis, L., Pang, S. C., and Bennett, B. M. (1993) Comparison of liposome fusion and electroporation for the intracellular delivery of nonpermeant molecules to cultured, adherent cells. *J. Pharmacol. Toxicol. Methods* **29**, 29–35.
10. Weaver, J. C. (1993) Electroporation: a general phenomenon for manipulating cells and tissues. *J. Cell. Biochem.* **51**, 426–435.
11. Chang, D. C., Chassy, B. M., Saunders, J. A., and Sowers, A. E. (1992) *Guide to Electroporation and Electrofusion.* Academic, San Diego, CA.
12. Tur-Kaspa, R., Teicher, L., Levine, B. J., Skoultchi, A. I., and Shafritz, D. A. (1986) Use of electroporation to introduce biologically active foreign genes into primary rat hepatocytes. *Mol. Cell Biol.* **6**, 716–718.
13. Matsumura, T., Konishi, R., and Nagai, Y. (1982) Culture substrate dependence of mouse fibroblasts survival at 4°C. *In Vitro* **18**, 510–514.
14. Kwee, S., Nielsen, H. V., and Celis, J. E. (1990) Electropermeabilization of human cultured cells grown in monolayers. Incorporation of monoclonal antibodies. *Bioelectrochem. Bioenerg.* **23**, 65–80.
15. Zheng, Q. and Chang, D. C. (1991) High-efficiency gene transfection by in situ electroporation of cultured cells. *Biochim. Biophys. Acta* **1088**, 104–110.
16. Frappier, L., Price, G. B., Martin, R. G., and Zannis-Hadjopoulos, M. (1987) Monoclonal antibodies to cruciform DNA structures. *J. Mol. Biol.* **193**, 751–758.
17. Frappier, L., Price, G. B., Martin, R. G., and Zannis-Hadjopoulos, M. (1989) Characterization of the binding specificity of two anticruciform DNA monoclonal antibodies. *J. Biol. Chem.* **264**, 334–341.
18. Raptis, L., Lamfrom, H., and Benjamin, T. L. (1985) Regulation of cellular phenotype and expression of polyomavirus middle T antigen in rat fibroblasts. *Mol. Cell. Biol.* **5**, 2476–2486.
19. Raptis, L. and Bolen, J. B. (1989) Polyoma virus transforms rat F111 and mouse NIH3T3 cells by different mechanisms. *J. Virol.* **63**, 753–758.
20. Lu, Y., Raptis, L., Anderson, S., et al. (1992) p21ras modulates commitment and maturation of 10T1/2 fibroblasts to adipocytes. *Biochem Cell. Biol.* **70**, 1249–1257.
21. Bahnson, A. B. and Boggs, S. S. (1990) Addition of serum to electroporated cells enhances survival and transfection efficiency. *Biochem Biophys. Res. Commun.* **171**, 752–757.
22. Zimmermann, U. (1982) Electric field mediated fusion and related electrical phenomena. *Biochim. Biophys. Acta* **694**, 227–277.
23. Raptis, L., Brownell, H. L., Firth, K. L., and MacKenzie, L. W. (1994) A novel technique for the study of intercellular, junctional communication; electroporation of adherent cells on a partly conductive slide. *DNA Cell Biol.* **13**, 963–975.
24. Raptis, L., Liu, S. K. W., Brownell, H., Firth, K. L., Stiles, C. D., and Alberta, J. A. (1995) Applications of electroporation of adherent cells in situ on a partly conductive slide. *Mol. Biotechnol.*, in press.

25. Raptis, L. H., Liu, S. K. W., Firth, K. L., Stiles, C. D., and Alberta, J. A. (1995) Electroporation of peptides into adherent cells in situ. *Biotechniques* **18,** 104–114.
26. Yu, C. L., Tsai, M. H., and Stacey, D. W. (1988) Cellular ras activity and phospholipid metabolism. *Cell* **52,** 63–71.
27. Blochlinger, K. and Diggelman, H. (1984) Hygromycin B phosphotransferase as a selectable marker for DNA transfer experiments with higher eucaryotic cells. *Mol. Cell. Biol.* **4,** 2929–2931.

CHAPTER 8

Electrotransformation of Chinese Hamster Ovary Cells

Danielle Gioioso Taghian and Jac A. Nickoloff

1. Introduction

Chinese hamster ovary (CHO) cells possess many advantages that make them among the most widely used and highly characterized mammalian cell lines. They are easy to culture, they grow rapidly without density dependence, and they do not have stringent culture medium or high serum growth requirements. CHO cell lines have a high plating efficiency, and many lines can grow either as a monolayer or in suspension. These features, and techniques for replica-plating CHO cells from culture dishes onto agar plates, have facilitated the isolation of mutant CHO strains, such as mutants that are UV-sensitive *(1,2)*, X-ray-sensitive *(1,3)*, DNA-repair-deficient *(1)*, glycosylation-deficient *(4,5)*, and have high sister chromatid exchange rates *(6)*. The isolation of mutants has led to the identification of human genes that complement the defective cell function and have been valuable to investigations of DNA damage and repair, recombination, cellular transformation, and protein compartmentalization. Many of these studies have utilized transfection technology to introduce selectable vectors and genes into CHO cell lines.

Although efficient transfection of CHO cells has been accomplished by both chemical methods and electroporation, these methods induce the uptake of DNA via different mechanisms. For example, with DEAE dextran and calcium phosphate coprecipitation (CPP), DNA-chemical aggregates enter the cells via endocytotic vesicles, with the DNA often becoming damaged and integrating as multiple copies in the genomic DNA. The electroporation mechanism is less understood, but because the DNA enters through transient holes formed in the cell membrane, packaging steps that cause DNA damage during chemical-mediated transfection are avoided, and the DNA may suffer less damage. Electroporation is not limited to DNA transfection; other macromolecules, such as RNA, regulatory proteins, and enzymes, have been electroporated into CHO cells.

A typical electrotransfection requires exponentially growing CHO cells to be washed once and harvested in prewarmed (37°C) phosphate-buffered saline (PBS). An appropriate type and amount of DNA is mixed with the resuspended cells in a cuvet where an electrical pulse is administered. The treated cells are plated in cell-culture dishes or wells, where a drug may be added to select for transformants. DNA type, transfection parameters, and selections will be discussed in Section 4.

2. Materials

1. CHO strains: Various CHO strains are commercially available through the American Type Culture Collection (ATCC) (Rockville, MD), including K1c and AA8 (*see* Note 1).
2. Plasmid DNA: Suitable plasmids with selectable markers include pMSG (Pharmacia, Piscataway, NJ), pSV2neo *(7)*, and pSV2gpt *(7)* (*see* Note 2).
3. Dulbecco's Ca^{2+}-, Mg^{2+}-free PBS (Gibco-BRL, Life Technologies, Inc., Grand Island, NY).
4. 0.15% Trypsin (Gibco-BRL).
5. Culture medium: αMEM (Gibco-BRL), supplemented with either 10% fetal or calf bovine serum (Sigma, St. Louis, MO) (*see* Note 3), 100 U/mL penicillin, and 100 μg/mL streptomycin.
6. Selective drugs: The specific drug to use depends on the marker carried by the vector DNA (*see* Notes 4 and 5 and Table 1). MAX stock solutions: 8.3 mg/mL mycophenolic acid (Sigma) in 95% ethanol, 5 mg/mL adenine (Sigma) in $0.1M$ HCl, 25 mg/mL xanthine (Sigma) in $0.025N$ NaOH to increase solubility. Store stock solutions at 4°C. Geneticin (G418) (Sigma) stock solution: 100 mg/mL in H_2O, stored at −20°C for up to 1 mo.
7. Dexamethasone (DEX) (Sigma) stock solution: 0.1 mM in H_2O (*see* Notes 2 and 3).

Table 1
Positively Selectable Markers in CHO-K1c Cells

Genetic marker	Expression vector	Selective drug	Drug concentration[a]
Aminoglycoside transferase	pSV2neo	G418	500 µg/mL
Xanthine-guanine phosphoribosyltransferase	pSV2gpt	Mycophenolic acid Adenine Xanthine	10 µg/mL 12.5 µg/mL 250 µg/mL
Hygromycin-B-phosphotransferase	pHyg	Hygromycin-B *(17)*	400 µg/mL

[a] Drug concentrations apply to CHO-K1c cells (*see* Note 4).

3. Methods

3.1. Cell Growth and Harvesting

1. Culture CHO cells for 2 d prior to harvest at a density that will yield 4×10^6 exponentially growing cells/electroporation. A 175-cm² culture flask normally yields 3×10^7 cells.
2. Aspirate growth media from 175-cm² culture flask.
3. Rinse cells once with 10 mL of prewarmed (37°C) PBS and aspirate.
4. Add 2.5 mL of trypsin, and incubate flask at 37°C for 2 min, making sure a thin layer of trypsin covers the entire bottom surface of the flask.
5. Detach cells by gentle tapping. Add 10 mL of prewarmed (37°C) culture media, and pipet cells into a single-cell suspension.
6. Transfer the suspension into a 50-mL centrifuge tube. Remove a 0.5-mL aliquot of the suspension, dilute appropriately with PBS, and count by hemocytometer or cell counter.
7. Centrifuge at 200*g* for 10 min at room temperature. Aspirate culture media from tube, without disturbing the cell pellet. Resuspend cells in 10 mL of PBS with gentle pipeting.
8. Centrifuge as above. Aspirate PBS, and then resuspend cells in PBS to a final concentration of 5.33×10^6 cells/mL.

3.2. Electroporation

1. The DNA to be electroporated should be suspended as a 50-µL aliquot in a 1.5-mL tube (*see* Notes 6 and 7).
2. To this tube, add 750 µL of the CHO cell suspension, and transfer the entire solution to a sterile electroporation cuvet (*see* Note 8). Cap the cuvet with a 4-mm electrode gap (*see* Note 9).

3. Place the cuvet in the electroporation chamber, and deliver an exponential decay pulse of 0.3 kV and 960-µF capacitance. These settings should produce a time constant of 12–14 ms.
4. Transfer the contents of the cuvet to a centrifuge tube containing 20 mL of prewarmed 37°C culture medium. Mix the cells by pipeting.
5. Remove a 50-µL aliquot, and add it to 450 µL of culture medium in a 1.5-mL tube. To determine cell survival after electroporation, add 50 µL from the 1.5-mL tube into a 100-mm dish containing 10 mL of culture medium (see Note 10).
6. Divide the remaining cells into two 100-mm dishes, 10 mL each.
7. Place all dishes into a 37°C humidified incubator containing 5% CO_2 and 95% air (see Notes 11 and 12).
8. Add selective drugs 16–24 h after plating. Colonies will form in 7 d under nonselective conditions and in 10–14 d under selective conditions.

4. Notes

1. CHO K1c cells grow as adherent monolayers, and slight movement of plated cells usually does not dislodge many cells. Because AA8 cells can grow as either a monolayer or in suspension, plated cells should not be moved for the duration of an experiment. Movement will cause colonies to become comet-shaped as the dislodged cells grow in their new position.
2. Stable transfectants carrying pSV2neo and pSV2gpt are selected by G418 and MAX, respectively (see Note 4). The SV40 promoter is expressed constitutively at high levels. The pMSG vector carries the DEX inducible MMTV promoter upstream of a polylinker into which different gene cassettes may be subcloned. pMSG also contains an SV40 promoter-driven bacterial xanthine-guanine phosphoribosyl transferase gene (gpt).
3. The MMTV promoter is inducible by the glucocorticoid hormone dexamethasone, is leaky under noninduced conditions, and can thus be used to express genes at high and low levels (8). We use fetal bovine serum to culture CHO strains with MMTV promoter constructs because it contains additional hormones not found in calf bovine serum. MMTV induction can be achieved within 30 min in CHO K1c by maintaining the cells in media supplemented with 10% fetal bovine serum and 1 μM DEX. Since the SV40 promoter is not inducible, we normally culture CHO strains containing SV40 promoter constructs in less expensive calf bovine serum.

4. Different CHO strains and cell lines have different sensitivities to drugs. A drug toxicity experiment should be performed for each new strain or line to determine the concentration that kills all nontransfected cells, but does not affect plating efficiencies of transfected cells.
5. One method to select for transfected CHO cells requires that the input vector gene complements a deficiency in the CHO strain. For example, CHO cells with defective thymidine kinase genes (tk^-) can be transfected with pTK2 *(9,10)*, a plasmid carrying a recombinant *tk* gene, and selected in medium containing hypoxanthine, aminopterin, and thymidine (HAT medium).

   CHO (DUKX B11) *(11)* cells have a defective dihydrofolate reductase gene, and will grow only in medium supplemented with thymidine, glycine, and purines. DHFR mutant cells *(11)* transfected with a *dhfr*-containing vector, such as pMT2 *(12)*, are able to grow without the nutrient-supplemented media. Transfection of dhfr⁻ cells with vectors containing *dhfr* linked to a gene of interest, followed by stepwise methotrexate selection, amplifies *dhfr* and the linked gene *(5)*.
6. Vector DNA is usually electroporated in a linear form to promote integration into genomic DNA. Expression vectors lacking a selectable marker may be cotransfected with a second selectable vector. If both plasmids are linearized, they will often become linked during electroporation *(13)*, with up to 90% of the selectable cells receiving both DNAs.
7. Carrier DNA may be added to enhance gene transfer, possibly by acting as a competitive inhibitor of cellular nucleases. Carrier DNA may consist of homologous DNA, heterologous human DNA, or salmon sperm DNA. A study using linear pSV2gptΔ DNA as a selectable vector electroporated with either circular or linear pSV2neo carrier DNA showed that carrier DNA type and form influence transfection efficiencies *(14)*. Although circular carrier plasmids stimulated electrotransfection, linear carrier plasmids had an inhibitory effect. In addition, transfection efficiency decreased with bacterial chromosome carrier DNA, but increased with calf thymus DNA and human DNA present in a cosmid.
8. Limiting dilutions of DNA are electroporated in order to obtain single-copy integrants, which are more likely to arise from low DNA concentrations. We select colonies from plates that contain 1–5 transfectants (*see* Fig. 1). Multiple-copy integration can be achieved by calcium phosphate coprecipitation, from which >40 copies/100 ng plasmid DNA and 10 µg carrier DNA have been reported for mouse LM cells *(15)*. Higher DNA concentrations must be used to obtain multiple-copy integration with electroporation. We have obtained 10–20 copies/40 µg plasmid DNA (unpublished results).

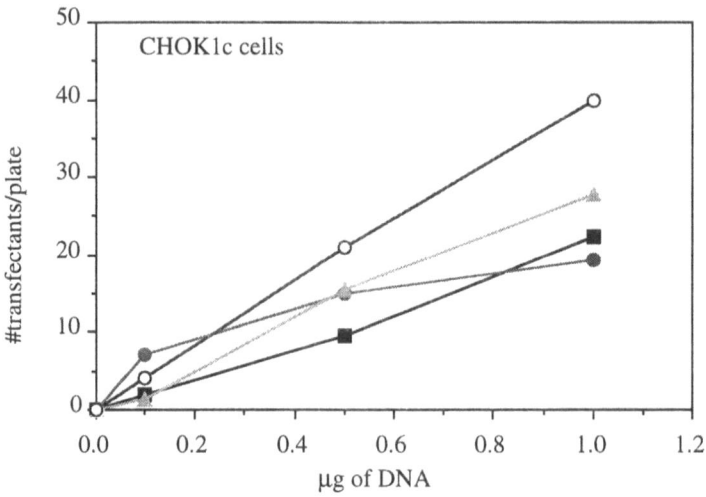

Fig. 1. The number of transfectants as a function of DNA concentration during electroporation. Four similar pMSGneo (16) constructs were each electroporated at three concentrations with a Bio-Rad (Hercules, CA) Gene Pulser. The number of transfectants increases linearly with the concentration of DNA. To obtain single-copy integrants, transfectants should be chosen from plates with the fewest number of colonies.

9. Seven hundred and fifty microliters of the final CHO cell dilution are mixed with a 50-µL aliquot of DNA, so that each electroporation cuvet contains 4×10^6 cells in a final volume of 0.8 mL.
10. The dilution in the 1.5-mL tube is plated to determine the number of viable cells after treatment. Approximately 50–70% CHO K1c cell death is expected from electroporation. As calculated, about 1000 cells are plated, and approx 300–500 colonies should arise within 10 d.
11. We perform additional controls that include:
    a. CHO cells, no DNA, and a pulse;
    b. CHO cells, DNA, and no pulse; and
    c. CHO cells, no DNA, and no pulse.
    Plate control cells in selective and nonselective media as described in Section 3.2. If any colonies arise, the drug concentration may be incorrect, and a cell-killing experiment should be performed (see Note 4).
12. We do not move or refeed plates for the duration of the experiment. If refeeding is necessary with a certain cell type or selective drug, move the plates in a manner that minimizes cell dislodgement (see Note 1).

References

1. Thompson, L. H., et al. (1987) Recent progress with the DNA repair mutants of Chinese hamster ovary cells. *J. Cell Sci. Supp.* **6,** 97–110.
2. Nairn, R. S., Humphrey, R. M., and Adair, G. M. (1988) Transformation of UV-hypersensitive Chinese hamster ovary cell mutants with UV-irradiated plasmids. *Intl. J. Radiat. Biol. & Relat. Stud. Phys., Chem. & Med.* **53,** 249–260.
3. Jeggo, P. A. and Kemp, L. M. (1983) X-ray-sensitive mutants of Chinese hamster ovary cell line. Isolation and cross-sensitivity to other DNA-damaging agents. *Mutat. Res.* **112,** 313–327.
4. Stanley, P. and Siminovitch, L. (1977) Complementation between mutants of CHO cells resistant to a variety of plant lectins. *Somatic Cell Genet.* **3,** 391–405.
5. Kaufman, R. J. and Sharp, P. A. (1982) Amplification and expression of sequences cotransfected with a modular dihydrofolate reductase complementary DNA gene. *J. Mol. Biol.* **129,** 611–618.
6. Hoy, C. A., Fuscoe, J. C., and Thompson, L. H. (1987) Recombination and ligation of transfected DNA in CHO mutant EM9, which has high levels of sister chromatid exchange. *Mol. Cell. Biol.* **7,** 2007–2011.
7. Southern, P. J. and Berg, P. (1982) Transformation of mammalian cells to antibiotic resistance with a bacterial gene under control of the SV40 early region promoter. *J. Mol. Applied Genet.* **1,** 327–342.
8. Beutti, E. and Diggelman, H. (1981) Cloned mouse mammary tumor virus DNA is bioloically active in transfected mouse cells and its expression is stimulated by glucocorticoid hormones. *Cell* **23,** 335–345.
9. Perucho, M., et al. (1980) Isolation of the chicken thymidine kinase gene by plasmid rescue. *Nature* **285,** 207–210.
10. Colbere-Garapin, F., et al. (1979) Cloning of the active thymidine kinase gene of herpes simplex virus type 1 in Escherichia coli K-12. *Proc. Natl. Acad. Sci. USA* **76,** 3755–3759.
11. Urlaub, G. and Chasin, L. A. (1980) Isolation of Chinese hamster cell mutants deficient in dihydrofolate reductase activity. *Proc. Natl. Acad. Sci. USA* **77,** 4216–4220.
12. Kaufman, R. J., et al. (1989) The phosphorylation state of eucaryotic initiation factor 2 alters translational efficiency of specific mRNAs. *Mol. Cell. Biol.* **9,** 946–958.
13. Perucho, M., Hanahan, D., and Wigler, M. (1980) Genetic and physical linkage of exogenous sequences in transformed cells. *Cell* **22,** 309–317.
14. Nickoloff, J. A. and Reynolds, R. J. (1992) Electroporation-mediated gene transfer efficiency is reduced by linear plasmid carrier DNAs. *Anal. Biochem.* **205,** 237–243.
15. Lin, F.-L. and Sternberg, N. (1984) Homologous recombination between overlapping thymidine kinase gene fragments stably inserted into a mouse cell genome. *Mol. Cell. Biol.* **4,** 852–861.
16. Nickoloff, J. A. (1992) Transcription enhances intrachromosomal homologous recombination in mammalian cells. *Mol. Cell. Biol.* **12,** 5311–5318.
17. Sugden, B., Marsh, K., and Yates, J. (1985) A vector that replicates as a plasmid and can be efficiently selected in B-lymphoblasts transformed by Epstein-Barr virus. *Mol. Cell. Biol.* **5,** 410–413.

CHAPTER 9

Electroporation of Rat Pituitary Cells

Ruth H. Paulssen, Eyvind J. Paulssen, and Kaare M. Gautvik

1. Introduction

Electroporation has become a useful tool for the introduction of biomolecules to a variety of cell types since the method was first described *(1,2)*. This technique has also been used for important functional studies in the different GH cell strains: rat pituitary tumor cells that produce prolactin or growth hormone. In most cases, transient-expression experiments have been used to study cellular processes, such as rat growth hormone gene expression *(3)*, prolactin promotor activity *(4–7)*, and the transcription regulatory properties of hormones, such as thyroid hormones and retinoic acid *(8–12)*.

Our long-term goal has been to analyze the intricacies of the cellular signal sequences activated by peptide hormones. As part of these studies, we have used antisense interference experiments for the inhibition of expression of intracellular proteins of importance for signal system activation. Thus, by introducing antisense RNA expression vectors, we have studied the influence of specific antisense RNA action on growth hormone mRNA expression and hormone secretion *(13)*, as well as on the expression and function of the stimulatory GTP binding protein, $G_s\alpha$, which conveys the actions of thyrotropin-releasing hormone and vasoactive intestinal peptide *(14)*.

For transfection studies, GH cells represent a relatively difficult system to work with, since they are small (spindle-shaped or spherical, with a diameter about 7 μm), with a high membrane electrical resistance. After testing different methods, it was possible to define an electroporation method and characterize the parameters that gave an efficient transfection of GH_12C_1 and GH_3 cells. We were also able to assess the general biological effects these handling procedures have on the cells (15). This method is described in detail in Section 3.3. Other methods applicable to these and other rat pituitary cell lines are also discussed.

2. Materials

1. GH_3 or GH_1 cells: American Type Culture Collection (Rockville, MD), cat. no. CCL 82.1 and CCL 82 (see Note 1).
2. Ham's F-10 cell culture medium (ICN Biomedicals Inc., Costa Mesa, CA) supplemented with 6.5% horse serum (Gibco-BRL, Gaithersburg, MD), 3% fetal calf serum (Gibco-BRL), 50 U/mL penicillin (Gibco-BRL), 50 μg/mL streptomycin (Gibco-BRL), and 2.5 μg/mL amphotericin B (ICN Biomedicals Inc.).
3. Trypsin (Gibco-BRL), 0.1% solution.
4. Sucrose (Merck, Darmstadt, Germany), $1M$ stock solution; filter-sterilize.
5. HEPES (N-hydroxyethylpiperazine-N'-2-ethane sulfonic acid) (Sigma, St. Louis, MO): $1M$ stock solution, pH 7.4; filter-sterilize and store at room temperature.
6. Phosphate-buffered saline (PBS), Dulbecco's formula (ICN Biomedicals Inc.): Dissolve one tablet in 100 mL H_2O, and autoclave.
7. Trypan blue solution: 0.4% in 0.85% saline (Gibco-BRL).
8. Electroporation buffer: 272 mM sucrose, 8 mM HEPES, pH 7.4.

3. Methods

3.1. Cell Culture

1. Grow GH cells as monolayer cultures in supplemented Ham's F-10 medium (16) at 37°C in a humidified atmosphere of 95% air and 5% CO_2 (17).
2. Change medium every 2–3 d and always 24 h before an experiment (see Notes 1 and 2).

3.2. Preparation of DNA

1. Purify plasmid DNA by two rounds of CsCl gradient centrifugation (see Note 4).
2. Dissolve DNA in sterile, deionized water; contaminating salts may interfere with the electroporating current (see Notes 4, 5, and 7).
3. Check the quality of the DNA by electrophoresis in a 0.8% agarose gel in a suitable buffer (e.g., TBE) followed by ethidium bromide staining.

A high ratio (>95%) of supercoiled to relaxed (or linearized) plasmid is crucial (*see* Note 4).
4. Determine DNA concentration by absorption at 260 nm (1 A_{260} unit = 50 μg/mL DNA). Dry the quantity to be used in the electroporation experiment in a sterile 1.5 mL microcentrifuge tube using a SpeedVac vacuum centrifuge, and dissolve it in 0.1 mL of electroporation buffer. Store the DNA sample on ice.

3.3. Electroporation

1. Wash the monolayer cells twice with cold PBS (*see also* Notes 1 and 2). Harvest the cells by brief trypsination (5–10 min at 37°C in 0.1% trypsin) *(18)*. Wash the cells twice by resuspending in ice-cold electroporation buffer and pelleting at 650–700g at 4°C. Resuspension of cell pellets at this stage must be done with great care to prevent disrupting the fragile cells.
2. Resuspend the cells at a concentration of $0.5 \times 10^6 - 1 \times 10^7$ cells/mL in ice-cold electroporation buffer.
3. Dissolve DNA in 0.1 mL ice-cold electroporation buffer. Optimal amounts of DNA are 50 μg for $GH_1 2C_1$ cells and 25 μg for GH_3 cells (*see* Notes 4, 5, and 7).
4. Combine 0.7 mL of the cell suspension with the 0.1 mL DNA solution, and resuspend carefully.
5. Transfer 0.8 mL of suspension to a sterile electroporation cuvet with a 0.4-cm electrode gap. Allow the suspension to reach room temperature before electroporation. Resuspend cells again, and mount the cuvet immediately into the electroporation chamber.
6. Electroporate cells at 240–300 V with a capacitance setting of 25 μF. The optimal parameters are 300 V for $GH_1 2C_1$ cells and 240 V for GH_3 cells.
7. Test cell viability by taking an aliquot (20–50 μL) of the electroporated cells, add an equal volume of trypan blue solution, and mix gently. Count the cells under a microscope in order to determine the ratio of dead to viable cells. Cells that do not take up the blue dye are counted as viable. A viability of approx 50% is optimal (*see* Note 3).
8. Resuspend the cell suspension carefully, divide into three equal portions, and add each to 8 mL of prewarmed culture medium in a 9-cm Petri dish.
9. Grow cells at 37°C, and allow viable cells to settle for 24 h. Change the culture medium every 24 h in order to discard floating cells and cell debris.
10. Cell parameters, such as transient gene expression or hormone production, can be assayed at any time-point. Suggested optimal time-points are 24–72 h after electroporation (*see* Notes 2, 6, and 8).

Table 1
Hormone Production in the Different GH Cell Lines[a]

Cell line	PRL production	GH production
GH_1	+	++
$GH_1 2C_1$	0	++
GH_3	+	+++
$GH_4 C_1$	+++	+
GC	0	+++

[a] The ability of the different rat pituitary tumor cell lines to produce and secrete the hormones prolactin (PRL) and growth hormone (GH) is indicated as none (0), poor (+), good (++), and very good (+++). The table is based on the work of others (19,20,22,23), as well as our own.

4. Notes

1. The GH cell lines were established by cloning of cells from a transplantable rat pituitary tumor (MtT/W5) (19,20), originally induced in a Wistar-Furth rat by irradiation (21). The clonal strains and subclones, GH_1, $GH_1 2C_1$, GH_3, and $GH_4 C_1$, and GC cells have preserved organ-specific functions with respect to regulated hormone secretion and synthesis (20,22). GH cells have for many years been an important model system for the study of regulation of pituitary hormone production (22–24). The GH cell strains differ in phenotype, and also produce and secrete varying levels of prolactin and/or growth hormone (Table 1), although generally, they respond to a variety of physiological signals similar to their normal pituitary counterparts. To our knowledge, there is no correlation between hormone production and the ability to express foreign genes. The cells have been characterized with respect to different surface and intracellular receptors for several hormones, such as the hypothalamic-releasing factors thyrotropin-releasing hormone and vasoactive intestinal peptide, as well as the inhibitory hormone somatostatin. In addition, there are receptors for thyroid hormone, corticosteroid hormone, and male and female sex steroid hormones (17,25–27).
2. All experiments should be carried out using cells in exponential phase of growth, i.e., passed (subcultivated) or harvested before reaching confluency. Hormone production (prolactin and/or growth hormone; see Table 1) can be checked by radioimmunoassay (RIA) of medium samples. More than 99% of the hormones produced during a 24-h period accumulate stably in the culture medium (28), thus reflecting the total hormone production.

3. An immediate parameter for the determination of the effect of the electroporation is cell viability, determined by the exclusion of trypan blue. Even small alterations in electroporation settings can cause unfavorable results. Recent reports have shown that effective electroporation seems to occur when the cell viability is about 20–50% *(29,30)*. In our hands, optimal DNA transfer into GH_12C_1 and GH_3 cells is achieved when cell viability after electroporation is approx 50%. However, we recommend that the cell viability optima at different voltage settings be determined individually for each cell line used in an experiment.
4. Introduction of DNA to GH cells takes place in the form of supercoiled plasmids containing the relevant gene or DNA fragment. In our experiments, we obtained a higher transient-expression efficiency when supercoiled, not linearized, DNA was used. For other cell systems, there are reports of both lower *(31–33)* and higher *(34)* transformation efficiency when using linearized DNA. The use of DNA purified with commercially available techniques is probably also valid.

    Changes in transfection efficiency will also result from variations in the quality of the electroporation buffer, which could vary with different batches of chemicals. We recommend that large batches of electroporation buffer be made and stored as aliquots at –20°C. Optimal electroporation settings will thus have to be determined only in the initial experiments. This will increase the reproducibility of the electroporation procedure.
5. Cell-line-specific differences in electroporation conditions are possible. Although GH_12C_1 cells and GH_3 cells can be transfected in a HEPES/sucrose buffer, GH_4C_1 cells show a cell viability of only 10% when transfected in the same buffer (data not shown). We have not explored these parameters on other GH cell strains.
6. It is important to note that electroporation itself may have an effect on the cell phenotype. We have observed a considerable transient decrease of growth hormone and prolactin production in GH_12C_1 and GH_3 cells. A maximum decrease (50–60%) was observed 24 h (GH_12C_1 cells) and 48 h (GH_3 cells) after electroporation. This effect was dependent on the voltage applied, and was similar in the presence or absence of DNA. However, protein and RNA synthesis levels remained stable. Others have also shown that electroporation can modulate cellular parameters. In human dermal fibroblasts, the expression of endogenous collagenase and collagen genes was attenuated *(35)*, as was DNA synthesis in electroporated plant protoplasts *(36)*. In the case of the fibroblasts, the effects were observed for up to 10 d after electroporation. Therefore, if cellular functions subsequent to transient gene expression are to be studied, it is imperative also to assess these functions in treated cells in the absence of DNA.

7. Several alternative electroporation procedures have been described for the transfection of GH cell lines, where different conditions for electroporation (cell density, cell viability, electroporation buffer, DNA concentration, voltage) have successfully been applied in gene transfer experiments. A change of electroporation buffer most likely implicates different ionic strength, and thus a requirement for different electroporation voltage and capacitance settings. GH_4C_1, GH_1, and GC cells (0.5–1 × 10^8 in 0.4 mL) were successfully transfected in PBS with 0.91 mg/mL glucose using voltage settings in the range of 0.25–2.0 kV *(8,10,37)*. Cell-culture medium (DMEM with serum) was used as a buffer in order to electroporate GH_3 and GH_4C_1 cells (3 × 10^6 cells in 0.1 mL medium, 250 V/250 µF) *(12)*. Similar conditions (0.2 mL DMEM with serum, 220 V/500 µF) have also been employed *(4,5)*. Our method describes electroporation in a low-ionic-strength buffer, which does not require high capacitance. However, the use of a different buffer, such as PBS, will necessitate a higher capacitance setting.
8. For studies of gene expression, whether transient or in stably transfected cell lines, the choice of the promoter upstream of the foreign gene can affect the outcome of the experiment. We have tested a number of different promoters regulating transcription of the bacterial chloramphenicol transferase gene *(15)*. The simian virus 40 (SV40) early promoter and the β-globin promoter were strong, whereas the glucocorticoid-inducible mouse mammary tumor virus long-terminal repeat (MMTV-LTR) promoter was silent. We have found that the metallothionein promoter also is suitable for expression studies *(14)*. In different studies, others have shown the usefulness of these promoters, as well as the promoter from the Rous sarcoma virus LTR *(7,37)*. Not surprisingly, endogenous promoters, such as the rat growth hormone promoter *(10,12)*, and the rat prolactin promoter *(6,38)*, and the related mouse TSHβ promoter *(12)*, are also effective in expression experiments.
9. As an indicator of gene-transfer efficiency, plasmids containing sequences for chloramphenicol acetyltransferase *(32)*, β-galactosidase *(39,40)*, or luciferase *(41,42)* can be introduced into cells. The reproducibility of the transfections is then measured by assaying the activity of the transiently expressed enzyme. Transfection efficiency varied between 10 and 20% for various GH cell strains under different electroporation conditions *(4,8,37,38)*.

References

1. Neumann, E. and Rosenheck, K. (1972) Permeability changes induced by electric impulses in vesicular membranes. *J. Membrane Biol.* **10**, 279–290.

2. Potter, H. (1988) Electroporation in biology: methods, applications, and instrumentation. *Anal. Biochem.* **174**, 361–373.
3. Ye, Z.-S., Forman, B. M., Aranda, A., Pascual, A., Park, H.-Y., Casanova, J., and Samuels, H. H. (1988) Rat growth hormone gene expression. *J. Biol. Chem.* **263**, 7821–7829.
4. Keech, C.A. and Gutierrez-Hartmann, A. (1991) Insulin activation of rat prolactin promoter activity. *Mol. Cell. Endocrinol.* **78**, 55–60.
5. Keech, C. A. and Gutierrez-Hartmann, A. (1989) Analysis of rat prolactin promoter sequences that mediate pituitary-specific and 3',5'-cyclic adenosine monophosphate-regulated gene expression in vivo. *Mol. Endocrinol.* **3**, 832–839.
6. Iverson, R. A., Day, K. H., d'Emden, M., Day, R. N., and Maurer, R. A. (1990) Clustered point mutation analysis of the rat prolactin promoter. *Mol. Endocrinol.* **4**, 1564–1571.
7. Day, R. N., Walder, J. A., and Maurer, R. A. (1989) A protein kinase inhibitor gene reduces both basal and multihormone-stimulated prolactin gene transcription. *J. Biol. Chem.* **264**, 431–436.
8. Forman, B. M., Yang, C., Au, M., Casanova, J., Ghysdael, J., and Samuels, H. H. (1989) A domain containing leucine-zipper-like motifs mediate novel in vivo interactions between the thyroid hormone and retinoic acid receptors. *Mol. Endocrinol.* **3**, 1610–1626.
9. Park, H.-Y., Davidson, D., Raaka, B. M., and Samuels, H. H. (1993) The herpes simplex virus thymidine kinase gene promoter contains a novel thyroid hormone response element. *Mol. Endocrinol.* **7**, 319–330.
10. Flug, F., Copp, R. P., Casanova, J., Horowitz, Z. D., Janocko, L., Plotnick, M., and Samuels, H. H. (1987) Cis-acting elements of the rat growth hormone gene which mediate basal and regulated expression by thyroid hormone. *J. Biol. Chem.* **262**, 6373–6382.
11. Forman, B. M., Yang, C., Stanley, F., Casanova, J., and Samuels, H. H. (1988) c-erbA protooncogenes mediate thyroid hormone-dependent and independent regulation of the rat growth hormone and prolactin genes. *Mol. Endocrinol.* **2**, 902–911.
12. Wood, W. M., Kao, M. Y., Gordon, D. F., and Ridgway, E. C. (1989) Thyroid hormone regulates the mouse thyrotropin β-subunit gene promoter in transfected primary thyrotropes. *J. Biol. Chem.* **264**, 14,840–14,847.
13. Paulssen, R. H., Paulssen, E. J., Aleström, P., Gordeladze, J. O., and Gautvik, K. M. (1990) Specific antisense RNA inhibition of growth hormone production in differentiated rat pituitary tumor cells. *Biochem. Biophys. Res. Commun.* **171**, 293–300.
14. Paulssen, R. H., Paulssen, E. J., Gautvik, K. M., and Gordeladze, J. O. (1992) The thyroliberin receptor interacts directly with a stimulatory guanine-nucleotide-binding protein in the activation of adenylyl cyclase in GH_3 rat pituitary tumor cells. *Eur. J. Biochem.* **204**, 413–418.
15. Paulssen, R. H., Paulssen, E. J., Aleström, P., and Gautvik, K. M. (1990) Electroporation of rat pituitary (GH) cell lines: optimal parameters and effects on endogenous hormone production. *Biochem. Biophys. Res. Commun.* **171**, 1029–1036.

16. Ham, R. G. (1963) An improved nutrient solution for diploid hamster and human cell lines. *Expt. Cell Res.* **29**, 515–526.
17. Naess, O., Haug, E., and Gautvik, K. M. (1980) Effects of glucocorticosteroids on prolactin and growth hormone production and characterization of the intracellular hormone receptors in rat pituitary tumor cells. *Acta Endocrinol. (Copenh.)* **95**, 319–327.
18. Bancroft, F. C. and Tashjian, A. H., Jr. (1971) Growth in suspension culture of rat pituitary cells which produce growth hormone and prolactin. *Exp. Cell Res.* **64**, 125.
19. Tashjian, A. H., Jr. (1979) Clonal strains of hormone producing pituitary cells. *Methods Enzymol.* **58**, 527–535.
20. Tashjian, A. H., Jr., Yasumura, Y., Levine, L., Sato, G. H., and Parker, M. L. (1968) Establishment of clonal strains of rat pituitary tumor cells that secrete growth hormone. *Endocrinology* **82**, 342–352.
21. Takemoto, H., Yokoro, K., Furth, J., and Cohen, A. I. (1962) Adrenotropic activity of mammosomatotropic tumors in rats and mice. I. Biological aspects. *Cancer Res.* **22**, 917–924.
22. Tashjian, A. H., Jr., Bancroft, F. C., and Levine, L. (1970) Production of both prolactin and growth hormone by clonal strains of rat pituitary tumor cells. *J. Cell Biol.* **47**, 61–70.
23. Bancroft, F. C. (1973) Measurement of growth hormone synthesis by rat pituitary cells in culture. *Endocrinology* **92**, 1014–1021.
24. Gautvik, K. M., Bjøro, T., Sletholt, K., Østberg, B. C., Sand, O., Torjesen, P., Gordeladze, J. O., Iversen, J.-G., and Haug, E. (1988) Regulation of prolactin secretion and synthesis by peptide hormones in cultured rat pituitary cells, in *Molecular Mechanisms in Secretion*. Alfred Benzon Symposium 25 (Thorn, N. A., Treiman, M., and Pedersen, O. H., eds.), Munksgaard, Copenhagen, pp. 211–227.
25. Haug, E., Naess, O., and Gautvik, K. M. (1978) Receptors for 17β-estradiol in prolactin-sectreting rat pituitary cells. *Mol. Cell. Endocrinol.* **12**, 81–95.
26. Davis, J. R. E., Belayew, A., and Sheppard, M. C. (1988) Prolactin and growth hormone. *Bailliere's Clin. Endocrinol. Metab.* **2**, 797–834.
27. Wehrenberg, W. B., Janowski, B. A., Piering, A. W., Culler, F., and Jones, K. L. (1990) Glucocorticoids: potent inhibitors and stimulators of growth hormone secretion. *Endocrinology* **126**, 3200–3203.
28. Gautvik, K. M. and Kriz, M. (1976) Measurements of prolactin and growth hormone synthesis and secretion by rat pituitary cells in culture. Endocrinology **98**, 344–351.
29. Andreason, G. L. and Evans, G. A. (1988) Introduction and expression of DNA molecules in eukaryotic cells by electroporation. *Biotechniques* **6**, 650–660.
30. Andreason, G. L. and Evans, G. A. (1989) Optimization of electroporation for transfection of mammalian cell lines. *Anal. Biochem.* **180**, 269–275.
31. Cann, A. J., Koyanagi, Y., and Chen, I. S. Y. (1988) High efficiency transfection of primary human lymphocytes and studies of gene expression. *Oncogene* **3**, 123–128.
32. Chu, G., Hayakawa, H., and Berg, P. (1987) Electroporation for the efficient transfection of mammalian cells with DNA. *Nucleic Acids Res.* **15**, 1311–1326.

33. Iannuzzi, M. C., Weber, J. L., Yankaskas, J., Boucher, R., and Collins, F. S. (1988) The introduction of biologically active foreign genes into human respiratory epithelial cells using electroporation. *Am. Rev. Respir. Dis.* **138,** 965–968.
34. Toneguzzo, F., Hayday, A. C., and Keating, A. (1986) Electric field-mediated DNA transfer: transient and stable gene expression in human and mouse lymphoid cells. *Mol. Cell. Biol.* **6,** 703–706.
35. Lambert, C. A., Lefebvre, P. Y., Nusgens, B. V., and Lapiere, C. M. (1993) Modulation of expression of endogenous collagenase and collagen genes by electroporation: possible involvement of Ca^{2+} and protein kinase C. *Biochem. J.* **290,** 135–138.
36. Rech, E. L., Ochatt, S. J., Chand, P. K., Davey, P. R., and Mulligan, B. J. (1988) Electroporation increases DNA synthesis in cultured plant protoplasts. *Biotechnology* **6,** 1091–1093.
37. Wu, J., Kovacic-Milivojevic, B., Lapointe, M. C., Nakamura, K., and Gardner, D. G. (1991) Cis-active determinants of cardiac-specific expression in the human atrial natriuretic peptide gene. *Mol. Endocrinol.* **5,** 1311–1322.
38. Jackson, S. M., Keech, C. A., Williamson, D. J., and Gutierrez-Hartmann, A. (1992) Interaction of basal positive and negative transcription elements controls repression of the proximal rat prolactin promoter in nonpituitary cells. *Mol. Cell. Biol.* **12,** 2708–2719.
39. Buonocore, V., Sgambati, O., De Rosa, M., Esposito, E., and Gambacorta, A. (1980) A constitutive ß-galactosidase from the extreme thermoacidophile archaebacterium *Caldariella acidophila:* properties of the enzyme in the free state and immobilized whole cells. *J. Appl. Biochem.* **2,** 390–397.
40. Jagota, S. K., Rao, M. V. R., and Dutta, S. M. (1981) Beta-galactosidase of streptococcus cremoris H. *J. Food Sci.* **46,** 161–168.
41. De Wet, J. R., Wood, K. V., De Luca, M., Helinski, D. R., and Subramani, S. (1987) Firefly luciferase gene: structure and expression in mammalian cells. *Mol. Cell. Biol.* **7,** 725–737.
42. Maxwell, I. H. and Maxwell, F. (1988) Electroporation of mammalian cells with a firefly luciferase expression plasmid: kinetics of transient expression differ markedly among cell types. *DNA* **7,** 557–562.

CHAPTER 10

Electroporation of Plasmid DNA into Normal Human Fibroblasts

F. Andrew Ray

1. Introduction

It is frequently desirable to use normal genetically stable human cells as recipients of exogenous genes. However, normal human cell strains have not often been used because of their finite life-span. Instead, immortal human cell lines have usually been used as genetic recipients. These cell lines derived from tumors or by transformation with SV40 virus have accumulated numerous genetic changes (1,2), and may in fact contain changes that will affect the gene or genes that are of interest to a particular study. Normal human fibroblast cell strains have a proliferative potential of from 50–75 cumulative population doublings (CPD) (3,4). Since expansion from one cell to a million cells requires approx 20 population doublings, human fibroblasts can readily be used for stable transfection experiments. However, before initiating an experiment, the life-span of the recipient cell strains should be characterized, and electroporation should not be performed on cells with insufficient proliferative potential remaining.

Electroporation is a method used to transfer macromolecules across the cell membrane and into a variety of different types of cells (5,6). DNA and proteins are commonly introduced into cells in this manner (5,7). By exposing cells to an electric pulse, electroporation is believed to open transmembrane pores that allow large molecules, such as DNA, to enter (8). When transferring genes via electroporation, a percentage of the DNA molecules reach the nucleus. A reduced percentage of these molecules recombine with the cellular DNA and become stably integrated into the genome *(9)*. Advantages of the technique include simplicity and reproducibility. Another significant advantage is that, typically, only one to a few copies of a gene transfected by electroporation become integrated at a single site within a cell *(10)*. This facilitates the identification of single-copy integrants if desired.

A simple, reproducible electroporation protocol to transfect plasmid DNA into normal human fibroblasts is presented here. Using this protocol, individual clones of normal human cells expressing desired proteins can be obtained. Such clones are expected to have a single stable integration site containing one to a few copies of the gene(s) of interest. The protocol described can be divided into four general steps: preparation of the DNA and cells, electroporation, biochemical selection, and clonal expansion. The protocol described here is a modification of the procedure described by Chu et al. *(11)*.

2. Materials
2.1. Plasmid Vectors

In order to identify cells successfully transfected with the gene of interest, it is generally preferable to use one of the dominant antibiotic resistance selections in human cells, either linked in a plasmid construct or linked in the cell after coelectroporation. Any gene that imparts a distinctive phenotypic change to the cells can be used with some minor technique modifications. The two most frequently used genes and selective agents are the neomycin resistance gene *(neo)*, which allows survival of mammalian cells in G418-sulfate (Geneticin, Gibco BRL, Gaithersburg, MD) and the hygromycin gene *(hyg)* which allows cells to survive in hygromycin B (Boehringer Mannheim, Indianapolis, IN) *(12,13)*. These genes are available commercially (e.g., LacSwitch Inducible Mammalian Expression System from Stratagene, La Jolla, CA includes both *neo* and *hyg* in separate plasmids) and from other investigators, providing a variety of different promoters and cloning sites. Approximately 100 µg of the chosen plasmid should be prepared and purified using the Plasmid Midi Kit (Qiagen, Chatsworth, CA).

2.2. Cell Strains

The salient feature regarding fibroblast strain selection is that the cells are at low passage and therefore have a high proliferative potential. It is difficult to purchase cell stains with sufficient proliferative potential to obtain an expanded population of cells. If many millions of cells are required for a given experiment, cells at the first or second passage should be used. If possible, characterize the proliferative potential of the cells prior to electroporation. Make sure to follow institutional guidelines if initiating cell strains. Foreskins from male newborns are readily obtainable from pediatric colleagues and make an excellent source of fibroblasts.

2.3. Solutions and Reagents for Electroporation

All solutions and reagents should be sterilized by autoclaving or sterile filtration.

1. TE: 10 mM Tris-HCl, pH 7.4, 1 mM EDTA, pH 8.0 *(14)*.
2. HEPES-buffered saline (HeBS): 20 mM HEPES, pH 7.05, 137 mM NaCl, 5 mM KCl, 0.7 mM Na$_2$HPO$_4$, and 6 mM dextrose *(11,15)*.
3. Phosphate-buffered saline (PBS): 137 mM NaCl, 2.7 mM KCl, 14.1 mM Na$_2$HPO$_4$, and 2.9 mM KH$_2$PO$_4$ *(16)*.

2.4. Solutions and Reagents for Human Fibroblast Culture

1. Trypsin/EDTA (Gibco-BRL).
2. Growth medium: αMEM (Gibco-BRL), 10% fetal bovine serum (FBS), 50 U/mL penicillin G sodium, 50 µg/mL streptomycin sulfate.
3. Collagenase solution: growth medium with 400 U/mL of collagenase.
4. G418 Sulfate (Geneticin), (Gibco BRL): Dissolve in αMEM without serum at 40 mg/mL to make a 100X stock solution. The media will turn quite yellow, but it is not necessary to adjust the pH. Sterilize by filtration, and store at 4°C for up to 1 mo.
5. Cloning cylinders (8-mm diameter, Bellco Biotechnology, Vineland, NJ) are placed into a 1–2-mm layer of vacuum grease on the bottom of a glass Petri dish and autoclaved.

3. Methods

3.1. Initiating a Cell Strain for Electroporation

1. Observing appropriate biosafety protocols for handling human tissue, immerse foreskin in 70% ethanol, and then rinse in growth medium.

2. Mince the tissue with sterile scissors, and digest with 10 mL of collagenase solution for 2–4 h at 37° C in a sealed centrifuge tube with gentle agitation *(17,18)*.
3. Pellet clumps by centrifugation at 100g for 1 s and plate supernatant with dispersed cells in growth medium.

3.2. Preparation of Plasmid DNA

1. On the day prior to electroporation, digest plasmid DNA with a restriction enzyme that linearizes the plasmid and does not cut within required genetic elements (such as the promoter or 3'-splice sites) of the gene to be studied (*see* Notes 1 and 2).
2. Remove a small aliquot of the DNA after the restriction endonuclease reaction and assay for complete digestion on an agarose gel (*see* Note 3).
3. Add 1/10 vol of 3*M* sodium acetate and 2 vol 100% ethanol to the remaining DNA sample, incubate at –20°C for at least 1 h, and pellet the precipitated DNA by centrifugation at 10,000g for 10 min. Pour off the supernatant, invert for 5 min to dry, and resuspend the DNA in 10–20 µL of sterile TE at pH 7.4 (*see* Note 4). Remove 2 µL of the DNA, and dilute in 500 µL TE for spectrophotometric determination of DNA concentration (*see* Note 5).

3.3. Electroporation Protocol

1. Usually on the day prior to electroporation, subculture the fibroblasts to a density such that a 50–80% confluent culture is obtained on the day of electroporation (*see* Note 6). The cells should have as much proliferative potential as necessary to accomplish the goals of the experiment (*see* Note 7).
2. Detach the cells with trypsin/EDTA. Add 5 mL of the trypsin solution to each 75-cm^2 flask, rock back and forth over surface of flask 5–10 times, and then aspirate the solution. This leaves a layer of trypsin over the cells that is sufficient for detachment. Monitor the detachment with an inverted microscope, and when the cells have rounded up, detach by gently striking the flask.
3. Add 10 mL of growth medium. Resuspend the cells by pipeting, transfer them to a 15-mL centrifuge tube, and count a 0.2-mL aliquot diluted to 10 mL with PBS using a Coulter counter or hemocytometer. Concomitantly, pellet the remaining cells by centrifugation at 200g for 10 min.
4. After centrifugation, aspirate the media completely from the cell pellet, and resuspend the cells at a concentration of 2.5×10^6 cells/mL in sterile HeBS. Add 0.8 mL cells to a 1.5-mL microcentrifuge tube containing the plasmid DNA(s), and gently pipet to mix (*see* Note 8). Transfer the cell–DNA mixture to an electroporation cuvet with a 0.4-cm electrode gap.

5. Electroporate the cells at 230 V and 960 µF, and record time constant (*see* Note 9). Remove the cells from the chamber using a Pasteur pipet, and add them to 100 mL of growth medium. Mix and plate 10 mL into each of ten 100-mm diameter tissue-culture dishes. Place the dishes into a CO_2 incubator at 37°C.
6. After 48 h, dilute the 100X stock solution of G418 to 2X in growth medium. Gently add 10 mL of 2X G418 to each 100-mm dish to a final concentration of 400 µg/mL (*see* Note 10). Media should be kept separate for each individual electroporation experiment to avoid the slight chance of cellular crosscontamination. Return the cells to the incubator.

3.4. Cloning and Expansion Protocol

1. Twelve days after electroporation, examine dishes for visible colonies by holding the dishes up to a light and tilting the dish slightly. If colonies are not visible, return the dishes to the incubator for 1–2 d. Colonies should have >500 cells, and mitotic cells should be observed on the periphery of the colony.
2. Mark the colonies to be cloned with a dot in the center and a ring around the periphery.
3. Aspirate the media from the dish, and surround the colony with a cloning cylinder (*see* Note 11).
4. Fill the cloning cylinder with trypsin/EDTA and monitor detachment. Remove the detached cells from a single colony with a sterile cotton-plugged Pasteur pipet, and place in a 25-cm^2 flask containing 5 mL of growth medium and G418.
5. Continue selective pressure at least until the cells have been expanded to a confluent 75-cm^2 flask.
6. If an antibody to the gene product of interest is available, at the first transfer (e.g., 25-cm^2 flask to 75-cm^2 flask), a small aliquot of cells can be placed in LabTek slides (Nunc, Naperville, IL) for detection by immunofluorescence assay (*see* Note 12).

4. Notes

1. If coelectroporation is performed, it is necessary to provide a 10-fold excess of the unselected plasmid compared to the selected plasmid *(19)*.
2. It is not known why linear DNA is more efficient in the electroporation protocol *(11,20)*, although it is likely that the DNA ends recombine more efficiently with chromosomal DNA than do circular DNAs. Avoid digesting DNA at low concentrations (<0.2 µg/µL), because it is more time-consuming to precipitate DNA from dilute solution *(14)*. At least 20% more DNA should be digested than anticipated for electroporation, since some DNA will be lost during the ethanol precipitation and quantitation steps.

3. If the DNA is not completely digested, add more buffer and restriction endonuclease. By having excess DNA at the onset and starting a day prior to electroporation, a second round of digestion can be performed if necessary.
4. The DNA sample is resuspended in a small volume of TE to avoid altering the ionic composition of the electroporation solution.
5. A reasonably accurate determination (and good crosscheck) of the DNA concentration can be made by comparing the ethidium bromide-stained plasmid band on the agarose gel to a band representing known marker DNA closest in size to the sample DNA.
6. Cells must be in logarithmic growth, or the transfection efficiency drops dramatically *(11,21)*. For example, transfecting 1 µg pRSVneo into stationary fibroblasts usually yields no transfectants, whereas 10–20 transfectants are commonly obtained using cycling cells. The condition of the cells is the first place to start when troubleshooting the protocol. The number of culture flasks of cells required depends on the number of experiments and the size of the flasks. A 75-cm^2 flask at 50% confluence will yield approx 2×10^6 human fibroblasts, enough for one electroporation experiment.
7. If concerned about the proliferative potential of the cell strain, include a control sample with 10 µg of a plasmid containing the SV40 T-antigen, such as pSV3neo *(22)*. The T-antigen will extend the life-span of the fibroblasts by 10–20 population doublings *(19)*. If colonies are obtained in this sample, but not from the experimental samples, then the proliferative potential was likely a limiting factor.
8. An alternative to mixing the selection plasmid (e.g., pRSVneo) with the experimental plasmid in the individual microcentrifuge tubes is to mix the selection plasmid with all of the cells at the desired concentration. Then add the premixed DNA and cells to the various experimental plasmids as described above. This minimizes pipeting variation, and the number of colonies obtained per electroporation cuvet will be comparable.
9. Using the Gene Pulser (Bio-Rad, Hercules, CA) with capacitance extender and 1 µg pRSVneo, we obtain 1–2 G418 resistant colonies/100-mm dish. This low number of colonies is optimal when transfectants having a single copy of an integrated vector is desired, and it minimizes potential colony crosscontamination. If this is not a consideration or if more colonies are necessary, the DNA concentration of the plasmids can be increased until the toxicity becomes a significant factor. Electroporation frequency increases linearly with DNA concentration over a wide range of DNA concentrations *(6,10,11,21)*. Sodium butyrate treatment after the voltage pulse has also been shown to increase the electroporation efficiency in human fibroblasts *(23)*. Time constant values should range from 12.7–15.1 ms.

10. Pipet onto side of dish slowly to avoid dislodging cells and starting satellite colonies. Nontransfected fibroblasts will begin to die and detach in 4 d. When plating cells using the conditions described, it is not necessary to change the media during the 2-wk selection period.
11. Using sterile forceps, twist cylinder slightly to seat over the colony. Make sure that vacuum grease does not occlude the opening prior to applying.
12. Use most of the colony for expansion. The narrow tip of the Pasteur pipet containing the remaining few drops of the cell suspension from the colony should have enough cells for two wells of a LabTek slide (control and experimental) for analysis by indirect immunofluorescence. To assay SV40 T-antigen, for example, grow the cells on a LabTek slide until they are ~50% confluent, and then rinse them with PBS. Separate the plastic chamber from the slide, and leave the silicon rubber gasket in place. Next, fix the cells for 3 min in 100% acetone, and let dry. Add 20 µL of antibody to T-antigen, such as mouse monoclonal Ab1 (Oncogene Science, Uniondale, NY) diluted 1:10 in PBS, to each appropriate well, cover each sample mixture with a plastic cover square (e.g., a piece from an autoclave bag), and incubate at room temperature for 1 h in a humidified chamber. Rinse the slide with PBS, and then wash the slide for 5 min in PBS with agitation. Remove the silicon gasket and apply 100 µL of secondary FITC-labeled goat antimouse antibody (CalTag, S. San Francisco, CA) diluted 1:10 in PBS to the entire slide, replace coverslip, and incubate at 37°C in humidified chamber in the dark for 20 min. Rinse the slide with PBS, and use fluorescence microscopy to determine if the protein is expressed in cells that were initially treated with primary antibody, but not in nontransfected control cells or in transfected control cells that were initially treated with PBS only.

References

1. Nowell, P. C. (1976) The clonal evolution of tumor cell populations. *Science* **194,** 23–28.
2. Ray, F. A. and Kraemer, P. M. (1992) Frequent deletions in nine newly immortal human cell lines. *Cancer Genet. Cytogenet.* **59,** 39–44.
3. Hayflick, L. and Moorehead, P. S. (1961) The serial cultivation of human diploid cell strains. *Exp. Cell Res.* **25,** 585–621.
4. Smith, J. R. and Whitney, R. G. (1980) Intraclonal variation in proliferative potential of human diploid fibroblasts: stochastic mechanism for cellular aging. *Science* **207,** 82–84.
5. Neumann, E., Schaefer-Ridder, M., Wang, Y., and Hofschneider, P. H. (1982) Gene transfer into mouse lyoma cells by electroporation in high electric fields. *EMBO J.* **1,** 841–845.
6. Fromm, M., Taylor, L. P., and Walbot, V. (1985) Expression of genes transferred into monocot and dicot plant cells by electroporation. *Proc. Natl. Acad. Sci. USA* **82,** 5824–5828.

7. Winegar, R. A., Phillips, J. W., Youngblom, J. H., and Morgan, W. F. (1989) Cell electroporation is a highly efficient method for introducing restriction endonucleases into cells. *Mut. Res.* **225,** 49–53.
8. Kinosita, K. and Tsong, T. Y. (1977) Voltage-induced pore formation and hemolysis of human erythrocytes. *Biochim. Biophys. Acta* **471,** 227–242.
9. Mayne, L. V., Jones, T., Dean, S. W., Harcourt, S. A., Lowe, J. E., Priestly, A., Steingrimmsdottir, H., Sykes, H., Green, M. H., and Lehmann, A. R. (1988) SV40-transformed normal and DNA-repair-deficient human fibroblasts can be transfected with high frequency but retain only limited amounts of integrated DNA. *Gene* **66,** 65–76.
10. Boggs, S. S., Gregg, R. G., Borenstein, N., and Smithies, O. (1986) Efficient transformation and frequent single-site, single-copy insertion of DNA can be obtained in mouse erythroleukemia cells transformed by electroporation. *Exp. Hematol.* **14,** 988–994.
11. Chu, G., Hayakawa, H., and Berg, P. (1987) Electroporation for the efficient transfection of mammalian cells with DNA. *Nucleic Acids Res.* **15,** 1311–1326.
12. Colbère-Garapin, F., Horodniceanu, F., Kourilsky, P., and Garrapin, A.-C. (1981) A new dominant hybrid selective marker for higher eukaryotic cells. *J. Mol. Biol.* **150,** 1–14.
13. Sugden, B., Marsh, K., and Yates, J. (1985) A vector that replicates as a plasmid and can be efficiently selected in B-lymphocytes transformed by Epstein-Barr virus. *Mol. Cell. Biol.* **5,** 410–413.
14. Maniatis, T., Fritsch, E. F., and Sambrook, J. (1982) *Molecular Cloning: A Laboratory Manual,* Cold Spring Harbor Laboratory, Cold Spring Harbor, NY, pp. 448–461.
15. Graham, F. L. and van der Eb, A. J. (1973) A new technique for the assay of infectivity of human adenovirus 5 DNA. *Virology* **52,** 456–467.
16. van Wezel, A. L. (1973) Microcarrier cultures of animal cells, in *Tissue Culture: Methods and Applications* (Kruse, P. F. and Patterson, M. K., eds.), Academic, London, p. 373.
17. Freshney, R. I. (1987) *Culture of Animal Cells: A Manual of Basic Culture.* Alan R. Liss, New York, pp. 122–124.
18. Ray, F. A., Bartholdi, M. F., Kraemer, P. M., and Cram, L. S. (1984) Chromosome polymorphism involving heterochromatic blocks in Chinese hamster chromosome 9. *Cytogenet. Cell Genet.* **38,** 257–264.
19. Ray, F. A., Peabody, D. S., Cooper, J. L., Cram, L. S., and Kraemer, P. M. (1990) SV40 T antigen *alone* drives karyotype instability that precedes neoplastic transformation of human diploid fibroblasts. *J. Cell. Biochem.* **42,** 13–31.
20. Potter, H., Weir, L., and Leder, P. (1984) Enhancer-dependent expression of human k immunoglobulin genes introduced into mouse pre-B lymphocytes by electroporation. *Proc. Natl Acad. Sci. USA* **81,** 7161–7165.
21. Andreason, G. L. and Evans, G. A. (1988) Introduction and expression of DNA molecules in eukaryotic cells by electroporation. *Biotechniques* **6,** 650–660.
22. Southern, P. J. and Berg, P. (1982) Transformation of mammalian cells to antibiotic resistance with a bacterial gene under control of the SV40 early region promoter. *J. Mol. Applied Genet.* **1,** 327–341.
23. Tatsuka, M., Orita, S., Yagi, T., and Kakunaga, T. (1988) An improved method of electroporation for introducing biologically active foreign genes into cultured mammalian cells. *Exp. Cell Res.* **178,** 154–162.

CHAPTER 11

Electroporation-Mediated Gene Transfer into Hepatocytes

Alphonse Le Cam

1. Introduction

The study of gene expression regulation relies on the introduction of foreign DNA into eukaryotic cells. A wide variety of DNA-transfer procedures have been developed that utilize retroviruses *(1)*, polycations *(2)*, liposomes *(3)*, chromosomes *(4)*, reconstituted viral envelopes *(5)* and other chemical reagents, such as calcium phosphate *(6)*, DEAE-dextran *(7)*, and lipopolyamines *(8)*. DNA also can be transferred into cells by physical means, such as microinjection *(9)*, laser beams *(10)*, and electroporation *(11)*. However, none of these methods work with high efficiency on every cell type, whether freshly isolated cells or established cell lines. Some of these techniques, such as microinjection and laser-mediated transfection, require sophisticated apparatus and are technically difficult, whereas others, such as protoplast fusion, retroviral vectors, and liposome fusion, require time-consuming biochemical manipulations. Electroporation, which circumvents most of these problems, has emerged as an effective tool for the transfection of eukaryotic cells in suspensions (for a review, *see* ref. *12*).

We are interested in genes specifically expressed in normal parenchymal liver cells (i.e., hepatocytes) and controlled by growth hormone (GH), glucocorticoids (GC), and cytokines. Because none of the hepatic-derived cell lines express their endogenous GH receptor, we chose the hepatocyte cell system to carry out gene-expression studies. Because hepatocytes have become a very attractive system to study gene expression and also for gene therapy, many investigators have targeted their efforts toward setting up efficient methods to introduce foreign DNA into those cells, both in vivo and in vitro (for a review, *see* ref. *13*). The electroporation method described here meets three requirements:

1. It is efficient and reproducible;
2. It does not appear to interfere with the sensitivity of hepatocytes to their hormonal effectors; and
3. It is fast and inexpensive and, thus, allows the comparison of a large number of parameters within a single experiment.

The procedure described in this chapter should allow an investigator to transfect normal rat hepatocytes and study the regulation of gene expression by a variety of effectors. It includes the conditions for hepatocyte isolation, electroporation, and culture, as well as the technique for measurement of the chloramphenicol-acetyltransferase (CAT) reporter gene activity.

2. Material
2.1. Sources of Enzymes, Hormones, and Chemicals

1. Collagenase A from *Clostridium histolyticum* (SA: 0.4–0.6 U/mg, *see* Note 4) (Boehringer Mannheim GmbH, Germany).
2. DMEM, Ham F12 medium, and gentamycin are available from Gibco-BRL (Cergy Pontoise, France).
3. Chloramphenicol, acetyl-CoA, fetal calf serum (FCS), adult bovine serum (ABS), pork insulin, dexamethasone (DEX), 5-bromo-4-chloro-3-indolyl β-D-galactopyranoside (X-Gal), and salmon sperm DNA are available from Sigma (St. Louis, MO).
4. Recombinant human GH is available from Serono Laboratories (Levallois-Perret, France). Recombinant human interleukin-6 (IL-6, specific activity: 10^7 U/μg) is available from Immunex (Seattle, WA).
5. [^{3}H]-acetyl-CoA (16 Ci/mmol) specific for CAT assays and Econofluor-2 are available from DuPont-New England Nuclear (Les Ulis, France).
6. Ion-exchange chromatography columns for DNA purification are available from Qiagen Inc. (Chatsworth, CA).

2.2. Plasmids

1. Amplify plasmids according to standard methods *(14)*, and purify the supercoiled form by ion-exchange chromatography (e.g., on Qiagen columns) or by CsCl gradient centrifugation.

2. Examples of plasmids used for electroporation (*see* Note 1):
   a. Homologous constructs containing promoters of two serine protease inhibitor (spi) genes *(15)* inserted upstream of the cat reporter gene, in the pEMBL vector *(16)*;
   b. Heterologous constructs containing fragments of the spi promoters inserted upstream of a minimal thymidine kinase *(tk)* promoter in the pBLCAT2 vector *(17)*, or the c-*fos* promoter in the Δ-56-c-*fos* plasmid *(18)*, linked to the cat reporter gene in both cases; and
   c. Homologous constructs consisting of spi promoter fragments placed upstream of the β-galactosidase reporter gene in the pLacF vector *(19)*. The RSV-β-Gal plasmid carrying the β-galactosidase gene driven by the strong Rous sarcoma virus promoter may be used for measuring transfection efficiencies. (*See* Note 4).

2.3. Solutions and Culture Medium

1. HEPES buffer: 10 mM HEPES, pH 7.8 at 37°C, 150 mM NaCl, 30 mM KCl. Autoclave and store at 25°C.
2. Collagenase solution: 0.5 mg/mL collagenase in HEPES buffer supplemented with 10 mM CaCl$_2$. Prepare extemporaneously, and filter-sterilize through a 0.22-μm membrane.
3. Culture medium: DMEM + Ham's (1:1, V/V) supplemented with 100 μg/mL gentamycin, 0.8 μM pork insulin, and 5% ABS or FCS.
4. Liver perfusion solution: HEPES buffer supplemented with 6 IU/mL heparin and 0.66 mM EGTA.
5. Electroporation medium: sterile phosphate-buffered saline (PBS) containing 5% ABS or FCS. Prepare extemporaneously using freshly filtered serum (through 0.22-μm membranes).
6. Sonicated salmon sperm DNA (SSS DNA): 10 mg/mL solution in 10 mM Tris-HCl, pH 7.5, 0.1 mM EDTA. Sonicate to give 200–300 bp DNA fragments and deproteinize by three to four phenol extractions. Store at –20°C.
7. Cell lysis buffer: 15 mM Tris-HCl, pH 7.8, 60 mM KCl, 15 mM NaCl, 2 mM EDTA. Add 0.15 mM spermine, 0.1 mM PMSF, and 1 mM DTT from frozen stock solutions before use *(20)*.
8. CAT assay reagent: A mixture containing 1.5 μL of 75 μM HCl, 1 μL of 150 μM unlabeled acetyl-CoA, 0.2 μL (0.2 μCi) of [^{3}H]-acetyl-CoA, 17.3 μL of 250 mM Tris-HCl, pH 7.8, 5 mM EDTA, and 20 μL of 5 mM chloramphenicol. Prepare extemporaneously from frozen stock solutions, except for HCl and Tris-HCl buffer.
9. Hepatocyte fixation mixture: 1% formaldehyde, 0.2% glutaraldehyde in PBS. Prepare just before use.
10. β-galactosidase staining solution: 4 mM potassium ferricyanide, 4 mM

potassium ferrocyanide, 4 mM MgCl$_2$, 0.4 mg/mL X-Gal (stock solution dissolved in DMSO or DMF) in PBS. Prepare extemporaneously from frozen stock solutions (protect potassium salts from light).

3. Methods

3.1. Preparation of Hepatocytes

1. Anesthetize adult male Wistar rats (200–300 g) with ip injection of pentobarbital and open the abdominal cavity.
2. Introduce a butterfly canula linked to tygon tubing connected through a peristaltic pump to a HEPES buffer reservoir placed in a 37°C water bath, into the portal vein, and ligate it tightly.
3. Cut the efferent vessels as quickly as possible with scissors, and turn on the peristaltic pump to perfuse the liver in situ with liver perfusion solution, at 40–50 mL/min for 2 min.
4. Continue the washing procedure for two more minutes using the same buffer devoid of heparin and EGTA and, concomitantly, remove the liver from the animal and transfer it onto a plastic grid (i.e., a perforated tissue-culture plastic dish) inserted in a funnel set on top of a 250-mL Erlenmeyer flask.
5. Recirculate 100 mL of a collagenase solution equilibrated at 37°C through the liver for 6–8 min. This treatment completely dissociates liver cells.
6. Disconnect the liver from the perfusion device, and transfer it into a plastic culture dish containing sterile HEPES buffer placed inside a tissue-culture hood.
7. Release hepatocytes into sterile HEPES buffer supplemented with gentamycin by tearing apart the liver capsule and swirling.
8. Filter the crude cell suspension (80–100 mL) through a nylon mesh (100 µm), and centrifuge it at 100g for 2 min at 25°C.
9. Resuspend each cell pellet in 40 mL of HEPES buffer containing gentamycin and recentrifuge as described in step 8; repeat this step once.
10. Resuspend pellets in 30 mL of culture medium, and centrifuge the suspensions at 500g for 2 min.
11. Resuspend hepatocytes in 30–40 mL of culture medium, and transfer them to a plastic culture flask (75 mL) for conservation before use.
12. Count the cells using a Nageotte cell; 400–500 × 10^6 hepatocytes (85–95% viable) are routinely obtained using this procedure, which takes 30–45 min to complete.

3.2. Electroporation and Culture Conditions

1. Centrifuge freshly isolated hepatocytes at 500g for 2 min, and resuspend in electroporation medium at 20–25 × 10^6 cells/mL.

2. Transfer 0.8 mL of cell suspension into a 15-mL sterile plastic tube containing 30 µg of the test plasmid DNA and 400 µg of SSS DNA (*see* Note 2).
3. Transfer the mixture into an electroporation cuvet and immediately apply the electrical shock (160 V, 960 µF). The time constant obtained under these conditions should be 20–25 ms.
4. Add 0.8 mL of culture medium to each electroporation cuvet, and transfer the cells to a sterile plastic tube.
5. Wash each cuvet twice with 0.8 mL of culture medium to recover all the cells.
6. Deliver 0.5-mL aliquots of cell suspension (2–3 × 10^6 hepatocytes) to 60-mm plates containing 3.5 mL of prewarmed culture medium (1–2 h at 37°C in the CO_2 incubator) supplemented with appropriate effectors (i.e., hormones or cytokines).
7. Gently swirl the culture plates to ensure good cell homogeneity, and incubate for 20–24 h. No medium change is necessary during this brief culture period.

3.3. Preparation of Cellular Extracts

1. Aspirate the culture medium, and carefully add 2.5 mL of cold PBS to each plate to wash hepatocyte monolayers. Repeat this step once.
2. Scrape out the cells in 1 mL of PBS using a rubber policeman, and transfer them into 1.5-mL tubes.
3. Centrifuge cell suspensions at 15,000g for 5 min at 25°C.
4. Resuspend pellets in 100 µL of cell lysis buffer.
5. Break down hepatocyte membranes by two freeze/thaw cycles using liquid nitrogen and a 37°C water bath. Centrifuge at 15,000g for 10 min at 25°C.
6. Recover supernatants (roughly 90 µL), and use them directly for β-galactosidase or luciferase assays. For CAT assays, the material must be heated for 7 min at 65°C to inactivate deacetylases and recentrifuged at 15,000g for 10 min.
7. Measure protein content by using the Coomassie blue protein staining assay (Bio-Rad, Richmond, CA).

3.4. CAT Assays

1. Incubate 60 µL (20–40 µg protein) of heated cell lysate with 40 µL of the CAT assay solution for 1 h at 37°C. Omit chloramphenicol in blank controls (include two or three controls per experimental series).
2. Stop the enzymatic reaction by heating the samples for 5 min at 80°C.
3. Transfer the reaction mixtures into a counting vial containing 5 mL of Econofluor-2, and shake thoroughly to allow partition of labeled-lipophilic metabolites (^{3}H-acetylated chloramphenicol) into the scintillation liquid *(21)*.
4. Incubate the vials for 10–15 min at 25°C, and determine the quantity of ^{3}H using a scintillation counter. Under these conditions, background values obtained in the absence of chloramphenicol range from 500–700 cpm (*see* Notes 3 and 5).

3.5. Transfection Efficiency

The transfection efficiency can be estimated by performing β-galactosidase histochemical assays (*see* Note 4).

1. Transfect hepatocytes with the pLacF-derived or RSV-β-Gal plasmids.
2. Transfer cells to 60-mm diameter dishes, and culture for 24 h.
3. Wash the cells twice with cold PBS, and incubate with the fixation mixture for 5 min at 25°C.
4. Wash the cells three times with cold PBS, and incubate for 24 h at 30°C with the β-galactosidase staining solution. Positive cells stain blue.

4. Notes

1. These procedures have been applied to the study of the regulation of spi promoter activities by hormones (22) and cytokines. Figure 1 shows the effects of GH and DEX on the transcriptional activity of the spi 2.1 and *tk* promoters. The spi 2.1 promoter (−175/+8) displays a fairly high level of basal activity in cultured hepatocytes (Fig. 1, experiments A, B, and C). GH and DEX enhance this activity by four- to sixfold and three- to fourfold, respectively. These two effectors had additive effects when combined, giving rise to an 8–10-fold activation ratio. Unlike the spi 2.1 promoter, the minimal *tk* promoter has low activity in the absence of GH (Fig. 1, BASAL). However, the presence of a single copy of the more distal spi 2.1 growth hormone response element (GHRE-II enhancer; 23) (experiments D, E, and F) confers GH sensitivity to the *tk* promoter. Thus, both the spi 2.1 and the GHRE-II-*tk* promoters are efficiently activated by GH in electroporated hepatocytes. Figure 2 shows that the spi 2.3 promoter also displays a high basal level of activity (experiments A, B, and C) and is activated by about threefold by both DEX and IL-6. The effects of these two effectors are additive. We tested the function of two copies of the spi 2.3 promoter region (−90/−70) containing a positive glucocorticoid response element (GRE) upstream of a minimal c-*fos* promoter. This construct has a low DEX-independent activity but becomes very active in the presence of DEX, since a 10–15-fold stimulation of CAT activity is observed (experiments D, E, and F). These results are quite representative of this biological system. They demonstrate:
    a. a high degree of reproducibility of the hormonal effects from one experiment to another (with various hormones and different promoters); and
    b. the possibility to reproduce largely in vitro with promoter constructs the effects seen with the endogenous genes in vivo.

Gene Transfer into Hepatocytes

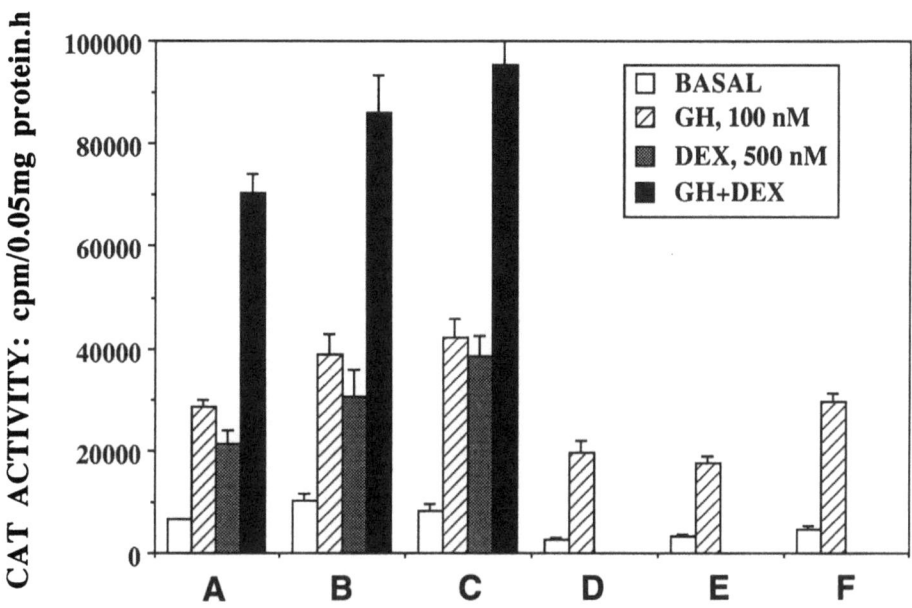

Fig. 1. Analysis of the basal and hormone-stimulated activities of the spi 2.1 and *tk* promoters in cultured hepatocytes. Cells prepared from three different livers were transfected with a pEMBL-derived plasmid carrying the −175/+8 spi 2.1 gene promoter fragment (**A–C**), or a pBLCAT2-derived plasmid bearing a single copy of the (−149/−104) spi 2.1 GHRE-II enhancer upstream of the *tk* promoter (**D–F**). CAT activities were measured in cell extracts prepared from control (BASAL) or hormone-treated hepatocytes.

2. The electroporation conditions (i.e., volume of cell suspension and DNA solutions) may be scaled down twofold.
3. The use of electroporation to transfect hepatocytes transiently for gene-expression studies has several major advantages.
    a. Because it is very easy to prepare several hundred million hepatocytes from a single rat liver, one can test a large number of experimental parameters (i.e., constructs, effectors, incubation conditions) with the same cell preparation (we routinely seed 70–80 dishes, 60-mm diameter).
    b. This procedure eliminates the need to use a cotransfected plasmid to account for variations in transfection efficiency when one wants to compare the influence of various effectors on the activity of any given construct, since the same pool of electroporated cells is used to test the various effectors.

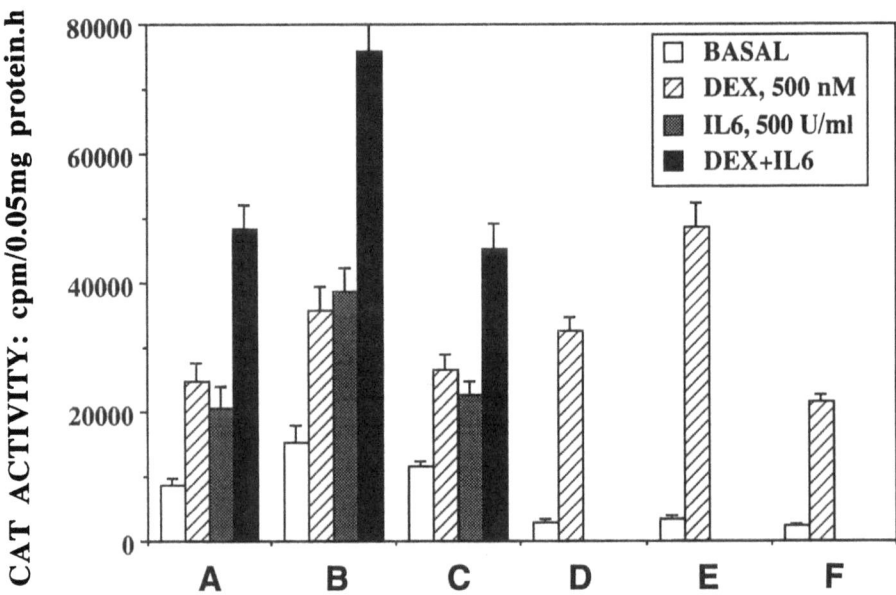

Fig. 2. Analysis of the basal and hormone-stimulated activities of the spi 2.3 and c-*fos* promoters in cultured hepatocytes. Cells prepared from three different livers were transfected with a pEMBL-derived plasmid carrying the −2500/+82 spi 2.3 gene promoter fragment (**A–C**), or a Δ-56-c-*fos*-derived plasmid bearing two copies of the (−90/−70) spi 2.3 GRE (**D–F**). CAT activities were measured in cell extracts prepared from control (BASAL) or hormone-treated hepatocytes.

    c. The procedure is fast and easy to perform. The entire protocol, from cell preparation to the CAT assay, takes roughly 30 h to complete. We have noted that CAT expression reaches a maximum after a 24-h culture period and declines by 50% after 48 h. However, it is still possible to test the effect of a given effector over longer periods, provided the culture medium is replaced every 24 h.
4. Based on β-galactosidase activity determinations, we estimate that 3–5% of the cells express the reporter gene under optimal promoter activity conditions (i.e., the β-galactosidase gene driven by the RSV promoter under basal conditions or by the spi promoters in the presence of hormones). Two parameters were found to be important:
    a. The dissociation of the liver with collagenase should not exceed 8–9 min. We have observed that cell survival, which normally is between

40 and 60%, as well as hormone responsiveness, is severely reduced with hepatocytes obtained after longer perfusion times (15–18 min). The use of a collagenase with a high specific activity (0.4–0.6 U/mg) to ensure rapid dissociation of the liver tissue is therefore important; and

b. The handling of the cells, including purification, electroporation, and returning to culture medium, should be performed as quickly as possible.

5. This procedure can easily be adapted to analyze a promoter using other reporter genes, such as the firefly luciferase gene.

References

1. Cone, R. D. and Mulligan, R. C. (1984) High efficiency gene transfer into mammalian cells: generation of helper-free recombinant retrovirus with broad mammalian host range. *Proc. Natl. Acad. Sci. USA* **81,** 6349–6353.
2. Bond, V. C. and Wold, B. (1987) Poly-L-ornithine-mediated transformation of mammalian cells. *Mol. Cell. Biol.* **7,** 2286–2293.
3. Schaeffer-Ridder, M., Wang, Y., and Hofschneider, P. H. (1982) Liposomes as gene carriers: efficient transformation of mouse L cells by thymidine kinase gene. *Science* **215,** 166–168.
4. Minden, M. D., Gusella, J. F., and Housman, D. (1984) Chromosome mediated transfer of the malignant phenotype by acute myelogenous leukemic cells. *Blood* **64,** 842–846.
5. Schaffner, W. (1980) Direct transfer of cloned genes from bacteria to mammalian cells. *Proc. Natl. Acad. Sci. USA* **77,** 2163–2167.
6. Graham, F. L. and Van Der Erb, A. J. (1973) Transformation of rat cells by DNA of human adenovirus 5. *Virology* **52,** 456–467.
7. McCutchan, J. H. and Pagano, J. S. (1968) Enhancement of the infectivity of simian virus 40 deoxyribonucleic acid with diethylaminoethyl-dextran. *J. Natl. Cancer. Inst.* **41,** 351–357.
8. Behr, J. P., Demeneix, B., Loeffler, J. P., and Perez-Mutul, J. (1989) Efficient gene transfer into mammalian primary endocrine cells with lipopolyamine-coated DNA. *Proc. Natl. Acad. Sci. USA* **86,** 6982–6986.
9. Yamamoto, F., Furusawa, W., Furusawa, I., and Obinata, M. (1982) A new efficient technique for mechanically introducing foreign DNA into the nuclei of cultured cells. *Exp. Cell Res.* **142,** 79–84.
10. Kurata, S., Tsukakoshi, M., Kasuyu, T., and Yakawa, Y. (1986) The laser method for efficient introduction of foreign DNA into cultured cells. *Exp. Cell Res.* **162,** 372–378.
11. Wong, T. and Neumann, E. (1982) Electric field mediated gene transfer. *Biochem. Biophys. Res. Commun.* **107,** 584–587.
12. Satyabhama, S. and Epstein, A. L. (1988) Laboratory methods. Short-term efficient expression of transfected DNA in human hematopoietic cells by electroporation: definition of parameters and use of chemical stimulators. *DNA* **7,** 203–209.

13. Parker Ponder, K., Dunbar, R. P., Wilson, D. R., Darlington, G. J., and Woo, S. L. C. (1991) Evaluation of relative promoter strength in primary hepatocytes using optimized lipofection. *Hum. Gene Ther.* **2,** 41–52.
14. Sambrook, J., Fritsch, E. F., and Maniatis, T. (1989) *Molecular Cloning: A Laboratory Manual, 2nd ed.* Cold Spring Harbor Laboratory, Cold Spring Harbor, NY.
15. Pagès, G., Rouayrenc, J. F., Rossi, V., Le Cam, G., Mariller, M., Szpirer, J., Szpirer, C., Levan, G., and Le Cam, A. (1990) Primary structure and assignment to chromosome 6 of three closely related rat genes which encode serine protease inhibitors expressed in liver. *Gene* **94,** 273–282.
16. Dente, L., Cesarini, G., and Cortese, R. (1983) pEMBL: a new family of single stranded plasmids. *Nucleic Acids Res.* **11,** 1645–1655.
17. Luckow, B. and Schütz, G. (1987) CAT constructions with multiple unique restriction sites for the functional analysis of eukaryotic promoters and regulatory elements. *Nucleic Acids Res.* **15,** 5490.
18. Pierce, J. W., Leonardo, M., and Baltimore, D. (1988) Oligonucleotide that binds nuclear factor NF-κB acts as a lymphoid-specific and inducible enhancer element. *Proc. Natl. Acad. Sci. USA* **85,** 1482–1486.
19. Mercer, E. H., Hoyle, G. W., Kapur, R. P., Brinster, R. L., and Palmiter, R. D. (1991) The dopamine β-hydroxylase gene promoter directs expression of E. coli LacZ to sympathetic and other neurons in adult transgenic mice. *Neuron* **7,** 703–716.
20. Pothier, F., Ouellet, M., Julien, J. P., and Guérin, S. L. (1992) An improved CAT assay for promoter analysis in either transgenic mice or tissue culture cells. *DNA and Cell Biol.* **11,** 83–90.
21. Paquereau, L. and Le Cam, A. (1992) Electroporation-mediated gene transfer into hepatocytes; preservation of a growth hormone response. *Anal. Biochem.* **204,** 147–151.
22. Paquereau, L., Vilarem, M. J., Rossi, V., Rouayrenc, J. F., and Le Cam, A. (1992) Regulation of two rat serine protease inhibitor gene promoters by somatotropin and glucocorticoids. Study with intact hepatocytes and cell-free systems. *Eur. J. Biochem.* **209,** 1053–1061.
23. Le Cam, A., Pantescu, V., Paquereau, L., Legraverend, C., Fauconnier, G., and Asins, G. (1994) *cis*-Acting elements controlling transcription from rat serine protease inhibitor 2.1 gene promoter. Characterization of two growth hormone response sites and a dominant purine-rich element. *J. Biol. Chem.* **269,** 21,532–21,539.

CHAPTER 12

Electroporation of Human Lymphoblastoid Cells

Fen Xia and Howard L. Liber

1. Introduction

There are many reasons to introduce new genes on recombinant vectors into mammalian cells. For example, *in situ* studies of the effects of particular gene products on DNA replication *(1–3)*, repair *(4–6)*, recombination *(7)*, or gene expression *(7–11)* can be performed. Alternatively, vectors that specifically respond to a particular enzymatic function, such as homologous recombination *(12,13)* or mismatch repair *(14)*, can be introduced to determine whether a cell type of interest is positive or negative for that function.

A wide range of organic molecules with different sizes and features, including proteins and nucleic acids, can be introduced into mammalian cells *(15,16)*. Two common ways to stably transfect cells are calcium phosphate transfection and liposome fusion. Both methods, which yield high levels of stable transfectants, are very suitable for anchorage-dependent cells, because the procedure can be easily performed in cell-culture dishes *(15–19)*. A third technique, electroporation, applies a high electric pulse to open reversible pores in the cell membrane through which macromolecules can enter. Electroporation is very efficient with many different cell types; it is a highly reproducible, fast, easy, and chemical-free technique, and it is especially suitable for suspension-grown cells, such as lymphoblasts *(15,16)*.

Since only a fraction of cells integrate and express the gene of interest, it is usually necessary to include a dominant selection scheme to isolate transfectants. Typically, this is an expressible drug resistance gene encoded within the recombinant vector. Common selectable markers include the aminoglycoside phosphotransferase *(neo)* gene, which confers resistance to G418 *(20,21)*, and dihydrofolate reductase, which confers resistance to methotrexate *(15)*.

We discuss in this chapter procedures for electroporating a plasmid with an expressible gene into human lymphoblastoid cells. Human B-lymphoblast cell lines have several advantages as an experimental system. They can be established from any human donor, are immortal, and many lines can be grown easily in suspension culture with rapid doubling times to very high cell numbers. Some lines, even after growth for many generations in vitro, have normal or near-normal karyotypes that are stably maintained.

2. Materials
2.1. Cells

Human lymphoblast cell lines from many different donors are available from the American Type Tissue Collection (Rockville, MD) or from the National Institutes of Health Human Genetic Mutant Cell Repository (Camden, NJ). Culture medium is RPMI-1640 medium (Gibco-BRL, Gaithersburg, MD) supplemented with 10–20% fetal bovine or horse serum, according to the requirements of the particular cell line. Serum is usually heat-inactivated at 56°C for 30 min.

2.2. Reagents

1. Plasmid DNA: Isolate plasmids using commercially available kits (e.g., Qiagen). Purify plasmid DNAs by standard phenol/chloroform extraction and ethanol-precipitation. Allow to air-dry, resuspend in sterilized ddH$_2$O, and store at 4°C for up to 8 wk, or –20°C for longer-term storage. Details of these standard procedures can be found in most laboratory manuals (e.g., refs. *15,16*) *(see also* Note 1).
2. Restriction enzymes: An appropriate enzyme will cleave at only one site, in a nonfunctional region.
3. Sepharose CL-6B (Pharamcia): Wash this 10 times with Na$^+$-free TE before use. Prepare simple spin columns as described in *(22)*, and store at 4°C.
4. PBS: Dulbecco's phosphate-buffered saline, free of Ca^{2+} and Mg^{2+}. It consists of 8.0 g/L NaCl, 0.2 g/L KCl, 1.15 g/L Na$_2$HPO$_4$, and 0.2 g/L KH$_2$PO$_4$. Store at 4°C.

5. G418 (Gibco-BRL). Prepare a 5 mg/mL G418 stock solution in 100 mM HEPES, pH 7.3, and store at –20°C.

3. Methods
3.1. Preparation of Plasmid DNA

1. Linearize purified plasmids by digestion with an appropriate restriction enzyme (e.g., *Eco*RI for pSV2neo), according to the supplier's recommendations.
2. After 3–4 h of digestion (*see* Note 2), stop the reaction by placing the vials in a 65°C water bath for 10 min, to heat-inactivate the restriction enzyme. Purify the digest by passage through a Sepharose CL-6B spin column *(18)*. Alternatively, use phenol/chloroform followed by ethanol precipitation to purify digested DNA *(15,16)*.
3. Quantify DNA on a spectrophotometer, by absorbance readings at 260 and 280 nm (1 absorbance unit at $A_{260} \approx 50$ µg/mL; $A_{260}/A_{280} = 1.8$ for maximal purity). Adjust the concentrations with sterile ddH$_2$O according to the amount of DNA needed for one transfection in a final volume of 10 µL.

3.2. Human Lymphoblast Cell Culture

1. Maintain human lymphoblast cell lines as exponentially growing cultures in RPMI-1640 medium supplemented with horse or fetal bovine serum (FBS). Incubate cultures in T-25, -75, or -150 flasks (kept horizontally to maximize surface area for growth) at 37°C in 5% CO$_2$ and 100% humidity. Maintain cultures at densities of $1–10 \times 10^5$ cells/mL by regular dilutions, depending on growth rate (*see* Note 3).
2. It is important to determine the plating efficiency (PE) of each cell line. To do this, seed small aliquots of cells into 96-well microtiter dishes. For example, seed two plates each at 1, 10, or 100 cells/well (0.2 mL/well), and incubate for 12–20 d. Count colonies and calculate PE with the Poisson distribution *(23)*. It is essential to observe both positive and negative wells. This method is most accurate when half the wells are empty and half are full.

$$PE = [-\ln(\text{fraction of negative wells})] \div \text{number of cells plated/well} \quad (1)$$

3.3. DNA Transfection

1. Harvest a total of 1×10^8 cells by centrifuging at 250g for 5 min at 4°C.
2. Wash cells once with 10 mL of ice-cold PBS.
3. Resuspend cells in 0.6 mL of ice-cold PBS. Mix well with 3–30 µg of digested plasmid (*see* Note 4), and transfer to a precooled cuvet with a 0.4-cm electrode gap. Incubate the filled cuvet on ice for 5 min (*see* Notes 5 and 6).

4. Expose the mixture of cells and DNA to an electric pulse of 270 V at 960 µF (*see* Note 7).
5. Incubate the shocked cells on ice for 10 min.
6. Add the cells to 15–30 mL of prewarmed growth medium in a T-75 flask (*see* Note 8), and incubate for 2 d at 37°C, to allow integration of the transfected gene into cellular DNA and subsequent expression of the new phenotype (*see* Note 9).
7. Determine the toxicity of the electric pulse by measuring the PE immediately after electroporation, using the protocol in Section 3.2., step 2.

$$\text{Toxicity} = 1 - (\text{PE}_{\text{electroporated cells}} / \text{PE}_{\text{controls}}) \qquad (2)$$

Mock electroporated cells (no plasmid DNA) are a necessary control if quantitative toxicity data are desired.

3.4. G418 Selection

1. After 2 d of incubation (*see* Note 10), seed cells into 24-well plates (*see* Note 11) at $1-10 \times 10^4$ cells/well in growth medium containing G418. The best cell concentration to use must be determined empirically, and will vary among lymphoblastoid lines. The final volume is 1.0 mL/well (*see* Note 12). The concentration of G418 must be optimized for each cell line; it must be high enough to kill the untransfected cells and low enough so as not to suppress the growth of the transfected cells. Appropriate concentrations ranging from 200–1500 µg/mL have been used (*see* Note 13). The number of plates to seed depends on the expected transfection frequency and the plating efficiency of the particular cell line, but seed at least 10 plates with the proper cell number per well if accurate transfection frequencies are required.
2. At the same time, determine PE by seeding the cells into 24-well plates at concentrations ranging from 1–100 cell/well (according to the predetermined PE for each different cell line) in nonselective growth medium. A minimum of eight plates should be plated for the PE determination. Calculate PE with the Poisson distribution, as in Section 3.2., step 2.
3. Refeed cells with fresh G418 every 4 d by adding 0.25 mL of growth medium that contains concentrated G418, so the final concentration will be restored to the original value (*see* Note 14).
4. Score G418 resistant colonies by microscopic examination at 4–10× after 16 d in selective medium. Positive colonies generally contain $0.2-2 \times 10^6$ cells, and are easily recognized against the background of cells killed by G418. Colonies generally have at least 50% more surface area than other wells, and if examined under higher magnification (100–400×), will contain many cells with a clear cytoplasm.

5. If desired, calculate transfection frequencies with the Poisson distribution. First, determine the number of colony forming units/well for cells plated in the presence of G418. As above, this is:

[–ln(fraction of negative wells)] / (cells plated/well)   (3)

Divide this value by the nonselective PE from this section, step 2 *(6)*.

4. Notes

1. Plasmid DNA of high purity is necessary to attain optimal transfection efficiency. After isolation, maintain the plasmid DNA in sterile solutions and containers to reduce the contamination to cell culture. The amount of DNA used for transfection can be varied. The more DNA used, the greater the number of transfectants produced. The common range for a single gene transfection in most cell lines is 1–10 µg.
2. Examine a sample of the digested DNA by gel electrophoresis to ensure that the DNA has been completely linearized before beginning the transfection. This is important because circular DNA does not integrate as well as linear DNA. If digestion is incomplete, prolong the incubation time by several hours, and/or add additional restriction enzyme and reincubate. If adding additional enzyme, be sure that the enzyme volume is ≤10% of the total volume of the reaction.
3. Maintain cells to be transfected in healthy exponential growth. It is better to dilute them 1 d prior to electroporation to an appropriate density (according to growth rate), so they will grow to 10×10^5 cells/mL just before transfection.
4. One-milliliter disposable serological plastic pipets are very convenient for mixing cells with DNA and transferring the cell suspension from tube to a cuvet or to a flask, because the narrow tip fits into the bottom of the 0.4-cm cuvet, and its length makes it easy to maintain sterility.
5. The final volume in the cuvet, and also condensation outside the cuvet, can influence electrical resistance. If quantitation is important, it is essential to control these small details. Measure volumes carefully, and dry the exterior of the cuvet with a paper towel immediately prior to administering the electric pulse.
6. The cuvet can be reused as long as there are no cracks on the wall of the electrode. Cuvets should be rinsed with ddH$_2$O thoroughly, soaked in 70% ethanol overnight, and UV-irradiated overnight before reuse (a cell-culture hood with a germicidal light works well for this).
7. Optimal electric shock pulse parameters vary for different cell types. Typically, the voltage should be 200–350 V, and the capacitance either 25 or 960 µF *(15,16)*.

8. Immediately after electroporation, plate cells at a density of about 5×10^5 cells/mL. Until selection, maintain them between 4 and 10×10^5 cells/mL.
9. To avoid characterizing a series of identical transfectants that are descendants of a single event, subdivide the transfection into different T-25 flasks immediately after transfection, and isolate no more than one colony from each flask.
10. The incubation time prior to selection can be longer than 2 d, but multiple copies of a particular transfected cell will be more prevalent (*see also* Note 9).
11. Use 24-well plates instead of 96-well plates because of volume considerations; refeeding every 4 d with fresh G418 quickly overfills the smaller wells.
12. The total number of cells plated and the number of cells per well depend on the survival of cells after electric shock, the plating efficiency, the transfection efficiency, recombination capacity, clonogenicity, and time and budgetary considerations. For quantitation of transfection frequency, the optimum is 12 resistant colonies/24-well plate. However, in these circumstances, there is a fairly high probability (28%) that any particular isolated colony is not monoclonal, but arose from two or more successfully transfected cells. One can achieve a higher probability of monoclonality if one plates at densities that yield 1 colony/plate (probability of monoclonality $\approx 98\%$) or less. In any case, if a monoclonal derivative is required, reclone at a low cell density (≤ 0.1 cell/well).
13. The G418 dose for selection is critical. Each cell line has a different sensitivity ranging from 200–1500 µg/mL *(15,16)*. Choose the optimal dose based on the dose–response relationship. The optimal dose should be 100–200 µg/mL higher than the dose at which no untransfected cells can survive. On the other hand, avoid using too high a dose. If quantitation is important, perform control experiments to make sure that the stable transfectants grow equally well in the presence and absence of the drug, and that there is no difference in plating efficiency between the cell lines in the presence and absence of G418. Another important point is that each lot of G418 has different biological activity, so recheck the toxicity when changing to a different lot.
14. If the original G418 concentration was 1 mg/mL, then the concentration of G418 required for the first feeding would be 5 mg/mL. Thus

    $$(0.25 \text{ mL} \times 5 \text{ mg/mL}) / (1.25 \text{ mL total volume}) = 1 \text{ mg/mL} \qquad (4)$$

    This calculation assumes that the effective concentration of G418 has been reduced to 0 after 4 d.
15. As an example we describe experiments to characterize homologous recombination in human lymphoblastoid cells. Homologous recombination between two DNA molecules may be an important event, involved in both physiological (during cell division) and pathological (to repair X-ray-

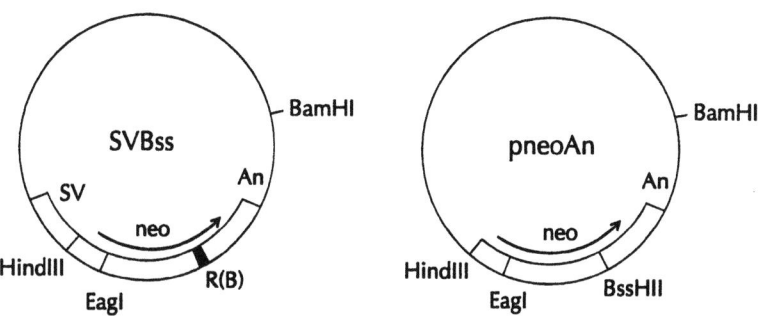

Fig. 1. Plasmids used for electroporation into TK6 and WTK1 human lymphoblastoid cells to monitor recombination (from ref. *12*). SVBss has a nonfunctional *neo* allele owing to the presence of a linker insertion, R(B). pneoAn has a transcriptionally silent *neo* allele.

induced damage, such as double-strand breaks) processes *(24)*. In mammalian cells, loss of heterozygosity, which may arise by homologous recombination, is often associated with the inactivation of tumor-suppressor genes. Therefore, it is important to understand the processes that mediate DNA homologous recombination in human cells. We have used the electroporation technique and the G418 selection system to analyze quantitatively interchromosomal homologous recombination capacities in two different human B-lymphoblast cell lines called TK6 and WTK1 *(25)*.

Plasmid pSV2neo *(26)* carrying a wild-type *neo* was utilized as a control; it carries the 1100-bp neomycin resistance coding region coupled to the SV40 promoter; after transfection, this *neo* gene can be transcribed to produce a protein that renders cells resistant to G418. Cells were mixed with 3 μg of *Eco*RI-digested pSV2neo and electroporated as described. This transfection served to (a) demonstrate that the *neo* gene could function in these cell lines and (b) compare the overall effectiveness of transfection. Since many factors (both known and unknown) can affect the process by which DNA molecules traverse the cell membrane and eventually integrate into a chromosome, it was essential to use the information from the pSV2neo control experiment to normalize the raw data from the homologous recombinational assay.

The plasmid pneoAn *(13)* was constructed by inserting a 2500-bp *Hind*III-*Bam*HI fragment from pSV2neo into pU19, which lacks the SV40 promoter. Thus pneoAn does not produce any active product of the *neo* gene. The plasmid SVBss is an inactive derivative of pSV2neo; it was made by inserting an *Eco*RI linker frameshift mutation into the *Bss*HII site (Fig. 1). When cotransfected, homologous recombination between pneoAn

Table 1
Recombination Capacity of TK6 and WTK1[a]

Cell line	Recombination frequency, $\times 10^6 \pm$ SD	
	Homologous	Nonhomologous
TK6	1.4 ± 0.8	3.9 ± 0.9
WTK1	51.3 ± 10.2	19.8 ± 2.3

[a]Homologous recombination frequencies were determined with 30 μg each of SVBss and pneoAn. Nonhomologous recombination (integration) frequencies were determined with 3 μg of pSV2neo. Values are averages of four determinations.

and SVBss can result in a product that will transcribe a functional *neo* message; subsequent integration of this recombinant into a mammalian chromosome can produce a stable G418 resistant transfectant.

Recombination assays were performed with 30 μg each of *Eco*RI-digested SVBss and *Hind*III-digested pneoAn. As a negative control, equal amounts of SVBss and the pneoAn were transfected separately, and no G418 resistant colonies were observed ($<5.8 \times 10^{-7}$ in both lines).

Transfection frequencies were calculated as the ratio of G418 resistant colonies to live cells plated. Frequencies in the two cell lines first were compared based on data obtained with the wild-type plasmid pSV2neo. Then, recombination frequencies were calculated based on transfection frequencies observed after cotransfection of SVBss and pneoAn.

Table 1 shows the transfection and recombination frequencies in TK6 and WTK1 cells. TK6 is less efficient at integrating the plasmid DNA. It also is less efficient at catalyzing intermolecular recombination, and it is interesting to speculate that both integration and recombination may be mediated by a single enzyme activity that is expressed at lower levels in TK6 than in WTK 1 *(25)*.

Acknowledgment

Development of these techniques was supported by grant CN-68a from the American Cancer Society.

References

1. Huberman, J. A. (1987) The in vivo replication origin of the yeast 2 μ plasmid. *Cell* **51**, 473–481.
2. Li, K., Smagula, C. S., Parsons, W. J., Richardson, J. A., Gonzalez, M., Hagler, H. K., and Williams, R. S. (1994) Subcellular partitioning of MRP RNA assessed by ultrastructural and biochemical analysis. *J. Cell Biol.* **124**, 871–882.

3. Nault, C., Veilleux, S., Delbecchi, L., Bourgauxramoisy, D., and Bourgaux, P. (1994) Intramolecular recombination in polyoma virus DNA is controlled by promoter elements. *Nucleic Acids Res.* **22,** 485–491.
4. Choi, I. S. and Park, S. D. (1991) Characterization and expression of RAD4 gene involved in nucleotide excision repair of UV-damaged Saccharomyces. *J. Toxicol. Sci.* **16 (Suppl. 1),** 75–82.
5. Gutman, P. D., Yao, H. L., and Minton, K. W. (1991) Partial complementation of the UV sensitivity of Deinococcus raddiodurans excision repair mutants by the cloned denV gene of bacteriophage T4. *Mutat. Res.* **254,** 207–215.
6. Kibitel, J. T., Yrr, V., and Yarosh, D. B. (1991) Enhancement of ultraviolet-DNA repair in denV gene transfectants and T4 endonuclease V-liposome recipients. *Photochem. Photobiol.* **54,** 753–760.
7. Sadofsky, M. J., Hesse, J. E., McBlane, J. F., and Gellert, M. (1993) Expression and V(D)J recombination activity of mutated RAG-1 proteins. *Nucleic Acids Res.* **21,** 5644–5650.
8. Hupp, T. R., Meek, D. W., Midgley, C. A., and Lane, D. P. (1992) Regulation of the specific DNA binding function of P53. *Cell* **71,** 875–886.
9. Kuerbitz, S. J., Plunkett, B. S, Walsh, W. V., and Kastan, M. B. (1992) Wild type P53 is a cell cycle checkpoint determinant following irradiation. *Proc. Natl. Acad. Sci. USA* **89,** 7491–7495.
10. Miura, M., Zhu, H., Rotello, R., Hartwieg, E. A., and Yuan, J. (1993) Induction of Apoptosis in fibroblasts by IL-1ß-converting enzyme, a mammalian homolog of the *C. elegans* cell death gene ced-3. *Cell* **75,** 653–660.
11. Tsujimoto, Y. (1989) Overexpression of human Bcl-2 gene product results in growth enhancement of Epstein-Barr virus-immortalized B cells. *Proc. Natl. Acad. Sci. USA* **86,** 1958–1962.
12. Nickoloff, J. A. (1992) Transcription enhances intrachromosomal homologous recombination in mammalian cells. *Mol. Cell Biol.* **12,** 5311–5318.
13. Nickoloff, J. A. and Reynolds, R. J. (1990) Transcription stimulates homologous recombination in mammalian cells. *Mol. Cell Biol.* **10,** 4837–4845.
14. Prolla, T. A., Pang, Q., Alani, E., Kolodner, R. D., and Liskay, R. M. (1994) MLH1, PMS1 and MSH2 interact during the initiation of DNA mismatch repair in yeast. *Science* **265,** 1091–1093.
15. Ausubel, F. M., Brent, R., Kingston, R. E., Moore, D. D., Seidman, J. G., Smith, J. A., and Struhl, K. (1989) *Current Protocols in Molecular Biology.* Green Publishing Associations and Wiley-Interscience, New York, pp. 9.1–9.5.
16. Sambrook, J., Fritsch, E. F., and Maniatis, T. (1989) *Molecular Cloning: A Laboratory Manual.* Cold Spring Harbor Laboratory, Cold Spring Harbor, NY, pp. 16.30–16.55.
17. Felgner, J. H., Kumar, R., Sridhar, C. N., Wheeler, C. J., Tsai, Y. J., Border, R., Ramsey, P., Martin, M., and Felgner, P. L. (1994) Enhanced gene delivery and mechanism studies with a novel series of cationic lipid formulations. *J. Biol. Chem.* **269,** 2550–2561.
18. Kamata, H., Yagisawa, H., Takahashi, S., and Hirata, H. (1994) Amphiphilic peptides enhance the efficiency of liposome-mediated DNA transfection. *Nucleic Acids Res.* **22,** 536–537.

19. Takeshita, S., Gal, D., Leclerc, G., Pickering, J. G., Riessen, R., Weir, L., and Isner, J. M. (1994) Increased gene expression after liposome-mediated arterial gene transfer associated with smooth muscle cell proliferation—in vitro and in vivo findings in a rabbit model of vascular injury. *J. Clin. Invest.* **93,** 652–661.
20. Colbere-Garapin, F., Horodniceanu, F., Kourilsky, P., and Garapin, A. C. (1981) A new dominant hybrid selective marker for higher eukaryotic cells. *J. Mol. Biol.* **150,** 1–14.
21. Jorgensen, R. A., Rothstein, S. J., and Reznikoff, W. S. (1979) A restriction enzyme cleavage map of Tn5 and location of a region encoding neomycin resistance. *Mol. Gen. Genet.* **177,** 65–72.
22. Nickoloff, J. A. (1994) Sepharose spin column chromatography: a fast, non-toxic replacement for phenol:chloroform extraction/ethanol precipitation. *Mol. Biotechnol.* **1,** 105–109.
23. Furth, E. E., Thilly, W. G., Penman, B. W., Liber, H. L., and Rand, W. M. (1981) Quantitative assay for mutation in diploid human lymphoblasts using microtiter plates. *Anal. Biochem.* **110,** 1–8.
24. Bollag, R. J., Waldman, A. S., and Liskay, R. M. (1989) Homologous recombination in mammalian cells. *Ann. Rev. Genet.* **23,** 199–225.
25. Xia, F., Amundson, S. A., Nickoloff, J. A., and Liber, H. L. (1994) Different capacities for recombination in closely related human lymphoblastoid cell lines with different mutational responses to x-irradiation. *Mol. Cell. Biol.* **14,** 5850–5857.
26. Southern, P. J. and Berg, P. (1982) Transformation of mammalian cells to antibiotic resistance with a bacterial gene under control of the SV40 early region promoter. *J. Mol. Appl. Genet.* **1,** 327–341.

CHAPTER 13

The Use of Electroporated Bovine Spermatozoa to Transfer Foreign DNA into Oocytes

Marc Gagné, François Pothier, and Marc-André Sirard

1. Introduction

The production of transgenic laboratory animals has rapidly gained importance as an experimental tool to study the factors that determine the tissue specificity of gene expression, to generate models for human diseases *(1,2)*, as well as to look at the consequences of oncogene expression *(3,4)*. Through this technique, recombinant DNA molecules of any type can be introduced into zygotes for use in producing transgenic domestic animals. Introduction of foreign genes into early embryos, particularly into bovine embryos, may be a powerful tool to obtain more valuable animals. Although pronuclear microinjection is the only proven route for gene transfer in livestock, the production of transgenic animals is still problematic and inefficient *(5)*, and importantly, microinjection may impair the development of the embryos *(6)*.

An alternative method that can be used to introduce new genetic information into bovine embryos employs electroporated sperm cells. Electroporation has previously been used both for obtaining stable transformants in eukaryotic cell lines *(7–9)* and to introduce plasmid constructs into bacteria *(10)*. Although it was reported that foreign DNA could be transferred into mouse oocytes by simply incubating the spermatozoa with plasmid DNA prior to in vitro fertilization *(11)*, this approach may be limited to specific, but as yet uncharacterized conditions, since so far no other groups have been able to produce transgenic animals using this method. By using electroporated sperm, it is possible to introduce a foreign DNA sequence into bovine oocytes during in vitro fertilization. The sperm electroporation procedures described below are based on methods originally described by Gagné and colleagues *(6)*.

2. Materials
2.1. Solutions and Reagents
For Sperm Preparation and Electroporation

1. Tyrode lactate sperm (TLS) *(12)*: 5.82 g/L NaCl, 230 mg/L KCl, 290 mg/L $CaCl_2 \cdot 2H_2O$, 310 mg/L $MgCl_2 \cdot 6H_2O$, 35 mg/L NaH_2PO_4, 2.38 g/L HEPES, 2.09 g/L $NaHCO_3$, 10 mg/L phenol red, 3.153 mL/L sodium lactate (60% syrup). Weigh and dissolve $CaCl_2 \cdot 2H_2O$ separately in ultrapure water before adding other reagents in solution. Adjust pH to 7.4 with $1.0M$ NaOH. Measure osmolarity (must be between 280 and 300 mosM; *see* Note 1 and Fig. 1). Filter-sterilize through a 0.22-μm membrane, aliquot (100 mL), and store at –20°C. Once thawed, store at 4°C (15 d maximum).
2. Gentamycin sulfate: 72.9 mg/mL (for an activity of 50 mg/mL) dissolved in a 0.9% NaCl aqueous solution. Stable when stored at 4°C for several months.
3. TALP *(12)*: 50 mL TLS, 300 mg bovine serum albumin, fatty-acid-free (BSA, Sigma, St. Louis, MO), 50 μL gentamycin (50 mg/mL). Filter-sterilize through a 0.22-μm membrane. Always prepare fresh.
4. M *(13)*: $0.3M$ mannitol, 0.1 mM $MgCl_2$, 0.05 mM $CaCl_2$, in distilled water. Filter sterilize through a 0.22-μm membrane. Store at 4°C for up to a year.
5. Fertilization medium stock (TLF-1) *(14)*: 6.66 g/L NaCl, 235 mg/L KCl, 300 mg/L $CaCl_2 \cdot 2H_2O$, 100 mg/L $MgCl_2 \cdot 6H_2O$, 2.104 g/L $NaHCO_3$, 41 mg/L NaH_2PO_4, 10 mg/L phenol red, 1.460 mL/L sodium lactate (60% syrup). Weigh and dissolve $CaCl_2 \cdot 2H_2O$ separately in ultrapure water before adding to other products in solution. Adjust pH to 7.4 with $1.0M$ NaOH. Measure osmolarity (must be between 280 and 300 mosM; *see* Note 1). Filter-sterilize through a 0.22-μm membrane, aliquot (100 mL), and store at –20°C. Once thawed, keep at 4°C (15 d maximum).
6. Heparin: Dissolve 1 mg of heparin in 1 mL of 0.9% NaCl aqueous solution, aliquot, and store at –20°C. Thaw just before use.
7. TLF-2: 10 mL of TLF-1, 60 mg of BSA, 10 μL of gentamycin (50 mg/mL), 2 μg/mL of heparin, 100 μL of 20 mM pyruvate (dissolve in distilled water).

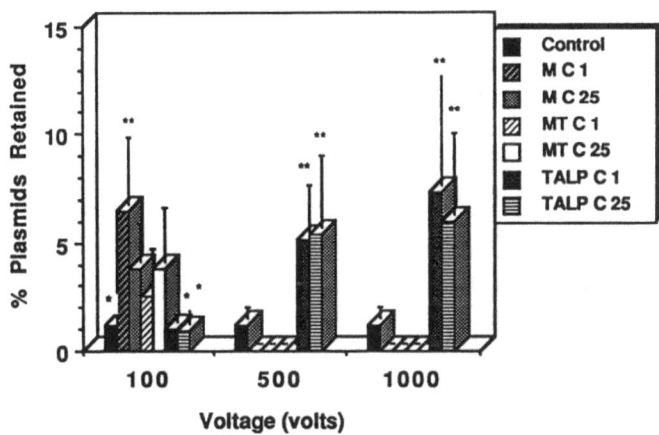

Fig. 1. Comparison of plasmid retaining rate by sperm cells after electroporation in three media and at a capacitance of 1 or 25 µF (C1 and C25). At 100 V, no difference ($p < 0.005$) was observed in sperm-plasmid attachment following electroporation in TALP (1.0% both for C 1 and C 25) compared to control (1.0%), whereas sperm treated in mannitol (M) and mannitol-TALP (MT) retained plasmids more efficiently. As sperm cells were killed (*see* Fig. 2) in MT and M over 100 V and in TALP over 1000 V, they were not assayed for the plasmid retention rates. Values of treatments with one asterisk were significantly lower ($p < 0.005$) than others in respective voltage classes. Values with two asterisks indicate treatments with significant differences from respective controls ($p < 0.005$, Fisher-Snedecor Test) *(14)*.

3. Methods

3.1. Preparation of the Semen

1. Use freshly ejaculated or frozen semen (*see* Note 2). For frozen semen, thaw the straws (0.05 mL) of frozen semen in a 35°C water bath for exactly 1 min.
2. Put the semen (1 mL) in a sterile 10-mL glass tube.
3. Add 4 mL/tube of TALP.
4. Wash the semen by centrifugation at 250*g* for 8 min at room temperature (between 20 and 25°C).
5. Remove as much supernatant as possible, taking care not to remove the spermatozoa in pellet.
6. Repeat steps 4 and 5.
7. Adjust the concentration of spermatozoa to 10^8 cells/mL.

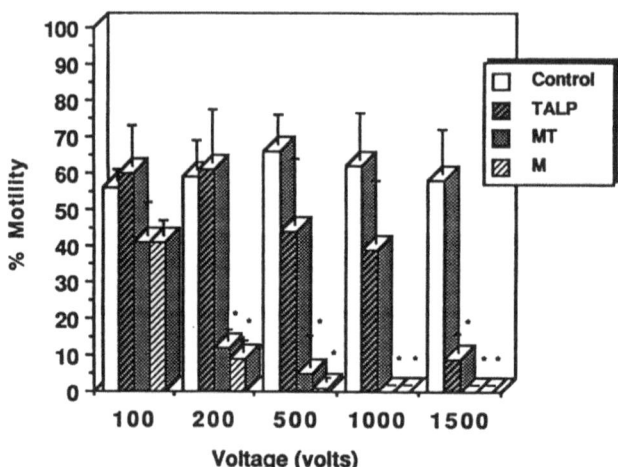

Fig. 2. Influence of electrical field on sperm cell motility. Motility was reduced by a 100-V pulse in mannitol (M) and in 50% mannitol–50% TALP (MT) and was essentially zero at 500 V. Treatment in TALP has a significant negative effect from 500 to 1500 V. Experiments were repeated in triplicate. Treatments with an asterisk differ from their corresponding controls, $p < 0.005$ (Chi-square test) (14).

3.2. Electroporation

1. Add 0.9 mL of TALP or 50% TALP-50% M (MT, v/v) or M only to a 1.4-mL cuvet with a 4-mm electrode gap.
2. Add 0.1 mL of the washed semen (10^8/mL).
3. Add 1 µg of transfectant DNA.
4. Pulse once at 100 V, 1- or 25-µF capacitance, for both M and MT medium (see Note 3), and 500 V for the TALP medium (see Note 4 for a description of the apparatus). The time constant should be about 0.4 ms (see Note 5 and Fig. 2). The electroporated spermatozoa are ready for fertilization (see Notes 6 and 7).

3.3. In Vitro Fertilization

1. Transfer 5 µL of electroporated spermatozoa (10^8 cells/mL) to oil-covered 50-µL drops of TLF-2 containing 5 matured oocytes (15).
2. Let fertilization occur for 24 h.

4. Notes

1. Under normal conditions, the osmolarity will range between 280 and 300 mosM. If not, the medium should be prepared again.

2. Motility should be >60% for both freshly ejaculated and frozen thawed semen.
3. Low voltage with nonconductive medium (M) is as efficient as high voltage with ionic medium (TLS-2) for retention of transfectant DNA after electroporation (see Fig. 1).
4. We use Gene Pulser cuvets (1.4-mL capacity with a distance of 4 mm between the electrodes, Bio-Rad, Richmond, CA), and the Gene Pulser Apparatus (Bio-Rad) to deliver electric pulses.
5. Because the intrinsic fertilization capacity and the motility of the spermatozoa vary among species, bulls, and even ejaculations of the same bull, it is suggested that these characteristics be assessed before and after the electroporation, and the electrical parameters be set to maximize the fertilization rate. However, the pulse should not exceed 1500 V for bovine spermatozoa (see Fig. 2).
6. Electroporation must be performed at room temperature (between 20 and 25°C) rapidly after the washing of the semen.
7. The final concentration of spermatozoa in fertilization drops must be adjusted to 10^6 cells/mL.

References

1. Léonard, J. M., Abramczuk, J. W., Pezen, D. S., Rutledge, R., Belcher, J. H., Hakim, F., Shearer, G., Lamperth, L., Travis, W., Fredrickson, T., Motkins, A. L., and Martin, M. A. (1988) Development of disease and virus recovery in transgenic mice containing HIV proviral DNA. *Science* **242**, 1665–1670.
2. Small, J. A., Scangos, G. A., Cork, L., Jay, G., and Khoury, G. (1986) The early region of hyman papovavirus JC induces dysmyelination in transgenic mice. *Cell* **46**, 13–18.
3. Andres, A. C., Schonenberger, C. A., Groner, B., Hennighousen, L., Lemeur, M., and Gerlinger, P. (1987) Ha-ras oncogene expression directed by a milk protein gene promoter: tissue specificity, hormonal regulation and tumor induction in transgenic mice. *Proc. Natl. Acad. Sci. USA* **84**, 1299–1303.
4. Sinn, E., Muller, W., Pattengale, P., Tepler, I., Wallace, R., and Leder, P. (1987) Coexpression of MMTV/V Ha-Ras and MMTV/c-myc genes in transgenic mice: synergistic action of oncogenes in vivo. *Cell* **49**, 465–475.
5. Clark, A. J., Simons, J. P., and Wilmut, I. (1992) Germ line manipulation: applications in agriculture and biotechnology, in *Transgenic Animals* (Grosveld, F. and Kollias, G., eds.), Academic, London, pp. 249–270.
6. Gagné, M., Pothier, F., and Sirard, M. A. (1990) Developmental potential of early bovine zygotes submitted to centrifugation and microinjection following in vitro maturation of oocytes. *Theriogenology* **34**, 417–425.
7. Knutson, J. C. and Yee, D. (1987) Electroporation: parameters affecting transfer of DNA into mammalian cells. *Anal. Biochem.* **164**, 44–52.
8. Neumann, E., Schaefer-Rider, M., Wang, Y., and Hofschneider, P. H. (1982) Gene transfer into mouse lyoma cells by electroporation in high electric field. *EMBO J.* **7**, 841–845.

9. Evans, G. A., Ingraham, H. A., Lewis, K., Cunningham, K., Seki, T., Moruichi, T., Chang, H. C., Silver, J., and Hyman, R. (1984) Expression of the thy-1-glycoprotein gene by DNA-mediated gene transfer. *Proc. Natl. Acad. Sci. USA* **81,** 5532–5536.
10. Miller, J. F., Dower, W. J., and Thompkins, L. S. (1988) High-voltage electroporation of bacteria: genetic transformation of campylobacter jejuni with plasmid DNA. *Proc. Natl. Acad. Sci.* **85,** 856–860.
11. Lavitrano, M., Canaioni, A., Fazio, V. M., Dolci, S., Farace, M. G., and Spadafora, C. (1989) Sperm cells as vectors for introducing foreign DNA into eggs: genetic transformation of mice. *Cell* **57,** 717–723.
12. Parrish, J. J., Susko-Parrish, J. L., and First, N. L. (1988) Capacitation of bovine sperm by heparin. *Biol. Reprod.* **38,** 1171–1180.
13. Zimmerman, U. and Vienken, J. (1982) Electric field-induced cell-to-cell fusion. *Membr. Biol.* **67,** 165–182.
14. Gagné, M., Pothier, F., and Sirard, M. A. (1991) Electroporation of bovine spermatozoa to carry foreign DNA in oocytes. *Mol. Reprod. Dev.* **29,** 6–15.
15. Sirard, M. A., Leibfried-Rutledge, L., Parrish, J., Ware, C., and First, N. L. (1988) The culture of bovine oocytes to obtain developmentally competent embryos. *Biol. Reprod.* **39,** 546–552.

Chapter 14

Electroporation of Embryonic Stem Cells for Generating Transgenic Mice and Studying In Vitro Differentiation

John S. Mudgett and Thomas J. Livelli

1. Introduction

The ability to transfect mammalian cell lines stably in culture has enabled countless scientific advances to be made in just a few decades. However, not until about 10 years ago did the true potential for the genetic manipulation of cells become a reality, with the ability to alter a mammalian gene precisely by targeted homologous recombination *(1–3)*. This powerful tool was first devised for gene therapy approaches, but was quickly adopted for the transfection *(4)* and genetic disruption of genes in embryonic stem (ES) cell lines to produce transgenic animals *(5,6)*. Currently, the most common cell line used is the mouse ES cell, but ES cells have recently been derived for other species (rabbit, rat, and hamster) *(7–9)*.

Murine ES cells are pluripotent cell lines isolated from the inner cell mass of d 3.5 postconception (pc) mouse blastocysts that can be maintained in culture, genetically manipulated by transfection with DNA vectors, and used either for producing transgenic mice or for investigating in vitro differentiation (reviewed in ref. *10*). Once reintroduced into murine blastocysts, the cells can contribute to all cell lineages and stages of embryonic development, including the germ line. Pluripotency must therefore be maintained before, during, and after transfection of the ES cells by using ES cell-qualified reagents and growth conditions, and coculturing the ES cells with feeder layers of inactivated mouse primary embryonic fibroblasts or murine STO *(10)* fibroblast cultures.

Electroporation is by far the most common and successfully used method to transfect ES cells, and targeted disruption or alteration of a murine gene is the most common reason for performing these transfections. There are several variables that may affect the efficiency of gene targeting. One that is easily controlled is the use of isogenic DNA in the construction of the targeting vectors *(11,12)*. The lengths of genomic DNA fragments used to construct the vector, the total homology of the vector to the targeted genomic region, and the final design of targeting vectors also influence the efficiency of targeted recombination. Another critical variable affecting recombination frequency is the genomic region being targeted. We have observed that making small changes in the genomic fragments included within similarly designed targeting vectors can alter recombination frequencies over a hundred-fold. These observations suggest that recombination-enhancing or inhibiting sequences may reside in the genomic fragments used to construct the targeting vectors. Therefore, one of the most straightforward approaches to improving gene targeting efficiencies may be to construct a different targeting vector.

The rationale and general approach to generating transgenic (knock-out) mice by use of ES cells have been reviewed *(13–17)*, as well as the more specialized use of ES cells for in vitro differentiation studies *(10,18–20)*. This chapter describes the electroporation of ES cell lines and the selection of stable transfectants for subsequent studies. There are many difficulties that can arise in the genetic manipulation of the mammalian genome and the generation of knock-out mice, but ES cell transfection should not be one of them.

2. Materials

2.1. Equipment

1. Multichannel pipeter.
2. Multichannel tissue-culture aspirator system (optional) (Inotech, Lansing, MI).
3. Inverted phase-contrast microscope with 10× and 20× objectives.

2.2. ES and Feeder Cells

The choice of ES cells depends on which cell lines are available for use in your laboratory. ES lines require coculturing with feeder cells to maintain pluripotency, and the feeder cell type required depends on which ES cell line is used (*see* Note 1). Primary mouse embryonic fibroblasts are most often used as feeder cells, but the mouse fibroblast line STO or STO derivatives are recommended for some ES cell lines. For purposes of example, this chapter will describe conditions that have proven successful for CCE *(21)*, E14 *(22)*, and J1 ES cells *(23)* with primary feeder cells, or D3 *(6)* and AB2.1 ES cells *(24)* with SNL feeder cells. SNL feeder cells are STO feeder cells transformed with neomycin resistance (neo^r) and murine leukemia inhibitory factor (LIF) genes *(24)*. For isolation of hygromycin resistant (hyg^r) ES clones, one may use an SNL feeder cell line transfected with a PGK-hygromycin construct *(25)*.

2.3. Solutions and Cell-Culture Reagents

We recommend using ES cell qualified reagents from Specialty Media, Inc. (Lavallette, NJ) or Gibco (Gaithersburg, MD), which are labeled below as "ESCQ." However, successful ES cell transfection is also possible using reagents purchased from other suppliers (e.g., Sigma, St. Louis, MO; J.R.H. Biosciences, Lenexa, KS) as described in references for the ES cell lines described in Section 2.2. Specific reagents that we have found to give optimal results with ES cells are marked "RECOMMENDED."

1. ESCQ fetal calf serum (FCS), RECOMMENDED: Store in the dark at −20°C for 1–18 mo or at 4°C for <1 mo. Heat-inactivate at 56°C for 30 min prior to use.
2. ESCQ DMEM (ES-DMEM): with 4500 mg/L glucose, L-Glu, and 2.25 g/L sodium bicarbonate, without sodium pyruvate. Store in the dark at 4°C up to 4 mo for growth of ES cells.
3. DMEM: with 4500 mg/L glucose and L-Glu, without sodium pyruvate. Store in the dark at 4°C for up to 4 mo for growth and preparation of feeder cells.
4. ESCQ 100X penicillin–streptomycin (Pen–Strep) solution (Gibco): Store in the dark at −20°C for up to 6 mo.
5. ESCQ 100X nucleosides (Specialty Media, Inc.) RECOMMENDED: Store in the dark at −20°C for up to 4 mo.
6. ESCQ 100X 2-mercaptoethanol for cell culture: Final concentration is 110 µM 2-mercaptoethanol (Specialty Media, Inc.). Store in the dark at 4°C for up to 1 yr.
7. ESCQ 100X nonessential amino acids (Specialty Media, Inc. or Gibco). Store in the dark at 4°C for up to 4 mo.

8. Complete ES medium, RECOMMENDED: ES-DMEM, 15% FCS, 1X Pen-Strep, 1X nonessential amino acids, 1X nucleosides, and 110 µM 2-mercaptoethanol.
9. Feeder cell medium: DMEM, 10% FCS, 1X Pen-Strep.
10. Leukemia-inhibitory factor (LIF) (ESGRO), RECOMMENDED (Gibco). Store at 4°C.
11. ESCQ 0.1% gelatin in sterile water, RECOMMENDED (Specialty Media, Inc.). Store at room temperature.
12. Trypsin-EDTA solution: 0.025% trypsin and 0.75 mM EDTA. Store at 4°C for <2 wk or at –20°C for up to 1 yr; 0.25% trypsin and 0.5 mM EDTA can also be used.
13. 1X ES cell-culture freezing medium, RECOMMENDED (Specialty Media, Inc.). Store at 4°C for <1 mo or at –20°C for up to 1 yr. This is for cryopreservation of feeder cells and ES cells (see Note 2).
14. 2X ES cell-freezing medium (80% FCS + 20% DMSO; filter-sterilized). Store at 4°C for <1 mo or at –20°C for up to 1 yr. This is for cryopreservation of ES cells in multiwell plates (see Note 2).
15. G418 (Geneticin) (Gibco): Dissolve 20–100 mg G418/mL sterile water for cell culture; filter-sterilize, and store aliquots at –20°C.
16. Hygromycin-B (Calbiochem): 20 mM in sterile water; filter-sterilize, and store at 4°C. **Note: Toxic by skin contact, ingestion, and inhalation.**
17. Sterile water for cell culture—ESCQ, RECOMMENDED. Store at room temperature.
18. ESCQ PBS, and PBS without Ca^{2+} and Mg^{2+} (Specialty Media, Inc. or Gibco): Store at room temperature.
19. Mitomycin C powder (Sigma): Dissolve 1 mg in 1 mL sterile water for cell culture, yielding a 100X stock solution. Store powder at 4°C; store stock solution at 4°C for up to 1 wk, and then discard of properly. **Note: Toxic by skin contact, ingestion, and inhalation.**
20. Lysis buffer: 100 mM Tris-HCl, pH 8.5, 5 mM EDTA, 0.2% SDS, 200 mM NaCl; add 100 µg/mL proteinase K just prior to use.
21. ESCQ electroporation buffer, RECOMMENDED (Specialty Media, Inc.). Store at 4°C (see Note 3 regarding alternative buffers).
22. TE: 10 mM Tris-HCl, pH 8, 1 mM EDTA.

2.4. Markers for ES Cell Transfection

Antibiotic resistance markers routinely used for the construction of targeting vectors include the neo^r and hyg^r genes. There are two versions of the neo^r gene in circulation, one of which contains a point mutation that reduces the resistance of transformants to antibiotic selection pressure *(26)*. The mutated version has been used for spontaneous double-recombinant selection in high G418 *(27,28)*. Select transfected *neo*^{r} ES cells in 150–300 µg/mL G418 (active wt/vol). The concentration of G418 required will vary among cell lines.

The *hyg*^{r} gene may be used, and both *neo*^{r} and *hyg*^{r} markers are sometimes used to create double knock-out ES cells by subsequent targeting of the second chromatid *(25,29)*. Hygromycin-B is used for selection of the *hyg*^{r} gene at 100 µg/mL. Greater hygromycin-B concentrations (150–500 µg/mL) must be used if the feeder cells are not *hyg*^{r}, but owing to the loss of the underlying feeder layer, many ES cell transformant colonies are lost as well.

Replacement vector approaches using counter selection have been described using negative-selection marker genes in the targeting vector, such as the herpes simplex virus thymidine kinase gene (HSV-tk) *(30)* or the diphtheria toxin-A gene *(31)*. If the HSV-tk counter selection approach is used, 1-(2-deoxy-2-fluoro-β–D-arabinofuranosyl)-5-iodouracil (FIAU) (fialuridine) (Bristol-Myers Squibb, Wallingford, CT) or gancyclovir (GANC) (Syntex, Palo Alto, CA) are included in the medium at concentrations of 0.2 µM (*see* Note 4).

2.5. Mice

Mice may be obtained from The Jackson Laboratory (Bar Harbor, ME), Taconic (Germantown, NY), or GenPharm International (Mountain View, CA).

3. Methods

3.1. Preparation of neo^r or hyg^r Primary Mouse Embryo Fibroblast Feeder Cells (PMEF)

Murine ES cells are cocultured with either primary or STO feeder cells prepared as described below.

1. Set up matings between a homozygous *neo*^{r} or *hyg*^{r} male mouse and several females.
2. On the following day, check the females for a copulation plug. Remove females with plugs and place them in a separate cage. This is d 1 for timing the embryos.
3. On d 13 or 14, add about 50 mL of PBS without Ca^{2+} or Mg^{2+} to three 150-mm (diameter) Petri dishes. Sacrifice the females, open the peritoneal cavity and transfer the uteri containing the embryos to one dish with PBS without Ca^{2+} or Mg^{2+}.
4. Dissect out the embryos, and remove the yolk sac, amnion, and placenta. Transfer the embryos to a fresh dish of PBS without Ca^{2+} or Mg^{2+} to wash and remove any blood.

5. Transfer the washed embryos into a fresh dish of PBS without Ca^{2+} or Mg^{2+}, and using a fine forceps, remove the head and liver (dark red tissue in center of embryo).
6. Place six of the prepared embryos into the barrel of a 5-mL syringe, and attach to the syringe an 18-gage needle. Add 3 mL of sterile PBS without Ca^{2+} or Mg^{2+} to the syringe, and replace the plunger.
7. Expel the syringe contents into a sterile 50-mL polypropylene conical centrifuge tube, and draw up the embryos through the needle into the syringe (this is one repetition). Repeat until the embryos are broken into a single-cell suspension containing small clumps of cells (usually five to six repetitions).
8. Add 25 mL of feeder cell medium.
9. Add approx 10 mL of cell suspension to each of three 175-cm^2 flasks containing 60 mL/flask of feeder cell medium. Place flask in a 37°C incubator with 5% CO_2. After 3–4 d the flasks should be confluent. The typical yield is $1–2 \times 10^7$ cells/embryo prior to expansion.
10. Split the cells 1:3 to expand the number of cells. To split cells, remove growth medium; wash cells with ~15 mL/flask PBS without Ca^{2+} or Mg^{2+}, discard wash; add 10 mL of trypsin-EDTA, rinse monolayer for 15–30 s, discard trypsin-EDTA solution, and place flasks in 37°C incubator for ~3–5 min until cells are "rounding-up" and dissociated from the flask surface; tap the flask to dislodge the cells completely, and immediately add 10 mL of feeder cell medium, resuspend the cells, and transfer to three 175-cm^2 flasks.
11. The PMEFs can be further expanded and cryopreserved at ~5×10^6 cells/mL/vial, or expanded for inactivation (as described in Sections 3.3. and 3.4.) and then cryopreserved. Note: PMEFs are only good for use as a feeder layer for 5–6 passages.

3.2. Preparation of STO or STO-Derived Feeder Cells

Grow STO or STO-derived (e.g., SNL) feeder cells in feeder cell medium to confluence (~$1–2 \times 10^5/cm^2$) in 225-cm^2 flasks. Wash, trypsinize, and expand for inactivation as described for PMEFs (Section 3.1., steps 10 and 11).

3.3. Mitomycin C Inactivation of PMEF, STO, or STO-Derived Feeder Cells

1. Grow PMEF, STO, or STO-derived (e.g., SNL) feeder cells in feeder cell medium to confluence (~$1–2 \times 10^5/cm^2$) in 225-cm^2 flasks.
2. Remove the growth medium from the culture, and replace with 30 mL/225-cm^2 flask of feeder cell medium supplemented with 10 µg/mL mitomycin C.

3. Incubate cells for 3 h at 37°C with 5% CO_2.
4. Remove medium, and wash cells twice with PBS and once with PBS without Ca^{2+} or Mg^{2+}. Then trypsinize cells as described in Section 3.1., step 10.
5. Count cells, centrifuge at $1000g$ for 5 min, and resuspend pellet in cold (4°C) cell culture freezing medium at a density of 5×10^6 cells/mL.
6. Cells in cryopreservation vials can be placed in dry ice and later transferred to liquid nitrogen. Frozen inactivated feeder cells can be stored for at least 6 mo, for later plating on gelatin-treated dishes.

3.4. γ Radiation Inactivation of PMEF, STO, or STO-Derived Feeder Cells

1. Grow PMEF, STO, or STO-derived (e.g., SNL) feeder cells in feeder cell medium to confluence ($\sim 1-2 \times 10^5/cm^2$) in 225-$cm^2$ flasks.
2. Remove medium, and wash cells twice with PBS and once with PBS without Ca^{2+} or Mg^{2+}. Trypsinize cells as described in Section 3.1., step 10. Pellet the cells at $1000g$ and resuspend in feeder medium to remove the trypsin-EDTA.
3. Irradiate with 70 gy of γ-rays.
4. Following irradiation, pellet cells by centrifugation at $1000g$ for 5 min, resuspend in freezing medium at 3.75×10^6 cells/mL, and freeze as described in Section 3.3., steps 5 and 6 (*see* Note 5).

3.5. Plating Mitomycin-Treated or γ-Irradiated Feeder Cells

1. At least 30 min prior to plating the inactivated feeder cells, gelatin-coat the cell-culture dishes with 3 mL of 0.1% gelatin solution/25-cm^2 to cover the bottom of the dish. Incubate in a cell-culture hood for at least 30 min.
2. Thaw cryopreserved mitomycin C-treated or γ-irradiated feeder cells at 37°C, and dilute the thawed cells into 10 mL/vial of feeder cell medium. Pellet cells at $1000g$ at 4°C for 5 min.
3. Resuspend cells in appropriate volume for plating feeder cells (Table 1). Remove 0.1% gelatin solution, and add diluted cells. Incubate cells at 37°C with 5% CO_2.
4. Feeder cells should be allowed to attach 4–8 h before adding ES cells. Inactivated feeder cells plated on gelatinized plates may be used for 12–14 d.

3.6. ES Cell Culture

ES cell cultures are passaged about every 3 d. Since passage number must be kept as low as possible for continued pluripotency, prepare as many cryopreserved vials at the lowest passages as possible, and only passage cells from frozen vials long enough to complete the transformation experiment.

Table 1
Growth Area, Volume of Media, and Feeder Cell Numbers
for Different Culture Dishes

T.C. dish	Volume, mL	Growth area, cm^2	Number of feeder cells/ flask or well
75-cm^2 flask	12	75	3.75×10^6
25-cm^2 flask	6	25	1.25×10^6
100-mm plate	10	56	2.8×10^6
60-mm plate	5	21	1.0×10^6
6-Well plate	4	9.5	4.75×10^5
12-Well plate	2	4	2×10^5
24-Well plate	1	2	1×10^5
96-Well plate	0.1	0.32	1.5×10^4

1. Thaw inactivated feeder cells, and plate on gelatin-treated dishes or flasks at a density of 1.25×10^6 cells/25 cm^2 as described in Section 3.5. Table 1 lists a variety of culture dishes, growth areas, and the approximate number of feeder cells needed. The dishes or flasks of feeder cells may be used as early as 6 h later, but are generally used the next day. Feeder cells can also be used for up to 14 d later, depending on how intact the feeder cell layer appears.
2. Thaw ES cells at 37°C, wash once by pelleting, resuspend in complete ES medium, and plate on the feeder cells at a density of $1-1.5 \times 10^6$ ES cells/ 25 cm^2 (d 0).
3. Change the growth medium (complete ES medium supplemented with 500 U/mL LIF) every day. Split the cells on d 3 using the following harvesting procedure: remove the growth medium; rinse the cells with 5–6 mL of PBS without Ca^{2+} or Mg^{2+} (25 cm^2); add trypsin-EDTA solution (0.5 mL/ 25 cm^2), and rinse the cells by tilting the plate back and forth for ~30 s; remove the trypsin-EDTA solution, and incubate the cells at 37°C for 4–5 min; add 5 mL complete ES medium (25 cm^2), and pipet the cells up and down with a transfer pipet until they are a single-cell suspension without any significant clumps of cells (>8 cells).
4. Count the cells and dilute in complete ES medium for passage at a density of $1-1.5 \times 10^6/25$ cm^2, or pellet cells at $1000g$ and resuspend in freezing medium at $1-1.5 \times 10^6$/mL as described in Section 3.7.

3.7. Cryopreservation of ES Cells
1. ES cells should be in log-phase growth at time of freezing. Feed cells fresh complete ES medium supplemented with 500 U/mL LIF 3–5 h prior to harvesting the cells for freezing.

2. Harvest the cells using the trypsinization procedure described in Section 3.6., step 3.
3. Resuspend the cells in complete ES medium, and count the cells.
4. Pellet the cells at $1000g$ and resuspend in cold (4°C) cell culture freezing medium at 6×10^6 cells/mL. Aliquot into cryopreservation vials, freeze in dry ice, and transfer to liquid nitrogen storage.
5. It is important to label each vial with the name of the ES cell line, the passage number, and the date frozen.

3.8. Culture of ES Cells for Electroporation

1. Prepare two or three 100-mm plates or 75-cm^2 flasks of inactivated feeder cells on gelatinized plates as described in Section 3.5. Harvest one or two 60-mm plates or 25-cm^2 culture flasks of ES cells (as described in Section 3.6., step 3), and dilute into 10 mL of complete ES medium supplemented with 500 U/mL LIF. Pellet cells at $1000g$ for 5 min at 4°C.
2. Resuspend ES cells in complete ES medium supplemented with LIF and plate ~3×10^6 cells/100-mm dish (or 75-cm^2 flask). Feed cells fresh medium every day or twice a day, since cells become more dense and begin to turn the medium more acidic in < 24 h (medium will appear more orange than the normal red-orange color; never allow the medium to become more yellow-orange). After 2–4 d, the ES cells will be ready for replating.
3. After the ES cells have been expanded for 2–4 d (but not overconfluent), they should be split 1:2 onto fresh feeder cells on the day before electroporation. A 100-mm plate or 75-cm^2 flask of ES cells should yield ~$5–10 \times 10^6$ cells or $10–20 \times 10^6$ cells, respectively.
4. Harvest the ES cells as described in Section 3.6., step 3. Plate ES cells on fresh 100-mm plates or 75-cm^2 flasks previously prepared with inactivated feeder cells, and culture overnight.
5. On the next day (the day of the electroporation), the ES cells should be fed fresh complete ES medium supplemented with LIF 3 h prior to harvesting the cells for electroporation.

3.9. Electroporation of ES Cells

There have been varied conditions and buffers described for transfection of ES cells by electroporation, so conditions are described that are optimum for a variety of ES lines and vectors (*see* Note 3).

1. Harvest ES cells for electroporation as described in Section 3.6., step 3. Pellet the cells at $1000g$, 4°C, for 5 min; wash cells once in PBS without Ca^{2+} or Mg^{2+}; resuspend cells in electroporation buffer at 5×10^6 ES cells/mL.

2. Linearize 10–20 µg of DNA, ethanol-precipitate, wash twice with 70% ethanol, dry in cell culture hood, and resuspend in electroporation buffer at a final concentration of 0.5–1.0 µg/µL. Add DNA to 0.85 mL of ES cells (in electroporation buffer), and transfer to an electroporation cuvet with a 0.4-cm electrode gap.
3. Electroporate cells with 300 V at 250 µF at room temperature. Allow the cells to recover for 5 min at room temperature. Disperse the cells by pipeting.
4. Plate in complete ES medium on 100-mm, 60-mm, or 6-well dishes previously prepared with feeder cells. The number of cells plated will depend on the desired density of colonies, and the specific vector and selection/counter selection approach used. In general, a typical yield is about 1000 G418 resistant colonies/800 µL of electroporated ES cells using the conditions described above. An initial transfection experiment will enable the investigator to determine the transfectant yield for their vector and selection conditions, allowing subsequent experiments to be designed accordingly. This is particularly important if a pooling approach is used (*see* Note 4), which requires a predictable yield of transformants per well.
5. Incubate the dishes overnight. Feed with complete ES medium supplemented with LIF (500 U/mL), and selective drugs (e.g., G418 or Hyg, GANC, or FIAU).
6. Replace complete ES medium supplemented with LIF and selective drug(s) daily until colonies appear, and are ready to pick (or pool). Colonies will appear in 6–12 d, depending on the ES cell line used.

3.10. Isolation of Single Clones Using a Micropipet

1. On the day before picking the ES colonies, prepare a 96-well flat-bottom plate with inactivated feeder cells in complete ES medium supplemented with LIF (500 U/mL).
2. Add 25 µL/well of trypsin-EDTA to each well of a round-bottom 96-well dish.
3. Wipe the barrel of a micropipet with 70% ethanol, and set to 10 µL vol.
4. Remove the growth medium from the dish of colonies, wash once with 10 mL of PBS without Ca^{2+} or Mg^{2+}, and add 2 mL of room temperature PBS without Ca^{2+} or Mg^{2+}. Place an inverted microscope in a cell-culture hood (optional). While viewing the colonies under the microscope or over a piece of black paper, carefully dislodge the colony from the dish while drawing it into the micropipet tip. Add the colony to a well of the 96-well dish containing trypsin-EDTA. Continue until each of the wells contain a colony (or as many as you can pick in 20–30 min).

5. Incubate the 96-well plate at 37°C for 4–5 min. Using a multichannel pipetor, add 75 µL of complete ES medium supplemented with LIF to each well. Dissociate the cells with the multichannel pipetor by pipeting up and down until the cells are a single-cell suspension (about five repetitions). Add the ~100 µL of cell suspension from each well to each well of a 96-well dish prepared with feeder cells. Replace complete ES medium supplemented with LIF and selective drug(s) daily.

3.11. Cryopreservation of and DNA Isolation from Selected ES Colonies

ES colonies should be ready to split into 24-well plates 2–3 d after picking colonies into 96-well plates. One set of 24-well plates is used for isolating DNA for Southern hybridization analysis, and a second set is used for cryopreservation of the colonies.

1. One day prior to splitting the colonies from the 96-well plate, gelatin-coat four 24-well plates, and prepare the plates with feeder cells; these plates will be used for cryopreservation of the ES colonies.
2. Examine the colonies in the 96-well plate for size and morphology (*see* Note 1). They should not be very different in size, and the morphology should be that of undifferentiated cells. If a colony appears to have differentiated, make a note of this observation, and do not use for subsequent blastocyst injections.
3. Aspirate the growth medium from the 96-well plate. Add 100 µL/well of PBS without Ca^{2+} or Mg^{2+} using a multichannel pipetor. Aspirate the PBS wash. Add 25 µL/well of trypsin-EDTA, and incubate at 37°C for 5 min.
4. Examine a few wells under the microscope to determine that the cells in the colonies have "rounded up." Using a multichannel pipetor, add 175 µL/well of room temperature complete ES medium with G418. Pipet cells up and down 10 times. Examine a few wells to determine that the colonies are now a single-cell suspension (clumps of two to five cells are not a problem, but large clumps can adversely affect the undifferentiated growth of the colonies).
5. Gelatin-treat and prepare four 24-well plates without feeder cells with 1 mL/well of complete ES medium with G418 (plates for DNA isolation). To the 24-well plates with feeder cells, add 1 mL/well of complete ES medium with G418 and 500 U/mL LIF (plates for cryopreservation). Transfer 100 µL cells into a well of the 24-well dish without feeder cells and 100 µL cells into a well of the 24-well dish with feeder cells. Incubate cells at 37°C with 5% CO_2.
6. Feed the cells daily with fresh medium.

7. After 2–3 d, the cells in the 24-well plate with feeder cells should be ready for cryopreservation. Aspirate the growth medium. Add 0.5 mL of PBS without Ca^{2+} or Mg^{2+}/well. Remove PBS wash, and add 0.2 mL trypsin-EDTA. Incubate cells at 37°C for 4 min. Tap the dish against the palm of your hand to loosen the cells, and immediately add 0.25 mL of room temperature complete ES medium. Pipet gently up and down two or three times to dissociate the cells. Add 0.5 mL of cold (4°C) 2X freezing medium to each well.
8. Wrap the plate with parafilm to seal it from evaporation, and place the plate on top of dry ice in an ice bucket with a lid. After the plate is frozen, transfer the dish to a styrofoam box (box should be precooled in the –80°C freezer) and store at –80°C. Plates can be stored for several months.
9. After 5–6 d, the cells in the 24-well plate without feeder cells should be ready for genomic DNA isolation. Aspirate the growth medium, and add 500 µL of lysis buffer/well. Incubate overnight at 37°C in a humidified incubator.
10. Place the plate on a gyrotory shaker for 15 min. Add 500 µL of isopropanol, and continue to swirl plate on shaker until all of the DNA has aggregated (about 15–20 min).
11. Recover the DNA by lifting the aggregate from the well using a sterile micropipet tip. Allow excess liquid to drip off, and place aggregate in a prelabeled sterile microcentrifuge tube.
12. Briefly centrifuge the sample to collect the DNA at the bottom of the tube. Dry the pellet in a vacuum centrifuge for 5 min, or incubate at 37°C for 1 h.
13. Add 100–150 µL of TE or sterile water, and dissolve overnight at 56°C. Store DNA in solution at 4°C or –20°C. This DNA is used for Southern or PCR analysis to identify targeted ES cell transfectants.

3.12. Pooled Colony Approach

If the homologous recombination frequency for a specific target locus is low (no targeted clones in several hundred assayed), it is advisable to screen pools or transfectants for the targeting event using a PCR-based detection assay. By this approach, several thousand transfectants can be screened in a relatively short amount of time.

1. Plate electroporated cells (Section 3.9., step 4) in six-well plates previously prepared with inactivated feeder cells. Allow colonies to grow until easily visible (6–10 d, depending on the cell type used), refeeding daily with complete ES medium supplemented with LIF (500 U/mL) and antibiotics.
2. Rinse dishes with 5 mL/well PBS without Ca^{2+} or Mg^{2+}, add 0.25 mL trypsin-EDTA, and incubate at 37°C for 8 min.
3. Add 1 mL of complete ES medium to each well, and then dissociate the

cells by pipeting.
4. Plate 0.5 mL of the cell suspension onto fresh six-well dishes for the isolation of genomic DNA (these dishes should not contain feeder cells).
5. Add 0.5 mL of cell suspension to 0.5 mL 2X freezing medium, mix, and store at –80°C or in liquid nitrogen.
6. Feed the dishes for genomic DNA every day with fresh complete ES medium with drug selection. Differentiation of the cells in the cultures for genomic DNA is not a concern. When the wells plated for isolation of genomic DNA are at high density, rinse the wells once with PBS without Ca^{2+} or Mg^{2+} and add 0.5 mL of lysis buffer. Incubate at 37°C overnight in a humidified incubator.
7. Extract the lysates once with phenol:chloroform:isoamyl alcohol (25:24:1), precipitate by adding 1 vol of isopropanol, wash with 70% ethanol, and resuspend in 500 µL TE overnight at room temperature while rocking.
8. Use 1 µL of the genomic DNA for PCR analysis of each ES cell transfectant pool. With some PCR primer pairs, there may be a need to titer the DNA concentration to determine the best PCR conditions.

3.13. Clone Expansion Using the Single-Clone Approach

The ES cell clones selected after Southern hybridization analysis (or PCR) are thawed from the cryopreserved 24-well dish using prewarmed complete ES medium as follows.

1. Remove the plate from the –80°C freezer, and place on dry ice in an ice bucket with a lid. Place the plate in the cell-culture hood, and immediately add 1 mL/well of 42°C complete ES medium to the appropriate wells. As soon as the well is mostly thawed, gently pipet up and down until it has thawed completely. Transfer the contents of the well to a 15-mL sterile, conical centrifuge tube containing 10 mL of room temperature complete ES medium. Pellet cells at 1000g for 5 min at 4°C.
2. Resuspend cells in complete ES medium supplemented with G418 and 500 U/mL LIF, and plate into a 60-mm dish or 25-cm^2 flask that was preseeded with feeder cells. Incubate cells at 37°C with 5% CO_2.
3. Feed every day with complete ES medium supplemented with LIF and antibiotics. In 2–3 d, split the cells 1:3 onto 60-mm dishes or 25-cm^2 flasks with feeder cells.
4. Feed cells with fresh complete ES medium supplemented with LIF and antibiotics every day. After 2–3 d, two of the dishes or flasks can be harvested for cryopreservation and blastocyst injection. The remaining dish or flask should be fed with fresh complete ES medium supplemented with LIF 3 h prior to harvesting the cells for blastocyst injection (*see* Note 6).

3.14. Clone Expansion Using the Pooled-Clone Approach

Once PCR-positive pools of ES cell transformants are identified, individual ES transfectant colonies must be grown, isolated, picked, and characterized as individual clones by Southern hybridization analysis, as described above.

1. Thaw the frozen vials of PCR-positive pool cells, and add aliquots of between 5 and 50 µL of cells directly to 100-mm dishes previously prepared with feeder cells.
2. Feed cells daily with complete ES medium supplemented with LIF (500 U/mL) and antibiotics until colonies are ready to pick (8–12 d).
3. Pick colonies, and split to freezer stock and genomic DNA plates as described above in Sections 3.10. and 3.11.
4. Isolate genomic DNAs, and analyze to identify targeted single clones as described in Section 3.11.
5. Positive clones are then thawed, expanded, and confirmed by Southern hybridization analysis, as described in Section 3.13.

3.15. Preparation of ES Cells for Blastocyst Injection

1. Using cells from Section 3.13., step 4, remove the growth medium; rinse the cells with a generous amount of PBS without Ca^{2+} or Mg^{2+}; add trypsin-EDTA solution, and rinse it over the cells by tilting the plate back and forth for ~30 s; remove the trypsin-EDTA solution, and incubate the cells at 37°C for 4–5 min; add complete ES medium to the cells, and pipet the cells with a transfer pipet until they are a single-cell suspension.
2. Remove 0.5 mL of cell suspension, and transfer to a sterile 35-mm tissue-culture dish that contains 2 mL of complete ES medium. Incubate cells at 37°C with 5% CO_2 in air for 20–30 min. The feeder cells carried over with the ES cells will attach to the dish and the ES cells still in suspension are transferred into a second 35-mm dish. Incubate cells at 37°C with 5% CO_2 for 15 min. Examine the cells under the microscope, and while gently swirling the dish, determine if there are a number of cells beginning to stick lightly to the surface. Continue to incubate, and check for lightly "sticking cells" until the number of cells appropriate for injection (15 cells/injection) are lightly sticking. Remove the medium containing the ES cells that are not sticking, and using a small volume of ES medium without serum, gently pipet the lightly sticking cells off from the surface. These "sticky" cells are then used for blastocyst injection.

Embryonic Stem Cells 181

4. Notes

1. ES cell morphology is often an indicator as to the general health of the cells, and proper colony morphologies need to be assessed for each cell line used. In general, a first-time user of ES cells should visit a laboratory that uses the cell line in question, or good photographs of totipotent and differentiated cells should be obtained. ES cultures are generally counted on hemocytometer counting slides, and the effectiveness of ES cell trypsinization and dispersion of the cells is assessed. It is desirable to count only the ES cells and not the feeder cells as judged by cell size (feeder cells in general are much larger) to obtain more accurate ES cell counts.
2. We have found that the traditional 1X freezing media of 10% FCS + 10% DMSO, or 2X freezing media of 20% FCS + 20% DMSO also work, but not as well.
3. We have found that electroporation buffer (Specialty Media, Inc.) reproducibly gives greater recovery of electroporated ES cells, but a variety of buffers have been described in literature for ES cell electroporation (*see* Section 2.2. for ES cell types and electroporation references). The choice of electroporation conditions described in the literature is also varied, but in general, we have found that colony yield is not as limiting as the time required to clone and assay transformants. With this in mind, the amount of DNA is usually reduced to about 10 µg linearized vector/cuvet in 0.85 mL cells (at 5×10^6 cells/mL electroporation buffer), and the capacitance is reduced to about 250 µF. This greatly reduces cellular lysis and clumping, thereby yielding a single-cell suspension after electroporation.
4. The choice of approach to use in the targeted alteration of a gene depends on the desired outcome of the experiment. To inactivate a gene, there are two general vector designs used: gene-replacement vectors and gene-insertion vectors. These vector designs have been reviewed previously, with advantages and disadvantages of each described *(30–32)*. In our experience, gene-replacement vectors work well and result in more stable genetic alterations, since they do not create duplications within the targeted site.

   Replacement vector design requires that two genomic DNA fragments ("arms") flank either side of the selectable marker gene (e.g., neo^r). Since targeted recombination frequencies are generally increased by increasing the length of the homologous arms, we suggest the use of a replacement vector with two relatively long (3–8 kb) vector arms to give greater targeting frequencies. Isolate single ES cell transformants as described in Section 3.10., and characterize the clones by Southern hybridization. If no targeted clones are detected within the first few hundred clones characterized, then the replacement vector can be modified so a pooled colony approach can be used (Section 3.14.). One of the vector arms needs to be relatively short (<2 kb) if PCR is going to be used to detect targeted recombinants, but >1 kb to reduce the chances of generating aberrant recombination events *(33)*. A short arm may decrease the frequency of targeted recombination, but a clone-pooling and PCR-analysis approach may be the quickest way to isolate targeted clones.

If more subtle mutations are desired, or if there is need to remove the neo^r marker in order not to disrupt the overall context of the target locus, then "hit-and-run" strategies may be used *(34,35)*. By this design, the neo^r and HSV-tk markers are adjacent to each other, and selecting against HSV-tk also selects for the loss of the neo^r gene by spontaneous intrachromosomal recombination. In a similar approach, "tag-and-exchange" procedures have been described, that allow any number of specific alterations to be made in the target gene while removing the neo^r gene, without having to perform sequential hit-and-run exchanges for each alteration *(36,37)*.

5. Do not γ irradiate in the freezing medium, since the DMSO will prevent adequate γ radiation-induced inactivation of the feeder cells.
6. Some ES cells from the Southern or PCR-positive clones are regrown at the time the clone is expanded from the 24-well plate to confirm the genotype of the targeted clone by Southern analysis. Putatively targeted ES cells from the isolated clones are cultured in the absence of feeders in complete ES medium containing LIF at 500 U/mL. The cells are lysed overnight in lysis buffer, and the next day extracted with phenol:chloroform, precipitated in 1 vol of isopropanol, washed with 70% ethanol, and resuspended in TE. Southern hybridization analyses are performed using several restriction endonucleases and unique 5'- and 3'-flanking probes, as well as probes derived from within the vector (e.g., *neo*) to confirm that the targeted allele has been altered as expected, and that the correct clones have been selected and expanded. Since this confirmation is subsequent to the expansion of clones from the 24-well freezer stock plates, these plates are refrozen and stored at –80°C until it is determined that the correct clones have been selected.

References

1. Smithies, O., Gregg, R. G., Boggs, S. S., Koralewski, M. A., and Kucherlapati, R. S. (1985) Insertion of DNA sequences into the human chromosomal ß-Golbin locus by homologous recombination. *Nature* **317,** 230–234.
2. Lin, F.-L., Sperle, K., and Sternberg, N. (1985) Recombination in mouse L cells between DNA introduced into cells and homologous chromosomal sequences. *Proc. Natl. Acad. Sci. USA* **82,** 1391–1395.
3. Thomas, K. R., Folger, K. R., and Capecchi, M. R. (1986) High frequency targeting of genes to specific sites in the mammalian genome. *Cell* **44,** 419–428.

4. Gossler, A., Doetschman, T., Korn, R., Serfling, E., and Kemler, R. (1986) Transgenesis by means of blastocyst-derived embryonic stem cell lines. *Proc. Natl. Acad. Sci. USA* **83,** 9065–9069.
5. Thomas, K. R. and Capecchi, M. R. (1987) Site-directed mutagenesis by gene targeting in mouse embryo-derived stem cells. *Cell* **51,** 503–512.
6. Doetschman, T., Maeda, N, and Smithies, O. (1988) Targeted mutation of the hprt gene in mouse embryonic stem cells. *Proc. Natl. Acad. Sci. USA* **85,** 8583–8587.
7. Doetschman, T., Williams, P., and Maeda, N. (1988) Establishment of hamster blastocyst-derived embryonic stem (ES) cells. *Devel. Biol.* **127,** 224–227.
8. Graves, K. H. and Moreadith, R. W. (1993) Derivation and characterization of putative pluripotential embryonic stem cells from preimplantation rabbit embryos. *Molec. Reprod. Dev.* **36,** 424–433.
9. Iannaccone, P. M., Taborn, G. U., Garton, R. L., Caplice, M. D., and Brenin, D. R. (1994) Pluripotent embryonic stem cells from the rat are capable of producing chimeras. *Devel. Biol.* **163,** 288–292.
10. Robertson, E. J. (1987) Embryo-derived stem cell lines, in *Teratocarcinomas and Embryonic Stem Cells: a Practical Approach.* (Robertson, E. J., ed.), IRL, Oxford, UK, pp. 71–112.
11. teRiele, H., Maandag, E. R., and Berns, A. (1992) Highly efficient gene targeting in embryonic stem cells through homologous recombination with isogenic DNA constructs. *Proc. Natl. Acad. Sci. USA* **89,** 5128–5132.
12. Deng, C. and Capecchi, M. R. (1992) Reexamination of gene targeting frequency as a function of the extent of homology between the targeting vector and the target locus. *Mol. Cell. Biol.* **12,** 3365–3371.
13. Capecchi, M. R. (1989) Altering the genome by homologous recombination. *Science* **244,** 1288–1292.
14. Robertson, E. J. (1991) Using embryonic stem cells to introduce mutations into the mouse germ line. *Biol. Reprod.* **44,** 238–245.
15. Pascoe, W. S., Kemler, R., and Wood, S. A. (1992) Genes and functions: trapping and targeting in embryonic stem cells. *Biochim. Biophys. Acta* **1114,** 209–221.
16. Bradley, A., Hasty, P., Davis, A., and Ramirez-Solis, R. (1992) Modifying the mouse: design and desire. *Bio/Technology* **10,** 534–539.
17. Robbins, J. (1993) Gene targeting: the precise manipulation of the mammalian genome. *Circ. Res.* **73,** 3–9.
18. Bradley, A. (1990) Embryonic stem cells: proliferation and differentiation. *Curr. Op. Cell Biol.* **2,** 1013–1017.
19. Brown, D. G., Willington, M. A., Findlay, I., and Muggleton-Harris, A. L. (1992) Criteria that optimize the potential of murine embryonic stem cells for in vitro and in vivo developmental studies. *In Vitro Cell. Dev. Biol.* **28A,** 773–778.
20. Chen, U. and Kosco, M. (1993) Differentiation of mouse embryonic stem cells in vitro: III. morphological evaluation of tissues developed after implantation of differentiated mouse embryoid bodies. *Devel. Dynamics* **197,** 217–226.
21. Robertson, E. J., Bradley, A., Kuehn, M., and Evans, M. (1986) Germline transmission of genes introduced into cultured pluripotent cells by retroviral vectors. *Nature* **323,** 445–447.

22. Hooper, M., Hardy, K., Handyside, A., Hunter, S., and Monk, M. (1987) HPRT-deficient (Lesch-Nyhan) mouse embryos derived form germline colonization by cultured cells. *Nature* **326,** 292–295.
23. Li, E., Bestor, T. H., and Jaenisch, R. (1992) Targeted mutation of the DNA methyltransferase gene results in embryonic lethality. *Cell* **69,** 915–926.
24. Soriano, P., Montgomery, C., Geske, R., and Bradley, A. (1991) Targeted disruption of the c-src proto-oncogene leads to osteopetrosis in mice. *Cell* **64,** 693–702.
25. Mortensen, R. M., Zubiaur, M., Neer, E., and Seidman, J. G. (1991) Embryonic stem cells lacking a functional inhibitory G-protein subunit (a_{i2}) produced by gene targeting of both alleles. *Proc. Natl. Acad. Sci. USA* **88,** 7036–7040.
26. Yenofsky, R. L., Fine, M., and Pellow, J. W. (1990) A mutant neomycin phosphotransferase II gene reduces the resistance of transformants to antibiotic selection pressure. *Proc. Natl. Acad. Sci. USA* **87,** 3435–3439.
27. Mortensen, R. M., Conner, D. A., Chao, S., Geisterfer-Lowrance, A. A. T., and Seidman, J. G. (1992) Production of homozygous mutant ES cells with a single targeting construct. *Mol. Cell. Biol.* **12,** 2391–2395.
28. Mortensen, R. M. (1993) Double knockouts: production of mutant cell lines in cardiovascular research. *Hypertension* **22,** 646–651.
29. teRiele, H., Maandag, E. R., Clarke, A., Hooper, M., and Berns, A. (1990) Consecutive inactivation of both alleles of the pim-1 proto-oncogene by homologous recombination in embryonic stem cells. *Nature* **348,** 649–651.
30. Mansour, S. L., Thomas, K. R., and Capecchi, M. R. (1988) Disruption of the proto-oncogene int-2 in mouse embryo-derived stem cells: a general strategy for targeting mutations to non-selectable genes. *Nature* **336,** 348–352.
31. McCarrick, J. W., Parnes, J. R., Seong, R. H., Solter, D., and Knowles, B. B. (1993) Positive-negative selection gene targeting with the diphtheria toxin A-chain gene in mouse embryonic stem cells. *Transgen. Res.* **2,** 183–190.
32. Hasty, P., Rivera-Perez, J., Chang, C., and Bradley, A. (1991) Target frequency and integration pattern for insertion and replacement vectors in embryonic stem cells. *Mol. Cell. Biol.* **11,** 4509–4517.
33. Thomas, K. R., Deng, C., and Capecchi, M. R. (1992) High-fidelity gene targeting in embryonic stem cells by using sequence replacement vectors. *Mol. Cell. Biol.* **12,** 2919–2923.
34. Hasty, P., Ramirez-Solis, R., Krumlauf, R., and Bradley, A. (1991) Introduction of a subtle mutation into the Hox-2.6 locus in embryonic stem cells. *Nature* **350,** 243–246.
35. Valancius, V. and Smithies, O. (1991) Testing an "in-out" targeting procedure for making subtle genomic modifications in mouse embryonic stem cells. *Mol. Cell. Biol.* **11,** 1402–1408.
36. Askew, G. R., Doetschman, T., and Lingpel, J. B. (1993) Site-directed point mutations in embryonic stem cells: a gene-targeting tag-and-exchange strategy. *Mol. Cell. Biol.* **13,** 4115–4124.
37. Wu, H., Liu, X., and Jaenisch, R. (1994) Double replacement: strategy for efficient introduction of subtle mutations into the murine *Col1a-1* gene by homologous recombination in embryonic stem cells. *Proc. Natl. Acad. Sci. USA* **91,** 2819–2823.

CHAPTER 15

Electrotransfection with "Intracellular" Buffer

Maurice J. B. van den Hoff, Vincent M. Christoffels, Wil T. Labruyère, Antoon F. M. Moorman, and Wouter H. Lamers

1. Introduction

Introduction of foreign molecules, such as DNA, RNA, or proteins, into living cells is a powerful means to test the biological functions of these molecules. One of the techniques by which foreign molecules can be introduced into living cells is electroporation (for reviews, *see* refs. *1,2*). Compared to other techniques for the introduction of molecules into living cells, electroporation does not seem to cause prolonged alterations in the biological structure and function of the target cells. Nevertheless, the pores that are induced by the electroporation technique *(3–5)* can be visualized by rapid-freezing electron microscopy techniques *(6)* and allow a direct contact between the cytoplasm of the cells and the electroporation medium. The components of the electroporation medium can, therefore, penetrate into the cell interior as long as the pores exist *(7)*. In this perspective, it is obvious that cells are exquisitely sensitive to the composition of the electroporation medium *(8–10)* and that an electroporation medium that resembles the composition of the cytoplasm would probably enhance the viability of the cells after electroporation.

A buffer that closely resembles the cytoplasmic concentration of most important ions was developed for metabolic studies in which the cell membrane of isolated cells was chemically permeabilized (for review, *see* ref. *11*). Such chemically permeabilized cells are successfully maintained in a buffer containing 120 mM KCl, 0.15 mM CaCl$_2$, 10 mM K$_2$HPO$_4$/KH$_2$PO$_4$, pH 7.6, 25 mM HEPES, pH 7.6, 2 mM EGTA, pH 7.6, 5 mM MgCl$_2$, 2 mM ATP, and 5 mM glutathione (*10*; and A. J. Meijer personal communication). Comparison of this intracellular buffer, termed "cytomix," with frequently used electroporation media, such as PBS (150 mM NaCl, 10 mM Na$_2$HPO$_4$/NaH$_2$PO$_4$, pH 7.4) or tissue-culture medium, shows that the concentration of calcium and magnesium in cytomix is low and buffered by EGTA. Furthermore, the potassium concentration is high, but the sodium concentration very low. Of the media used, the composition of cytomix best reflects the intracellular ionic conditions of eukaryotic cells.

Comparison of PBS, tissue-culture medium, and cytomix in electroporation experiments showed that cytomix substantially increases the viability of the cells after delivery of the electric pulse (Fig. 1A). The protective effect of cytomix can be understood from the effect of the medium composition on cellular organelles. When isolated mitochondria are suspended in isotonic sodium solutions, they rapidly swell and eventually rupture, whereas they behave as perfect osmometers when suspended in potassium-based media of different osmolarities (for review, *see* ref. *12*). Because cytomix is a slightly hypertonic potassium-based medium, the mitochondria will shrink a little and probably survive the electric pulse more readily than in culture medium or PBS, which are both isotonic and sodium-based media in which mitochondria swell and easily rupture. These observations of mitochondria can probably also be extended to other cell organelles and explain the increased viability of the cells when electroporated in cytomix.

In addition to sodium and potassium ions, the intracellular concentration of Ca^{2+} is strictly regulated (for review, *see* ref. *13*). Intracellular free calcium functions as a second messenger. The Ca^{2+} concentration is very low, but it can rapidly change after extracellular stimulation of cells. Changes in Ca^{2+} concentration will, therefore, have far-reaching consequences. An increase in the cytoplasmic Ca^{2+} concentration activates many enzymes, as well as the production of ATP by the mitochondria. However, the intracellular Ca^{2+} concentration, even after hormonal stimulation, for example, is very low compared to that in tissue-culture medium. Thus, when isolated mitochondria are suspended in medium containing Ca^{2+} at the concentration present in tissue-culture medium,

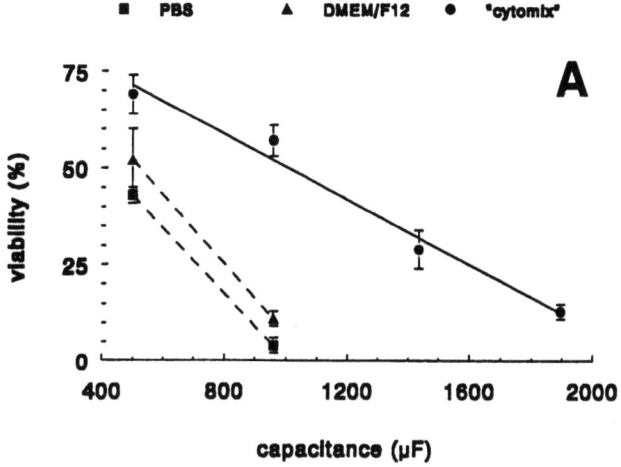

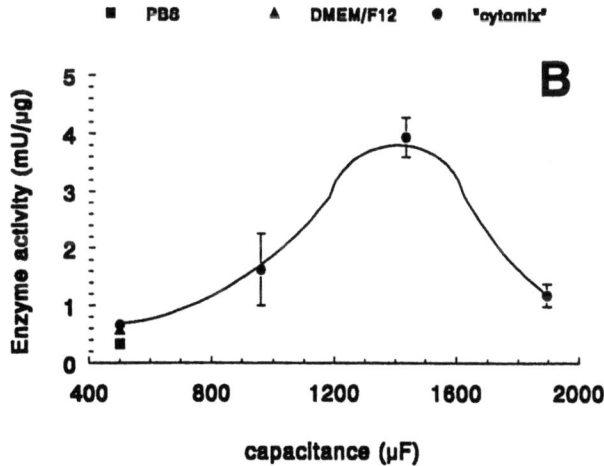

Fig. 1. Viability (**A**) and transient expression of a reporter gene (**B**) in different electroporation media as a function of capacitance. Using a Bio-Rad electroporation apparatus with two capacitance extenders, the EF was kept constant at 1500 V/cm and the capacitance was varied. The viability is expressed as the ratio of protein content per well of electroporated and control cells. All values are the mean ± SEM of three determinations (adapted from ref. *10*).

they rupture instantaneously. Since comparable effects are observed for all other cell organelles, the beneficial effects of cytomix over either PBS or tissue-culture medium most likely reside in its protection of the cellular organelles.

Electroporation experiments are performed to introduce molecules into living cells that are subsequently tested for their function. Therefore, it is desirable to combine high cell survival with high transfection efficiency. Figure 1B shows the expression of a promoter-reporter construct when introduced into FTO2B rat hepatoma cells using either cytomix, PBS, or tissue-culture medium. From these results, it can be concluded that electroporation of cells in cytomix yields high survival rates with high transfection efficiency. The observed differences in the transfection efficiency are owing to the differences in the ionic composition rather than to differences in the electrical conduction properties of the different electroporation media, because time constants were similar in all three media (data not shown).

Because cytomix increases cell viability, transient expression can be optimized by increasing the strength of the electric pulse. However, increasing the electric field (EF) to 2250 V/cm results in arcing and cell death. Increasing the capacitance prolongs the electric pulse, and the pores become larger until they no longer reseal. Figure 1A shows that the viability of the cells gradually declines with increasing capacitance, as expected. However, transient expression shows an optimum at 1435 µF (Fig. 1B). The initial increase of transient expression could be the result of a larger number of transfected cells with an equal amount DNA per cell or a constant number of transfected cells, each with a larger load of DNA molecules. To discriminate between these possibilities, the percentage of transfected cells has to be determined. In Section 4., we discuss experiments that support the first possibility. The mechanism underlying the drop in transient expression when capacitance settings >1435 µF are used is not yet understood. Results comparable to those with FTO2B rat hepatoma cells were observed with other cell lines, such as the hepatoma cell lines MH5123, MH1C1, and HepG2, the fibroblast cell lines Rat-1 and NIH3T3, the colon cell line CaCo-2, and the kidney cell line LLCPK. In addition, we have successfully transfected freshly isolated fetal and adult hepatocytes (data not shown). The protocols for the preparation of cytomix and its use in electrotransfection are described below.

2. Materials

1. Sterilize each of the following stock solutions a–f by autoclaving:
   a. 2.4M KCl: 17.9 g KCl/100 mL.
   b. 30 mM CaCl$_2$: 0.44 g CaCl$_2 \cdot$ 2H$_2$O/100 mL.
   c. 200 mM K$_2$HPO$_4$/KH$_2$PO$_4$, pH 7.6: 4.56 g K$_2$HPO$_4 \cdot$ 3H$_2$O/100 mL (solution 1) and 2.72 g KH$_2$PO$_4$/100 mL (solution 2). Titrate solution 1 with solution 2 to pH 7.6.
   d. 0.5M HEPES, pH 7.6: 11.92 g HEPES/100 mL; adjust pH with 1M KOH.
   e. 80 mM EGTA, pH 7.6: 3.04 g EGTA/100 mL; adjust pH with 1M KOH.
   f. 100 mM MgCl$_2$: 2.0 g MgCl$_2 \cdot$ 6H$_2$O/100 mL.
2. 50 mM ATP, pH 7.6: 0.28 g/100 mL; adjust pH with 5M KOH, filter-sterilize, and store at –20°C.
3. Glutathione powder: Store desiccated at 4°C to maintain its reduced state, which is essential to protect cells against free radical damage after exposure to the electric pulse.
4. 2X Cytomix without glutathione and ATP. Mix the stock solutions a–f above in the following order: 27 mL H$_2$O + 5 mL a + 0.5 mL b + 5 mL c + 5 mL d + 2.5 mL e + 5 mL f. This solution can be stored for several weeks at room temperature.

3. Method

Below is a step-by-step protocol for the transfection of FTO2B hepatoma cells by electroporation in cytomix using cuvets with a 2-mm electrode gap and with a final volume of 500 µL. An extensive discussion of the successive steps follows in the next section, and some comments are made on the interpretation of the results obtained in transient-expression assays:

1. Harvest the cells, which should be growing exponentially, and wash the cells at least twice by pelleting (50g for 10 min) and resuspending them in a large volume (e.g., 50 mL) of 1X cytomix without ATP and glutathione at room temperature to remove contaminating trypsin, tissue-culture medium, and fetal calf serum (FCS) (*see* Note 1).
2. Count the cells after the second wash, and pellet aliquots of 2 × 10^7 cells in 1.5 mL tubes at 50g for 10 min (*see* Note 2).
3. Prepare the electroporation medium by mixing 250 µL of 2X cytomix, 20 µL of 50 mM ATP, pH 7.6, 0.8 mg glutathione, and the DNA. Add water to a final volume of 500 µL (*see* Note 3).
4. Resuspend the cells in the electroporation medium, and transfer to the cuvet.
5. Incubate for 10 min at room temperature.
6. Pulse the cells at 260 V and 1500 µF (*see* Note 4).
7. Incubate the cell suspension in the cuvet at room temperature for 10 min (*see* Note 5).

8. Transfer the suspension to a sterile tube with a Pasteur pipet, and rinse the cuvet with 500 µL of tissue-culture medium.
9. Transfer the electroporated cells to a dish or dishes containing prewarmed culture medium with serum (e.g., 400 µL/well of a 6-well dish or 800 µL/10-cm dish).
10. Refresh the culture medium after overnight culture.
11. Harvest the cells 40–48 h after transfection for transient-expression assays (*see* Note 6 and Table 1).

4. Notes

1. Cell culture: Each specific cell line or primary cell culture will have its own specific culture and media requirements. However, these differences in culture conditions are not likely to influence the protocol for electrotransfection in cytomix. Cells must be growing exponentially to obtain the highest transient expression of a reporter construct. Exponentially growing cells are obtained by seeding the cells the day before transfection, such that on the day of transfection, the surface of the flasks or dishes is 70–80% confluent.

   Furthermore, it is important that the cells be free of infection with bacteria, yeast, and viruses, as well as mycoplasm. We observed that the expression of reporter genes driven by the Rous sarcoma virus longterminal repeat (RSV-LTR) or the simian virus 40 (SV40) early promoter was not affected in mycoplasma-infected cells, but the activity of the carbamoylphosphate synthetase I (CPS I) promoter was reduced 10-fold. We test for mycoplasma every 6 wk. Cells should be discarded, and a new batch of cells should be used, if infections or alterations in the morphological appearance or growth characteristics are observed.

2. Cell number and transfection efficiency: Optimal transfection results are obtained using cell concentrations varying between 1×10^6 and 1×10^8 cells/mL, partly depending on the cell type used (data not shown). The concentration of the cells in the electroporation medium should be carefully controlled to obtain reproducible transfection efficiencies. In this respect, we observed that transient expression can differ simply owing to differences in cell concentration.

   Care should be taken to assure that the osmolarity of the electroporation medium is 300 mosM. During transfections of several different cell lines, it was observed that the expression of the reporter gene and the viability of the cells were very sensitive to differences in the volume of DNA solution added, despite a constant setting of voltage and capacitance during electroporation. Electrotransfection of cells in electroporation media of which the osmolarity was varied revealed that the expression of reporter genes showed a sharp optimum when the medium was isotonic. Transient

Table 1
Transient-Expression Levels with TK, RSV,
and CPS I Promoters in Hepatoma Cells and Fibroblasts[a]

	FTO2B hepatoma	Rat-1 fibroblasts	Ratio Rat-1/FTO2B
TK promoter			
10^6 RLU/µg	0.05 ± 0.02	3.3 ± 0.5	66
% Transfected	1.9 ± 0.3	17.5 ± 7.3	9
Norm. 10^6 RLU/µg	2.7 ± 1.3	18.9 ± 2.9	7
TK = 1	1	1	1
n	4	5	
RSV promoter (31 µg)			
10^6 RLU/µg	39.6 ± 7.3	1824 ± 268	46
% Transfected	1.4 ± 0.2	37.5 ± 9.3	27
Norm. 10^6 RLU/µg	2840 ± 522	4860 ± 714	2
TK = 1	1054	246	0.2
n	2	2	
RSV promoter (5 µg)			
10^6 RLU/µg	1.0 ± 0.2	164 ± 44	164
% Transfected	0.6 ± 0.3	21.1 ± 9.8	35
Norm. 10^6 RLU/µg	164 ± 30	780 ± 207	5
TK = 1	61	71	1
n	4	5	
CPS promoter			
10^6 RLU/µg	3.5 ± 1.7	46.5 ± 7.9	13
% Transfected	2.2 ± 0.4	19.1 ± 7.0	9
Norm. 10^6 RLU/µg	164 ± 78	243 ± 41	1
TK = 1	61	14	0.2
n	4	5	

[a]FTO2B rat hepatoma cells and Rat-1 fibroblasts were transfected at 1500 µF and 1300 V/cm with 31 µg pT81luc (TK promoter), with 31 or 5 µg pRSVcat (RSV promoter), or with 31 µg pLCPSMP (CPS I promoter) in combination with 25 µg pRSV-n-*LacZ*. After electrotransfection, the cell suspension was divided into two equal parts. After 44 h of culture, one part was fixed and stained histochemically for *LacZ* expression, and the other part was lysed and the luciferase activity driven by the indicated promoter determined in the extract. The luciferase activity is given as relative light units (RLU)/µg protein, its expression level per transfected cell being calculated by normalizing for the fraction of transfected cells (norm. RLU/µg). The percentage transfected cells is the ratio of the number of *LacZ*-expressing cells to the total number of cells. The average transfection efficiency of all experiments is 1.5 ± 0.2 (n = 14) for FTO2B hepatoma cells and 21.4 ± 5.3 (n = 16) for Rat-1 fibroblasts. Promoter strength is also expressed relative to the TK promoter (TK = 1). Data are given as mean ± SEM (n = number of independent transfections).

expression decreased sharply when hypotonic media were used, probably a result of swelling and rupture of the organelles. Only a gradual decrease was observed when media of increasing ionic strength were used *(8,9)*.

It is important to note that several groups observed that the addition of carrier DNA can increase the expression of the introduced construct *(8,14,15)*. However, when testing various types of carrier DNA, including supercoiled or linear plasmid, or herring sperm DNA, we did not observe a major influence on the transient expression of a reporter gene introduced into hepatoma cells (data not shown), indicating cell-line-specific effects.

3. Purity of DNA and expression of transfected reporter genes: For most cell types, optimal transient expression is observed when the plasmids are introduced in a supercoiled conformation *(8,14,16)*. The best-quality supercoiled plasmid is obtained when purified using ethidium-bromide-CsCl gradient ultracentrifugation *(17)*. Even then, twofold differences in transient-expression levels may be observed between different batches of a plasmid (data not shown). After purification, the DNA must be dissolved in sterile H_2O, and not in TE (10 mM Tris, 1 mM EDTA, pH 8.0), because EDTA influences the equilibrium of EGTA with Ca^{2+} and Mg^{2+}.

It is important to note that if DNA is diluted in H_2O to determine the concentration of the DNA with a UV spectrophotometer, erratic readings may be obtained. We normally dilute the DNA in TE or a dilute salt solution to determine the concentration.

Column purification of DNA, e.g., by QIAGEN columns, may result in DNA of a lesser quality. In our hands, transient expression of DNA purified in this way results, on average, in a twofold lower expression compared to CsCl gradient-purified DNA (data not shown). Furthermore, we have observed up to 10-fold differences between two batches of the same construct when using column purification.

4. Electroporation conditions: Optimal electrotransfection is not solely dependent on the purity of the DNA and the exponential growth of the cells. Each cell line has it own specific requirements for capacitance and EF for creating resealable membrane pores *(8,18–20)*, and the combination resulting in optimal transient expression must be determined empirically. The results presented in Fig. 1 were obtained by switching two Bio-Rad (Richmond, CA) capacitance extenders in parallel. However, Bio-Rad strongly advises against this procedure. The Eurogentec Easyject *plus* allows such settings. To optimize the electroporation conditions for FTO2B rat hepatoma cells, 50 µg of a plasmid carrying a modified *Lac*Z gene, such that the proteins are translocated to the nucleus, and driven by the RSV promoter was introduced using 1500 V/cm, an infinite shunt resistance, and different capacitance settings. The advantage of the nuclear *Lac*Z reporter gene is that the percentage of transfected cells can be determined by counting blue nuclei and that the enzymatic activity in the culture can be determined *(17)*. Division of the enzyme activity by the fraction of trans-

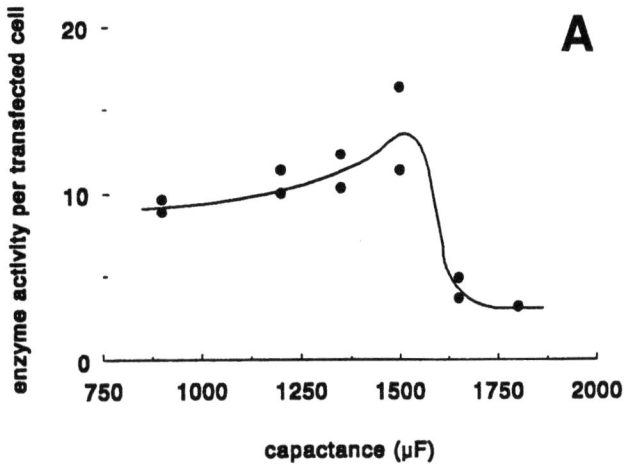

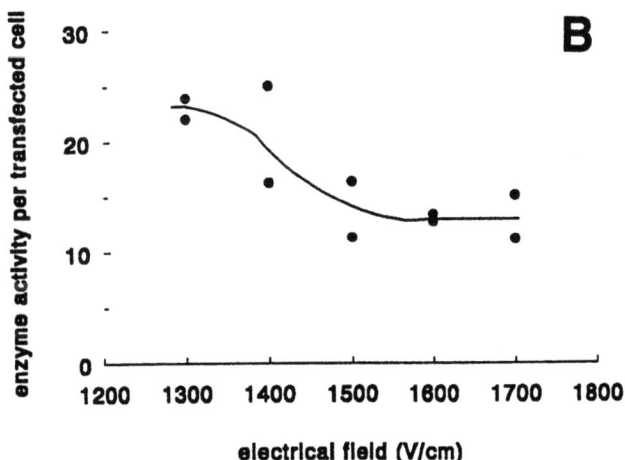

Fig. 2. Optimization of electroporation conditions for FTO2B hepatoma cells. (A) *LacZ* enzyme activity divided by the fraction of *LacZ*-expressing cells as a function of capacitance, using the Eurogentec Easyject *plus*. The EF was constant (1500 V/cm) and the shunt resistance infinite. (B) *LacZ* enzyme activity divided by the fraction of *LacZ*-expressing cells as a function of the strength of the electrical field. The capacitance was constant (1500 µF) and the shunt resistance infinite.

fected cells is a measure of the expression of the reporter per cell. In Fig. 2A, enzyme activity per transfected cell is plotted against capacitance. This figure shows that the expression per cell increases up to 1500 µF and drops sharply beyond, confirming our previous results (Fig. 1). The optimal EF was determined at a capacitance of 1500 µF and infinite shunt resistance. Transient expression is highest at 1300 V/cm and then decreases to reach a plateau until arcing occurs (Fig. 2B). Below 1300 V/cm, transient expression also gradually decreases (Fig. 1B).

5. Temperature during and after transfection: There have been conflicting reports of temperature effects on the transient expressing of reporter genes *(8,14,16)*. Performing the entire electroporation procedure at room temperature revealed a 25% higher transient-expression level than at 4°C when hepatoma cells were used (data not shown). However, a large increase in cell viability and transfection efficiency of Chinese hamster ovary cells was reported when a prepulse temperature of 4°C was combined with a postpulse temperature of 37°C *(21)*. We have not yet evaluated this modification for electroporation in cytomix. These data indicate that the optimal pre- and postpulse temperatures must be determined empirically.

6. Comparison between expression constructs: The technique of electroporation is often used to search for stretches of DNA that comprise the promoter and/or enhancers of a gene. Frequently, in such studies, deletion constructs are tested to determine the minimal DNA sequence that regulates the expression of a reporter gene. When testing constructs of different lengths, the question immediately rises of what should be kept constant: the amount of DNA or the number of DNA molecules. From the experiment shown in Fig. 3, it is clear that equimolar amounts of DNA should be used: When more DNA molecules are present in the transfection medium, more cells become transfected (Fig. 3A), but the expression per cell remains more or less constant (Fig. 3D). The interpretation of transfection results within one cell line is difficult, but becomes even more compli-

---

Fig. 3. *(Facing page)* Relationship between the concentration of transfected DNA and the expression of the reporter gene. Different concentrations of the pRSV-n-*LacZ* construct were transfected at 1500 µF, 1300 V/cm, and infinite shunt resistance. After transfection, the cells were divided into two equal parts, and after 44 h of culture, one part was fixed and the other was lysed. The fixed cells were stained histochemically for β-galactosidase activity. The percentage of expressing and nonexpressing cells was calculated (**A**). (**B**) Cell viability, calculated as described in the legend of Fig. 1. (**C**) *LacZ* activity as mol/µg protein × min. (**D**) *LacZ* activity per transfected cell calculated by dividing the *LacZ* activity by the fraction transfected cells.

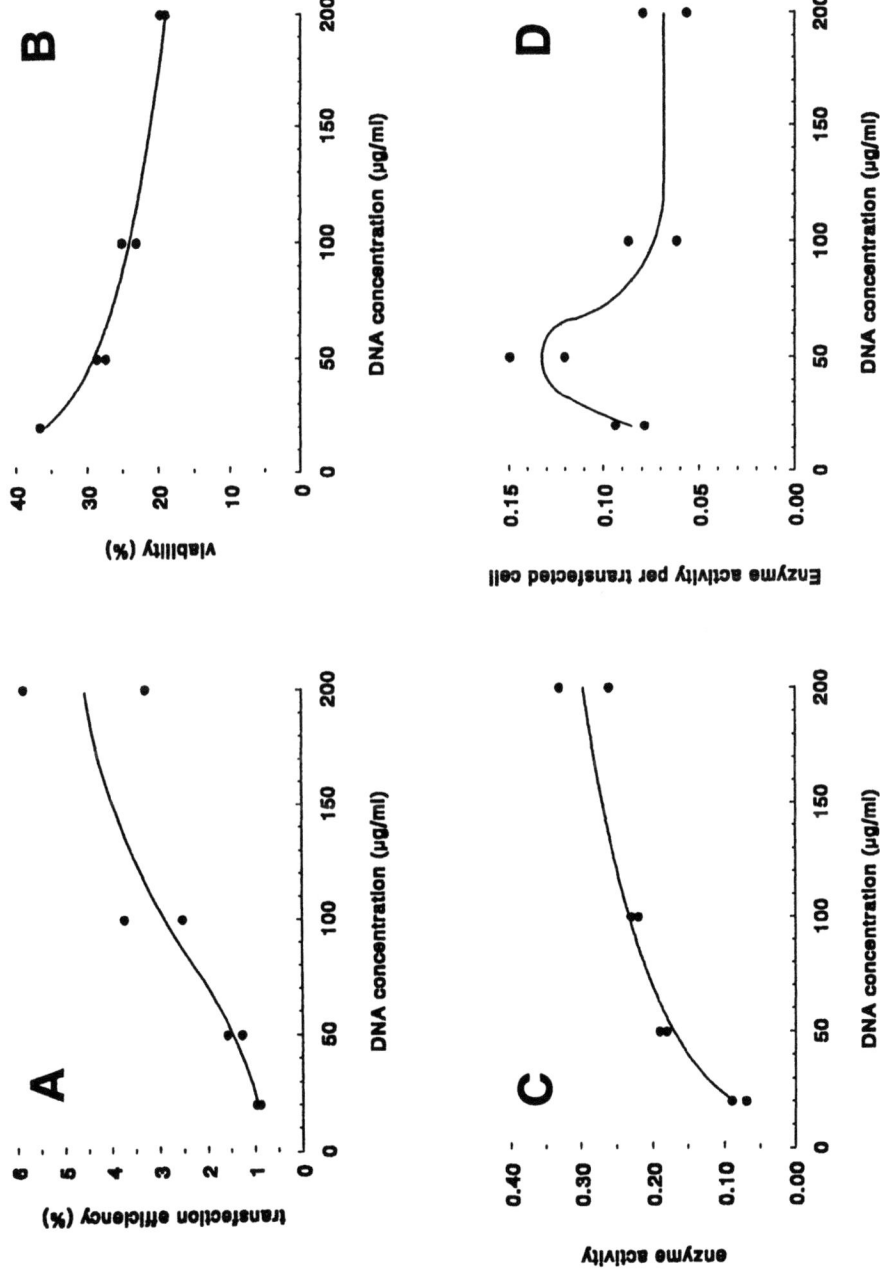

cated when different cell lines, different promoters, and different hormonal stimuli have to be compared. The complexity of such a comparison is immediately clear if one realizes that each cell line has it own transfection efficiency and that the transcriptional and translational capacity can vary several-fold between different cell lines.

An elegant and simple experiment can be performed in which the promoter of interest is transfected in combination with an internal standard. When different cell lines have to be compared, the internal standard should be chosen in such a way that it is possible to determine the percentage of transfected cells independent of the strength of its promoter (*see* ref. 2 for review on reporter genes). For this purpose, a *LacZ* reporter gene construct with a nuclear translocation signal can be used, since the cells can be fixed with 0.5% glutaraldehyde in PBS and stained for galactosidase activity *(17)*. By counting the number of blue cells, which contain galactosidase activity, and the total number of cells, the transfection efficiency can be calculated. The ratio of the activity of the reporter enzyme driven by the promoter of interest and the fraction of transfected cells is a measure for the activity of the promoter per transfected cell. Such normalized values can be compared directly. Table 1 summarizes the values for reporter enzyme activity before and after normalization for the frequently used thymidine kinase (TK) and RSV promoters, and also for the rat CPS I promoters after electrotransfection into FTO2B rat hepatoma cells and Rat-1 fibroblasts. These experiments revealed that the transient expression of the reporter gene (*see* RLU/µg in Table 1) is much higher in Rat-1 fibroblasts than in FTO-2B hepatoma cells. However, these differences are mainly the result of differences in transfection efficiency (*see* percent transfection in Table 1) and, to a lesser extent, differences in expression of each of these constructs in a transfected cell (compare RLU/µg and normalized RLU/µg in Table 1). In general, this analysis reveals that the transfection efficiency differs markedly between different cell lines. In our view, the transfection efficiency must be assayed with a reporter gene that can be visualized to determine the number of transfected cells.

References

1. Chang, D. C., Chassy, B. M., Saunders, J. A., and Sowers, A. E. (1992) *Guide to Electroporation and Electrofusion.* Academic, San Diego.
2. Alam, J. and Cook, J. C. (1990) Reporter genes: application to the study of mammalian gene transcription. *Anal. Biochem.* **188,** 245–254.
3. Saulis, G., Venslauskas, M. S., and Naktinis, J. (1991) Kinetics of pore resealing in cell membranes after electroporation. *Bioeletroch. Bioenerg.* **26,** 1–13.

4. Chernomordik, L. V., Sukharev, S. I., Popov, S. V., Pastushenko, V. F., Sokirko, A. V., Abidor, I. G., and Chizmadzhev, Y. A. (1987) The electrical breakdown of cell and lipid membranes: the similarity of phenomenologies. *Biochim. Biophys. Acta* **902**, 360–373.
5. Tsong, T. Y. (1991) Electroporation of cell membranes. *Biophys. J.* **60**, 297–306.
6. Chang, D. C. and Reese, T. S. (1990) Changes in membrane structure induced by electroporation as revealed by rapid-freezing electron microscopy. *Biophys. J.* **58**, 1–12.
7. Tekle, E., Astumian, R. D., and Chock, P. N. (1991) Electroporation by using bipolar oscillating electric field: an improved method for DNA transfection of NIH 3T3 cells. *Proc. Natl. Acad. Sci. USA* **88**, 4230–4234.
8. Chu, G., Hayakawa, H., and Berg, P. (1987) Electroporation for the efficient transfection of mammalian cells with DNA. *Nucleic Acids Res.* **15**, 1311–1326.
9. van den Hoff, M. J. B., Labruyère, W. T., Moorman, A. F. M., and Lamers, W. H. (1990) The osmolarity of the electroporation medium affects the transient expression of genes. *Nucleic Acids Res.* **18**, 6464.
10. van den Hoff, M. J. B., Moorman, A. F. M., and Lamers, W. H. (1992) Electroporation in "intracellular" buffer increases cell survival. *Nucleic Acids Res.* **20**, 2902.
11. Knight, D. E. and Scrutton, M. C. (1986) Gaining access to the cytosol: the technique and some applications of electropermeabilization. *Biochem. J.* **234**, 497–506.
12. Scarpa, A. (1979) Transport across mitochondrial membranes, in *Membrane Transport in Biology* (Giebisch, G., Tosteson, D. C., and Ussing, H. H., eds.), Springer-Verlag, Berlin, pp. 263–355.
13. McCormack, J. G. and Denton, R. M. (1986) Ca2+ as a second messenger within mitochondria. *TIBS* **11**, 258–262.
14. Andreason, G. L. and Evans, G. A. (1988) Introduction and expression of DNA molecules in eukaryotic cells by electroporation. *BioTechniques* **6**, 650–660.
15. Nickoloff, J. A. and Reynolds, R. J. (1992) Electroporation-mediated gene transfer efficiency is reduced by linear plasmid carrier DNA. *Anal. Biochem.* **205**, 237–243.
16. Potter, H. (1988) Electroporation in biology: methods, applications and instrumentation. *Anal. Biochem.* **174**, 361–373.
17. Sambrook, J., Fritsch, E. F., and Maniatis, T. (1989) *Molecular Cloning: A Laboratory Manual*, 2nd ed., Cold Spring Harbor Laboratory, Cold Spring Harbor, NY.
18. Rols, M. P. and Teissie, J. (1990) Electropermeabilization of mammalian cells. *Biophys. J.* **58**, 1089–1098.
19. Xie, T. D., Sun, L., and Tsong, T. Y. (1990) Study of mechanisms of electric field-induced DNA transfection. I. DNA entry by surface binding and diffusion through membrane pores. *Biophys. J.* **58**, 13–19.
20. Knutson, J. C. and Yee, D. (1987) Electroporation: parameters affecting transfer of DNA into mammalian cells. *Anal. Biochem.* **164**, 44–52.
21. Rols, M. P., Delteil, C., Serin, G., and Teissie, J. (1994) Temperature effects on electrotransfection of mammalian cells. *Nucleic Acids Res.* **22**, 540.

CHAPTER 16

Effect of Cis-Located Human Satellite DNA on Electroporation Efficiency

Djenann Saint-Dic and Michael S. DuBow

1. Introduction

The introduction of exogenous DNA into mammalian cells can ultimately result in the integration of the transfected DNA into the host genome *(1,2)*. Any genes containing appropriate expression signals will, in most cases, be expressed both prior to and after integration. However, stable expression of the transfected genes is not always observed *(3,4)*. One reason this is believed to occur is because the site of integration of the foreign DNA is virtually random with respect to both the exogenous DNA and the particular human chromosome *(5)*. Therefore, it can be expected that not only the expression, but also the stability and copy number of the transfected gene will depend on the integration sites and chromosomal flanking sequences.

The effect of flanking sequences on the stable expression of transfected genes has been reported in a number of studies. Butner and Lo *(3)* have shown that a herpes simplex virus type 1 (HSV-1) thymidine kinase gene *(tk)* was subject to transcriptional modulation when integrated into centromeric satellite DNA. The unexpressed *tk* genes were found to be heavily methylated, whereas derepression of the *tk* gene was accompanied by rearrangements in the flanking DNA. These observations suggest that the flanking heterochromatic DNA sequences may exert cis-effects on *tk* gene expression, mimicking position-effect variegation (PEV). In PEV, first documented in *Drosophila,* chromosomal rearrangements place euchromatic genes next to heterochromatin. The rearranged genes are expressed to different extents in different cell lineages, thus giving rise to a variegated animal *(6)*. In another study, it was noted that integration of transfected genes in centromeric or heterochromatic regions could be favored by the presence of satellite sequences in the transfected DNA *(2)*. Satellite DNAs, which represent at least 10% of the human genome *(7)*, are a family of short, highly repeated, tandemly arrayed DNA sequences, generally located in the α-heterochromatin near the centromere. This region of chromatin is highly compacted and transcriptionally inactive. It was observed that the integration events may preferentially occur via the satellite sequences when they are present in the exogenous DNA *(8)*.

The role of satellite DNA sequences in PEV has not been completely characterized. However, it is known that satellite DNA can affect gene expression, since it has been reported that cotransfection of satellite DNA sequences with a gene encoding a selectable phenotype leads to a decrease in the recovery of stable transfectants *(2)*. This observation is supported by results obtained in electroporation experiments performed using satellite DNA sequences in cis on a plasmid containing selectable genes in human cells. Having the satellite DNA and a selectable gene on the same plasmid provides a more direct way to study the effect of these satellite sequences. Plasmid constructs carrying both the HSV-1 *tk* and a human 1.797-kb *Eco*RI satellite II DNA *(9)* were generated from plasmid pSV2*neo* *(10)*. These plasmids were electroporated because of the high efficiency provided by this transfection method *(11)*. Electroporation of these constructs demonstrates that the satellite DNA can affect the electroporation efficiency of the gene in a plasmid location and/or orientation manner.

2. Materials

2.1. Plasmids

A systematic way to assess the effect of satellite DNA on electroporation efficiency is to have the satellite DNA sequence in cis with the gene of interest. This strategy is practical because it allows the variation of the orientation and/or location of the satellite sequence in relation to the selectable gene. Examples of such plasmids are depicted in Fig. 1. In this case, the genes of interest are the HSV-1 *tk* gene and the prokaryotic *neo* gene (expressed from the SV40 early promoter) encoding resistance to G418 (*see* Note 6).

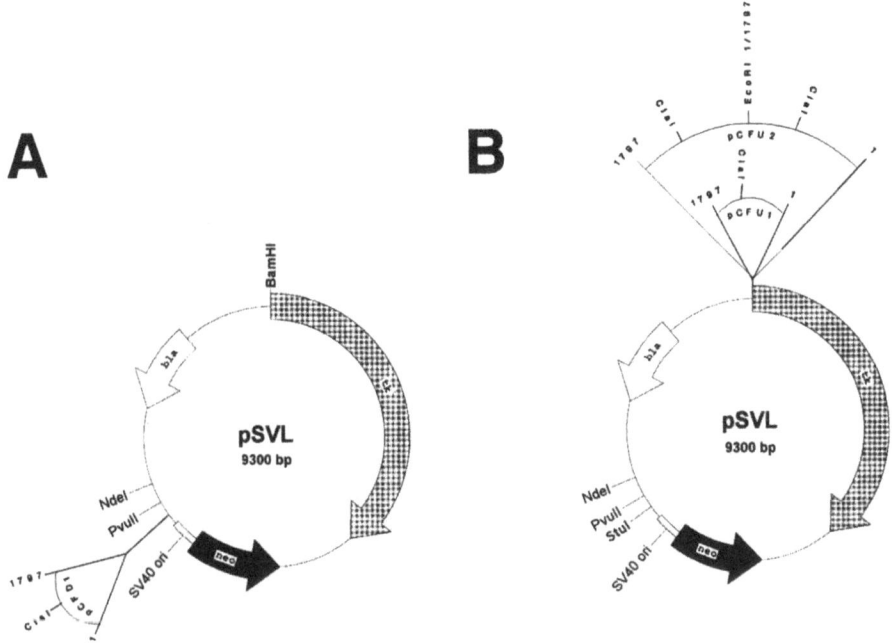

Fig. 1. Diagram of satellite DNA-containing plasmids used for electroporation. These plasmids are derived from pSVL *(9)*, a derivative of pSV2neoKT *(12)*. The different locations of the 1.797-kbp *Eco*RI satellite II DNA are shown, and their orientations are indicated with respect to their DNA sequence *(13)*. These plasmids contain the ColE1 origin and ampicillin-resistance gene *(bla)* from pBR322, the HSV-1 thymidine kinase gene *(tk)*, the SV40 origin of replication (SV40 ori), and the G418-resistance gene *(neo)*. A selected number of restriction enzyme sites are displayed.

2.2. Cell Lines

In the case of electroporation with plasmids carrying the HSV-1 *tk* gene, thymidine kinase-deficient cells need to be used. The transformed human fibroblastic cell line AK143B (ATCC CRL 8303) serves this purpose well. AK143B is a *tk*⁻ derivative of the murine sarcoma virus transformed cell line R970-5 *(14)*. However, the requirements for the cell line in electroporation experiments will vary depending on the selectable markers carried by the plasmids.

2.3. Solutions and Reagents for Cell Culture

1. Culture the AK143B cell line in Dulbecco's Modified Eagle Medium (DMEM) supplemented with final concentrations of 10% (v/v) fetal bovine serum (FBS), 100 U/mL penicillin, 100 µg/mL streptomycin, and 2.5 µg/mL amphotericin B (fungizone). Penicillin and streptomycin may be purchased combined (Pen-Strep) from Gibco/BRL (Burlington, Ontario, Canada). Stock solutions of 500 µg/mL (in deionized water [deIH$_2$O]) of fungizone (Squibb, Montreal, Quebec, Canada) can be prepared. Keep aliquots of FBS, Pen-Strep, and fungizone at –20°C for long-term storage. Most cell lines can be cultured in DMEM, but the required supplements may vary depending on the cell line.
2. Phosphate-buffered saline (PBS): 137 mM NaCl, 3 mM KCl, 8 mM Na$_2$HPO$_4$, 1 mM KH$_2$PO$_4$. Adjust the pH to 7.3 with HCl. Autoclave the solution and store at 4°C.
3. Trypsin (0.06%): 25 mL trypsin stock solution (2.5 mg/mL) plus 75 mL versene. Versene is comprised of 0.2 mg/mL EDTA in PBS. Autoclave the versene solution and store at 4°C. Filter-sterilize the trypsin stock solution through a 0.22-µm membrane filter, and store at –20°C. Store the 0.06% final trypsin solution at 4°C.
4. Buffered 50% DMSO: 50% dimethyl sulfoxide (DMSO) (v/v), 500 mM Trizma base, 73 mM dextrose, 174.9 mM citric acid. Adjust the pH to 6.7 with HCl. Filter-sterilize and store at 4°C. DMSO should be handled with caution, since it is a powerful solvent and may carry toxic substances through the skin *(15)*.

2.4. Solutions and Reagents for Electroporation

1. HEPES-buffered saline (HeBS): 140 mM NaCl, 5 mM KCl, 0.75 mM Na$_2$HPO$_4$ · 2H$_2$O, 6 mM dextrose, 25 mM HEPES. Adjust the pH to 7.05 with 0.5N NaOH. Filter-sterilize the final solution and store aliquots at –20°C.
2. TE: 10 mM Tris-HCl, pH 7.5, 1.25 mM EDTA. Autoclave and store at 4°C.
3. Plasmid DNA: Purify the plasmid DNA from *Escherichia coli* hosts on cesium chloride/ethidium bromide density gradients, dialyze vs TE, and store at 4°C *(16)*. Linear plasmids are obtained by digestion with an appropriate restriction endonuclease. Extract the digested DNA with phenol and chloroform, precipitate with 2.5 vol of ethanol *(17)*, and store at –20°C in TE.

2.5. Drug Selection and Colony Staining

1. HAT (hypoxanthine, aminopterin, thymidine) (Sigma, St. Louis, MO) is used at a final concentration of 0.1 mM hypoxanthine, 0.4 µM aminopterin, and 16 µM thymidine. These concentrations are achieved by directly

adding HAT (0.018 mg/mL final concentration) to the DMEM culture medium. Filter-sterilize the final solution. Caution should be taken with HAT, since it is highly toxic (18).
2. G418 (Geneticin) (Gibco/BRL or Sigma) is used at a final concentration of 400 µg/mL in DMEM. Store aliquots of 40 mg/mL in deIH$_2$O at –20°C after filter-sterilization. The potency of the product (i.e., the actual amount of G418 per mg of powder) should be taken into account in making the stock solution. Because of its toxicity, G418 should also be handled with care.
3. Fixative: 10% buffered formalin phosphate, pH 6.9–7.1 (4% [v/v] formaldehyde, 0.4% [w/v] NaH$_2$PO$_4$ · H$_2$O, 0.65% [w/v] Na$_2$HPO$_4$, 1.5% [v/v] methanol). Filter-sterilize and store at room temperature.
4. Giemsa stain (1%): 1 g Giemsa in a 100 mL glycerol/methanol solution (1:1 v/v). Filter-sterilize and store at room temperature.

3. Methods

3.1. Electroporation Protocol

1. Transfect 3×10^6 cells with each type of plasmid. One day prior to transfection, an average of 1.0×10^6 cells/flask are seeded (for cells that are relatively rapidly growing) in order to attain the desired confluency. It is very important that the cells do not reach more than 50–75% confluency (see Note 1). Seed cells into tissue-culture flasks of 80 cm^2/260 mL, using at least two flasks for each plasmid to be transfected. Incubate the cells overnight in nonselective medium (DMEM supplemented with FBS, Pen-Strep, and fungizone), in a 37°C incubator with a 5% CO$_2$ atmosphere.
2. On the next day, remove the culture medium by aspiration. Rinse the cells with 10 mL of cold PBS to wash away any trace of serum that may inhibit digestion by trypsin. To each flask, add 1.3 mL of trypsin to detach the adherent cells. After the cells are detached (usually 1–2 min), inactivate the trypsin by adding serum-containing DMEM. Resuspend the cells, and collect them by centrifugation for 5 min at 500g in a bench-top centrifuge (at the resuspension stage, the cells from 2 flasks [80 cm^2] may be combined).
3. After removal of the supernatant fluid, wash the cells twice with 7 mL of ice-cold HeBS by resuspension, and centrifugation as above. After the second wash, count an aliquot of the cells using a hemocytometer to determine the total number of cells. Subject the volume corresponding to 3×10^6 cells to centrifugation for 5 min as above. Resuspend the cells in 0.5 mL of ice-cold HeBS.
4. Transfer 0.5 mL of cell suspension to an ice-cold electroporation cuvet with a 0.4-cm electrode gap. Add 1.5 pmol of plasmid DNA (circular or

linear) to the cell suspension (*see* Note 2). Mix the DNA/cell suspension by tapping the bottom of the cuvet, and incubate for 10 min at room temperature.
5. Set the electroporation apparatus to 250 V and 500 µF; no additional resistance is needed (*see* Note 4). Pulse the cells in the cuvet once at these settings, at room temperature (*see* Note 3). Note the time constant; it should be between 10 and 12.4 ms (*see* Note 4).
6. Immediately following the electric pulse, add 1.4 mL of serum-containing DMEM to the cuvets to allow the cells to recover (*see* Note 5). Divide the contents of each cuvet into five 10-cm plates (or two 15-cm plates). The important factor here is to have a concentration of 10^5 cells/mL in each plate. Each plate should contain the appropriate volume of nonselective medium to give the desired cell concentration; 6 mL of medium are sufficient for a 10-cm plate. Incubate the plates in a 37°C incubator with a 5% CO_2 atmosphere.

3.2. Drug Selection Protocol

1. Twenty-four hours after transfection, change the nonselective medium for fresh medium, and incubate at 37°C with a 5% CO_2 atmosphere. Forty-eight hours after electroporation, change the nonselective medium to medium containing the selective drugs (HAT and/or G418 at the final concentrations specified in Section 2.5.) (*see* Note 6). Change the selective medium for fresh selective medium every 3 d.
2. Colonies will appear after approx 10 d. Fourteen days after electroporation, stain the colonies with Giemsa and count. Before staining, aspirate the selective medium from the plates. Wash each plate with 10 mL of PBS. Remove the PBS, add 10 mL of fixative/small plate, and then incubate for 1–12 h at room temperature. Remove the fixative, and add 5 mL/plate of Giemsa stain diluted just before use (1:50 [v/v] with $deIH_2O$). Then incubate for 1–12 h at room temperature. Remove the stain, rinse the plates with $deIH_2O$, and count the colonies (*see* Note 7).

4. Notes

1. In most electroporation experiments selecting for permanent transfectants, the actual number of cells transfected varies between 1.5×10^6 and 5×10^6 cells/0.5 mL of buffer. The concentration range of cells that can be used is, however, much larger. Chu et al. *(19)* tested a number of cell densities, ranging from 2.5×10^6 to 4.0×10^7/mL. A sample of these electroporated cells was assayed for transient expression, and the results were found to be independent of the density of cells, even with a fixed amount of DNA. Transient-expression experiments may require cell densities up to 8×10^7 cells/mL, depending on the promoter and gene product to be assayed *(11)*.

On the other hand, the extent of growth of the cells, before they are harvested for electroporation, is critical. The best results are obtained with cells that are approx 50% confluent *(19)*. Depending on the cell line used, there may be up to a 10-fold decrease in electroporation efficiency if fully confluent cells are used.

2. The electroporation efficiency is also affected by the quality and quantity of the plasmid DNA. The efficiency has been shown to increase linearly with DNA concentration *(19)* and, depending on the cell line, higher electroporation efficiencies may be obtained with linear DNA. In the case of the AK143B cell line, there may be up to a fivefold increase in transfection efficiency with linear plasmids (Table 1). Therefore, it may be preferable to determine the efficiency vs the geometry of the DNA to be electroporated.

In experiments where the effect of satellite DNA sequences is being assayed, it may be judicious to use plasmids of varying sizes devoid of satellite DNA sequences. In this way, it can be assessed whether it is plasmid size, as opposed to the satellite sequences, that is affecting transfection efficiency. Fouquet and DuBow *(9)* found that larger plasmids tend to decrease transfection efficiency (keeping the molar amounts constant), but to a much lesser extent than equivalent-sized plasmids containing satellite DNA sequences. Different sizes of DNA fragments from bacteriophage λ, for example, can be used to construct control plasmids of varying sizes.

The location and copy number of these satellite sequences in the plasmid can also be varied *(9)*. Heartlein et al. *(1)* have shown that alphoid satellite DNA copy number exerts a dose-dependent effect on chromosomal aberrations, such as sister chromatid exchange and dicentric or ring chromosomes.

3. Some electroporation experiments were shown to be more effective if performed at 4°C (on ice) rather than at room temperature *(20)*. The temperature setting may vary depending on the cell line *(19)*. It is always possible to compare the efficiency at both temperatures. In the case of electroporation at 4°C, the mixture of cells and plasmid DNA has to be incubated on ice for 5 min before and, if desired, after the electric pulse.

4. The most critical factors to control in order to obtain optimal electroporation efficiencies are the parameters of the electroporation apparatus. It is not necessary to use additional electrical resistance when buffers of high ionic strength, such as PBS or HeBS, are employed. Buffers of high salt concentration have less resistance and therefore require a shorter electric pulse. Electroporation experiments performed using HeBS have shown that this *isotonic* buffer leads to higher viabilities and higher transfection efficiencies among buffers of high salt concentration *(19)*.

Table 1
Electroporation Efficiency of Satellite DNA-Containing Plasmids

Plasmid form	Plasmid	Number of colonies in HAT + G418, %[a]	Number of colonies in HAT, %	Number of colonies in G418, %	Number of colonies in G418/HAT, %[d]
Circular	pSVL	1.8×10^2 (100)	3.9×10^2 (100)	9.9×10^2 (100)	4.9×10^1 (100)
	pCFU1	1.0×10^2 (56 vs 37)[c]	—[b]	—	—
	pCFU2	2.0×10^1 (11 vs 9)[c]	—	—	—
	pCFD1	0 (0)	0 (0)	7.65×10^2 (77)	1.8×10^0 (3.6)
Linear	pSVL	4.8×10^2 (100)	9.15×10^2 (100)	1.11×10^3 (100)	1.22×10^2 (100)
	pCFD1	5.6×10^0 (1)	2.4×10^2 (26)	2.7×10^2 (24)	3.0×10^0 (2.4)

[a]Transfection efficiency. The efficiency obtained with pSVL is designated as 100%.
[b]"—": Not done.
[c]The "vs % number" is the efficiency obtained with control plasmids of identical size containing a λ DNA insert instead of the satellite DNA.
[d]Number of transfectants obtained in a stepwise selection using G418 first, followed by G418 + HAT.

One of the most important parameters to follow is the duration of the electric pulse, which is determined by the capacitance of the electroporation apparatus and the buffer resistance *(11)*. Therefore, if excessive cell death results from the electroporation *(see* Note 5), the capacitance should be decreased. It may be necessary to test a few capacitance values in order to find the one that best suits a specific cell line. In certain cases, cell viability appears to depend more on total charge than capacitance alone *(19)*. Since the electric charge is dependent on both the capacitance and the voltage, it is also important to determine the proper voltage. Voltage dependence is inversely proportional to the cell size; therefore, different cell lines will have different voltage requirements. In general, the voltage dependence varies between 200 and 275 V for mammalian cells.

5. In electroporation experiments, 40–80% cell viability is optimal after the electric pulse *(11)*. Electroporations performed with the plasmid pSV2*neo* have yielded viabilities ranging between 25 and 50% *(19)*. Colonies with nonintegrated plasmids first grow and then die. The colonies appearing after 10–14 d contain integrated plasmids. To determine the cell viability or plating efficiency, a subset of cells (500, for example) are plated in nonselective medium following transfection, and counted after 2 or 3 d.

   In certain experiments, cells are incubated for a few minutes at room temperature after the electric pulse, before being plated *(11)*. However, this treatment has been shown in some cases to result in a very low viability. This can be corrected by incubating the cells in a small volume of nonselective medium before plating or by directly plating them immediately following electroporation.

6. The combination of both HAT and G418 selects for plasmid integration sites that are not within the *tk* and *neo* genes, and for chromosomal locations where both genes are expressed. However, it is possible to select for cells that have taken up and integrated the plasmid DNA by only using one drug (HAT or G418 in the case of the plasmids described here). In this way, only the expression of one of the genes is followed. Selection with each drug independently helps to assess more accurately the effect of satellite DNA on the expression of a particular gene (Table 1). Different marker genes can be used depending on the cell line or their availability. In general, *neo* is used because it is a dominant selectable marker in all mammalian cell lines *(11)*, as is the histidinol resistance gene *(21)*.

   It should be noted that the *neo* gene was found to act as a typical silencer by exerting a cis-acting negative effect on expression from promoters in plasmid or retroviral vectors transfected into mammalian cell lines *(22)*. Therefore, using another gene as a selectable marker may prove useful in eliminating the potential negative effects of the *neo* gene.

In experiments where the cotransfection of the *tk* gene with satellite DNA sequences was performed, it has been shown that revertants may be obtained by selecting for bromo-deoxyuridine (BUdR) resistant clones *(2)*. The phenotypic switching was attributed to the influence of the satellite DNA on the gene, since no modulation of gene expression was observed in cell lines transfected with the *tk* gene only. By re-exposing the cells to HAT medium, it is possible to obtain re-revertants to the tk^+ phenotype. In the same way, null tk^- mutants can be selected using the antiviral drugs acyclovir, trifluorothymidine, and ganciclovir in combination *(23)*.

7. There may be differences in the sizes of the colonies on a given plate. Tiny colonies should not be counted, because they could be caused by cells having detached from larger colonies. For fibroblasts, the frequency of permanent transfectants is generally expected to be $1/10^3$–10^4 cells *(11)*. This value may vary depending on the experiment, and is likely to decrease up to 40-fold or even to zero when plasmids carrying satellite DNA sequences are used *(9)*.

Acknowledgments

We would like to thank Claire Fouquet for helpful discussions, and Peter Ulycznyj, Felix Sieder, and Caroline Diorio for their help in the preparation of the manuscript. We also thank Gina Macintyre and Kirsty Salmon for their comments and suggestions. M. S. D. is a research scholar (Exceptional Merit) of the Fonds de la Recherche en Santé du Québec (FRSQ). The work in our laboratory is supported by an operating grant (OGP0003222) from the Natural Sciences and Engineering Research Council of Canada (NSERC) to M. S. D.

References

1. Heartlein, M. W., Knoll, J. H. M., and Latt, S. A. (1988) Chromosome instability associated with human alphoid DNA transfected into the Chinese hamster genome. *Mol. Cell. Biol.* **8,** 3611–3618.
2. Talarico, D., Peverali, A. F., Ginelli, E., Meneveri, R., Mondello, C., and Della Valle, G. (1988) Satellite DNA induces unstable expression of the adjacent herpes simplex virus *tk* gene cotransfected in mouse cells. *Mol. Cell. Biol.* **8,** 1336–1344.
3. Butner, K. and Lo, C. W. (1986) Modulation of *tk* expression in mouse pericentromeric heterochromatin. *Mol. Cell. Biol.* **6,** 4440–4449.
4. Butner, K. A. and Lo, C. W. (1986) High frequency DNA rearrangements associated with mouse centromeric satellite DNA. *J. Mol. Biol.* **187,** 547–556.
5. Murnane, J. P., Yezzi, M. J., and Young, B. R. (1990) Recombination events during integration of transfected DNA into normal human cells. *Nucleic Acids Res.* **18,** 2733–2738.

6. Grigliatti, T. (1991) Position-effect variegation—an assay for nonhistone chromosomal proteins and chromatin assembly and modifying factors. *Methods in Cell Biol.* **35,** 587–627.
7. Beridze, T. (1986) *Satellite DNA.* Springer-Verlag, Berlin, Germany.
8. Wallenburg, J. C., Nepveu, A., and Chartrand, P. (1987) Integration of a vector containing rodent repetitive elements in the rat genome. *Nucleic Acids Res.* **15,** 7849–7863.
9. Fouquet, C. and DuBow, M. S. (1992) Effect of *cis*-located human satellite DNA on the electroporation efficiency of *neo* and HSV-1 *tk* containing plasmids. *Mut. Res.* **284,** 321–328.
10. Mulligan, R. C. and Berg, P. (1980) Expression of a bacterial gene in mammalian cells. *Science* **209,** 1422–1427.
11. Potter, H., Selden, R. F., and Kingston, R. E. (1987) Transfection by electroporation, in *Current Protocols in Molecular Biology* (Ausubel, F. M., Brent, R., Kingston, R. E., Moore, D. D., Seidman, J. G., Smith, J. A., and Struhl, K., eds.), Greene Publishing Associates and Wiley-Interscience, Brooklyn, NY, pp. 9.3.1–9.5.6.
12. Goring, D. R. and DuBow, M. S. (1985) A cytotoxic effect associated with 9-(1,3-dihydroxy-2 propoxymethyl) guanine is observed during the selection for drug resistant human cells containing a single Herpesvirus thymidine kinase gene. *Biochem. Biophys. Res. Commun.* **133,** 195–201.
13. Sol, K., Lapointe, M., Macleod, M., Nadeau, C., and DuBow, M. S. (1986) A cloned fragment of Hela DNA containing consensus sequences of satellite II and III DNA hybridizes with the *Drosophila* P-element and with the 1.8 kb family of human *Kpn*I fragments. *Biochim. Biophys. Acta* **868,** 128–135.
14. Rhim, J. S., Cho, H. Y., and Huebner, R. J. (1975) Non-producer human cells induced by murine sarcoma virus. *Int. J. Cancer* **15,** 23–29.
15. Horita, A. and Weber, L. J. (1964) Skin penetrating property of drugs dissolved in dimethylsulfoxide (DMSO) and other vehicles. *Life Sci.* **3,** 1389–1395.
16. Sambrook, J., Fritsch, E. F., and Maniatis, T. (1989) Plasmid vectors, in *Molecular Cloning, a Laboratory Manual,* vol. 1, Cold Spring Harbor Laboratory, Cold Spring Harbor, NY, pp. 1.42–1.46.
17. Tolias, P. P. and DuBow, M. S. (1987) The amino terminus of the bacteriophage D108 transposase protein contains a two-component, sequence-specific, DNA-binding domain. *Virology* **157,** 117–126.
18. Stubblefield, E. (1968) Synchronization methods for mammalian cell cultures, in *Methods in Cell Physiology* (Prescott, D. M., ed.), Academic, New York, pp. 25–43.
19. Chu, G., Hayakawa H., and Berg P. (1987) Electroporation for the efficient transfection of mammalian cells with DNA. *Nucleic Acids Res.* **15,** 1311–1326.
20. Potter, H., Weir, L., and Leder, P. (1984) Enhancer-dependent expression of human κ immunoglobulin genes introduced into mouse pre-B lymphocytes by electroporation. *Proc. Natl. Acad. Sci. USA* **81,** 7161–7165.
21. Hartman, S. C. and Mulligan, R. C. (1988) Two dominant-acting selectable markers for gene transfer studies in mammalian cells. *Proc. Natl. Acad. Sci. USA* **85,** 8047–8051.

22. Artlet, P., Grannemann, R., Friel, J., Bartsch, J., and Hauser, H. (1991) The prokaryotic neomycin-resistance-encoding gene acts as a transcriptional silencer in eukaryotic cells. *Gene* **99,** 249–254.
23. Brisebois, J. J. and DuBow, M. S. (1993) Selection for spontaneous null mutations in a chromosomally integrated HSV-1 thymidine kinase gene yields deletions and a mutation caused by intragenic illegitimate recombination. *Mut. Res.* **287,** 191–205.

CHAPTER 17

Quantitation of Transient Gene Expression

Michael K. Showe and Louise C. Showe

1. Introduction

In many situations, it is important to quantitate accurately expression of a transiently transfected gene. This is most often the case when mechanisms controlling gene expression are under study, and the roles of several possible promoter and enhancer elements and their interacting factors are to be assessed. Accurate quantitation of expression driven by these different elements requires a gene-transfer procedure that will efficiently and reproducibly deliver the reporter construct into a relatively high percentage of the cells being transfected *(1,2)*. In addition, the reporter assay should give high sensitivity, reproducibility, and linearity of response to increasing concentration of construct, and have a wide dynamic range. The combination of transfection by electroporation with a colorimetric or photometric reporter gene assay meets these requirements, and provides a rapid and sensitive method for the study of genetic regulatory elements in many types of cells.

Electroporation is particularly suited to quantitation of gene expression in vitro, because cells can be transfected synchronously and essentially instantaneously *(3)*. Since there need be little manipulation of cells after the electric pulse, there is good reproducibility from sample to sample. A variety of cell types, especially those grown in suspension, which are poorly transfected by other means, can be successfully transfected by electroporation *(1,2,4–6)*.

A number of new sensitive reporter gene assays that depend on bioluminescence or chemiluminescence have recently become available *(7)*. Two reporters for which vectors are already widely used are the β-galactosidase (β-gal) gene of *Escherichia coli (8,9)* and the luciferase (luc) gene from the firefly *Photinus pyralis (10)*. Both can be conveniently used to optimize conditions for quantitative measurements of gene expression in many cell lines and a number of primary cell types. The luciferase assay has the advantage of a very wide dynamic range and higher sensitivity than β-gal assayed with o-nitrophenyl-β-D-galactoside (ONPG), but the latter assay is less expensive and does not require a luminometer or other photon-counting equipment (*see* Note 1). We describe below conditions that allow measurement of these two reporters from the same cell extract, allowing the use of one reporter for the experimental construct and the other as an internal control.

A disadvantage of using electroporation for transfection to quantitate gene expression has been the relatively large amounts of DNA required. Cells transfected by electroporation must be treated with 5–10 times the amount of DNA required by other techniques, such as lipofection or transfection using DEAE dextran *(11)*. This is inconvenient when large numbers of different constructs must be prepared. DNA yields for reporter constructs are sometimes low, and DNA quality can differ from one preparation to another, confounding direct comparisons among related constructs.

A number of conditions have been found to increase the efficiency of transient expression of electroporated DNA. Most of these have been tested in only a few cell types, and the effects of some are known to be cell-type-dependent. Exposure to 5-mM sodium butyrate for a 14-h period after electroporation has been reported to increase transiently expressed chloramphenicol acetyl transferase *(1)*. DEAE-dextran has been found to increase expression from microgram concentrations of Rous sarcoma virus (RSV)-luc from 7- to 150-fold in various cell lines *(12)*.

DNA from various sources and in various forms have been tested as carriers to increase efficiency for both transient and stable transfection. Chu et al. *(13)* found a twofold increase in transient expression using 500 µg/mL of salmon sperm DNA. McNally et al. *(14)*, using relatively high amounts of DNA (120 µg/mL), found that stable transfectants of hematopoietic cell lines were produced on average five times more efficiently when plasmid was linearized than when it was supercoiled, but transient assays of K562 cells were reported not to be influenced by the plasmid form. When rates of stable transfection of Chinese hamster ovary cells by microgram amounts of a linearized plasmid were determined, addition of supercoiled carrier (at 25 µg/mL) increased rates three- to fivefold, and linear eukaryotic DNA was also stimulatory; however, linearized plasmid carrier was found to diminish efficiency *(15)*. The observed variations of transfection efficiency with cell line, DNA form, and source on stable transfectants may reflect the actions of nucleases and recombination enzymes whose effects are not relevant when transient expression from related constructs is being measured.

Using transient expression of an electroporated luciferase reporter gene as an assay, a comparison of a variety of DNAs (including Bluescript [BL] plasmid, bacteriophage, *E. coli,* and salmon sperm DNA at various sizes) showed that BL or similar bacterial plasmid DNA was 10–50 times more effective as a carrier than the other DNAs tested *(16)*. The best carrier was DNA of low complexity and from 0.4 kb to a few kilobases in size; linearized carrier was consistently twice as efficient as circular.

The mechanism by which carrier DNA functions has not been established, but its utility is illustrated by the fact that 2 µg of a luciferase construct electroporated with 50 µg of carrier can have nearly the same activity as 50 µg of construct alone. Above a few tenths of a picomole of construct, most of the added DNA is not being expressed, but only increasing the efficiency of the electroporation. It has been suggested that DNA may play a role in pore formation during electroporation *(17)*.

Since transient expression depends on DNA concentration and purity, the variability inherent in comparing different reporter constructs can be minimized by making the carrier the principal DNA present during electroporation. The use of a large excess of carrier provides a range over which reporter gene activity is directly proportional to moles of construct electroporated, independent of construct size.

We have adopted 0.5 pmol of reporter construct brought to 50 µg total DNA with plasmid carrier as a standard assay condition that provides good activity and is reasonably sparing of material. Including a constant amount of a second reporter plasmid (in the range of 0.1–0.5 pmol) allows correction for variations in electroporation efficiency in individual samples without affecting expression of the experimental construct *(16)*.

2. Materials

1. Plasmids: RSV-luc, RSV-β-gal, and SV-β-gal denote the plasmids pRSVluc *(10)*, pTB1 *(9)*, and pSVβgal *(8)*, respectively. These are control plasmids that are used to monitor electroporation efficiencies. The promoterless and enhancerless luciferase vectors pXPI and pXP2, which contain multiple cloning sites, as well as psLuc, which contains the simian virus 40 early region minimal promoter, may be purchased from ATCC, Rockville, MD. BL (KSII+) plasmid, which is used as carrier DNA, may be obtained from Stratagene (La Jolla, CA). All plasmids are grown in *E. coli* JM 109 (Stratagene or Promega, Madison, WI).
2. Terrific broth (TB) *(18)*:
   a. Nutrients: Add 12 g of tryptone, 24 g of yeast extract, and 4 mL of glycerol to a final volume of 900 mL of H_2O.
   b. Buffer: Add 12.5 g of K_2HPO_4 and 2.3 g of KH_2PO_4 to a final volume of 100 mL of H_2O.
   Autoclave nutrient and buffer solutions separately, and mix when cool.
3. Tris-EDTA (TE): 10 mM Tris-HCl, pH 8.0, 1 mM disodium EDTA.
4. Ethidium bromide extraction buffer: Mix 80 mL of isopropanol with 10 mL of 50 mM Tris-HCl, 1 mM EDTA, pH 8.0, and then add 10 mL of 5M NaCl.
5. Phenol/$CHCl_3$: Just before using, mix 1 vol of $CHCl_3$ with an equal volume of ultrapure phenol (Gibco-BRL, Gaithersburg, MD or United States Biochemical, Cleveland, OH) that has been equilibrated with 10 mM Tris-HCl, pH 8.0.
6. Mammalian cells: K562, a human erythroleukemia cell line, is available from ATCC. This cell line is easily transfected and can be used to test the electroporation protocols.
7. Cell lysis buffer: 0.1M K_2HPO_4, 0.6% Triton X-100, 1 mM dithiothreitol, pH 7.8. Just before using, add 1 µL of 1M dithiothreitol (1000X stored frozen) and 20 µL of 30% (v/v) Triton X-100 (50X) to each 1 mL of 0.1M K_2HPO_4, pH 7.8.
8. Luciferase assay buffer: 0.023M final conc. glycyl-glycine, 0.015M $MgSO_4$, 0.005M ATP, pH 7.8. Just before use, mix 9.35 mL of 0.025M glycyl-glycine, pH 7.8, 0.15 mL 1M $MgSO_4$, and 0.5 mL 0.1M ATP. Prepare ATP stock solution by dissolving 600 mg disodium ATP in 10 mL distilled water. Titrate to pH 7.0 with approx 160 µL of 10N NaOH. Store frozen at –70°C as 500-µL aliquots. It may be stored for several weeks at –20°C.
9. Luciferin stock solution: Prepare 0.1 mM luciferin (Analytical Luminesence, San Diego, CA) by disolving 50 mg luciferin in 30 mL H_2O *(see* Note 2).

10. β-gal assay buffer: $0.1M$ Na_2HPO_4, 1 mM $MgCl_2$, 0.05 mM β-mercaptoethanol, 1 mg/mL ONPG, pH 7.3. Just before using, add 100 µL of Mg-β ME stock solution ($0.1M$ $MgCl_2$, 5 mM β-mercaptoethanol, stored frozen) and 200 µL of 50 mg/mL ONPG (stored frozen) to 9.7 mL of $0.1M$ of sodium phosphate buffer, pH 7.3.
11. Luminometer: Monolight 2010 luminometer (Analytical Luminescence).
12. Luminometer cuvets: (Analytical Luminescence).

3. Methods
3.1. Preparation of DNA for Transfection
1. Inoculate 250 mL of TB in a 2-L flask from a frozen bacterial stock or from an overnight culture.
2. Grow inoculated cultures with rapid rotation (300–400 rpm) for good aeration at 37°C for 16–60 h (see Note 3).
3. Isolate plasmid DNA by alkaline lysis (19) (see Note 4).
4. Purify large-scale DNA preparations by two rounds of CsCl equilibrium centrifugation (using, for example, a Beckman NVT65 rotor), and collect the plasmid bands with a syringe (19, see Note 5).
5. Add the collected plasmid immediately to a tube containing 2–3 vol of ethidium bromide extraction buffer, mix, and allow the phases to separate. Remove the upper layer containing the ethidium bromide with a Pasteur pipet and discard. Repeat two to three times until all the color from ethidium bromide is removed.
6. Transfer the DNA solution to a dialysis bag. Dialyze at room temperature against a 200–500-fold excess of TE for at least 2 h with one buffer change after 1 h (see Note 6).
7. Add SDS (UltraPure, Gibco-BRL) to dialyzed DNA to a final concentration of 0.5%, extract once with an equal volume of phenol/$CHCl_3$, and precipitate twice with 2.5 vol of 90% ethanol. Dry the DNA precipitate under vacuum to remove residual ethanol, and resuspend in TE at either a concentration of 0.5–2.5 µg/µL (luciferase-construct DNA) or 5 µg/µL (carrier DNA) (see Note 7).
8. Determine DNA concentrations by measuring the optical density (OD) at 260 nm (see Note 8). Store working stock of DNA at 4°C in TE. Backup stocks may be stored frozen.
9. Compare DNA preparations by agarose gel electrophoresis before using them for transfection (see Notes 9 and 10).

3.2. Preparation of Cells for Electroporation
1. Split exponentially growing suspension cells into fresh medium the day before the electroporation.

2. Determine cell viability, which should be >95% for good transfection efficiencies.
3. Determine the concentration of the cells to be transfected using a hemocytometer.
4. Harvest cells just before using by centrifugation at 300g for 5 min at room temperature.
5. Remove the old medium completely, and gently resuspend in fresh complete medium (*see* Note 11) at room temperature at $1-2 \times 10^7$ cells/mL.
6. Carry out all manipulations of cells using sterile technique. Grow adherent cells that are to be electroporated to 60–70% confluence. Remove cells from the growth substrate by whatever method is normally used in propagation of the cell line, and process for transfection as described in steps 4–6.

3.3. Determination of Optimum Voltage for Electroporation (see Note 12)

1. Mix on ice 0.5 pmol RSV-luc (2.2 µg) or RSV-β-gal (2.6 µg) with BL carrier to bring total DNA to 50 µg/transfection. The volume should be <25 µL.
2. Pipet 1.7 mL of cell-culture medium for each electroporation into a well of a 12-well tissue-culture plate, and place in the incubator to equilibrate temperature and CO_2.
3. Centrifuge 5×10^6 cells/electroporation as described above (Section 3.2. step 4), and gently resuspend in 0.3 mL of fresh culture medium (*see* Note 11).
4. Add the DNA mixture to the resuspended cells, and pipet 300 µL of the mixture into each 0.4-cm electrode gap cuvet (*see* Note 13).
5. Set capacitance level of the electroporator at the maximum following the manufacturers' instructions. For the Gene-Pulser (Bio-Rad, Richmond, CA), this is 960 µF.
6. Electroporate the contents of the first cuvet at 50 V, tapping the cuvet to mix the contents on removal from the electroporator (*see* Note 14), and record the time constant (*see* Note 15).
7. Increase the voltage by 50, and repeat with the next cuvet, up to 350 V (*see* Note 16) recording the time constant for each voltage.
8. Transfer the contents of each cuvet to one of the wells of the tray of prewarmed medium (*see* Note 17). This operation may be carried out with a Pasteur pipet. Rinse the cuvet once with the diluted cells to ensure complete transfer (*see* Note 18 for a modification using cells that grow attached rather than in suspension).
9. Incubate the tray of electroporated samples 16–24 h (overnight) in a CO_2 incubator (*see* Note 19).

3.4. Preparation of Cell Extracts for Luciferase and/or β-gal Assays

1. Resuspend transfected cells by pipet, and transfer completely into 2.2-mL microcentrifuge tubes.
2. Centrifuge at 12,000g for 5 min and carefully and completely remove the supernatant by aspiration (*see* Note 20).
3. Add 100 µL of cold cell lysis buffer to the cell pellet, and resuspend by pipeting. Lysis is evident when cell debris makes it difficult to pipet the extract (*see* Note 21).
4. Clear the cell extract by centrifugation for 5 min at 4°C at 12,000g. Extracts are routinely assayed for luc immediately, but may be frozen and assayed at a later time (*see* Note 22).

3.5. Luciferase Assay

1. Measure the luciferase activity *(10)* with a luminometer set to integrate light output over a 10-s interval.
2. Add 350 µL of luciferase assay buffer, equilibrated at room temperature, to a separate luminometer cuvet for each sample to be measured.
3. Load the luciferin solution, also at room temperature, into the luminometer.
4. Add 50 µL of cell lysate to each luminometer cuvet just before placing it in the luminometer (*see* Note 23). The luciferin is automatically injected and the reading recorded (*see* Note 24). The results of typical voltage curves at two electroporation volumes are shown in Fig. 1.

3.6. β-gal Assay

1. Add 10–50 µL of cell extract to enough β-gal assay buffer on ice to total 300 µL. If samples begin to turn yellow while on ice, set up a second reaction with less extract.
2. Start the reaction by transferring the tubes simultaneously to 37°C. Incubate for at least 30 min (*see* Note 25).
3. Monitor the color development by eye until a clearly visible yellow color is observed (*see* Note 26).
4. Stop the reactions by transferring the tubes to a 60°C water bath for 10 min (*see* Note 27).
5. Centrifuge to clear each sample at 12,000g for 5 min, and measure the OD in a spectrophotometer at 410 nm (*see* Note 28).

3.7. Comparison of Expression from Different Reporter Plasmids

When several reporter plasmids are to be compared for their abilities to drive transcription of the luciferase reporter, it is important that all

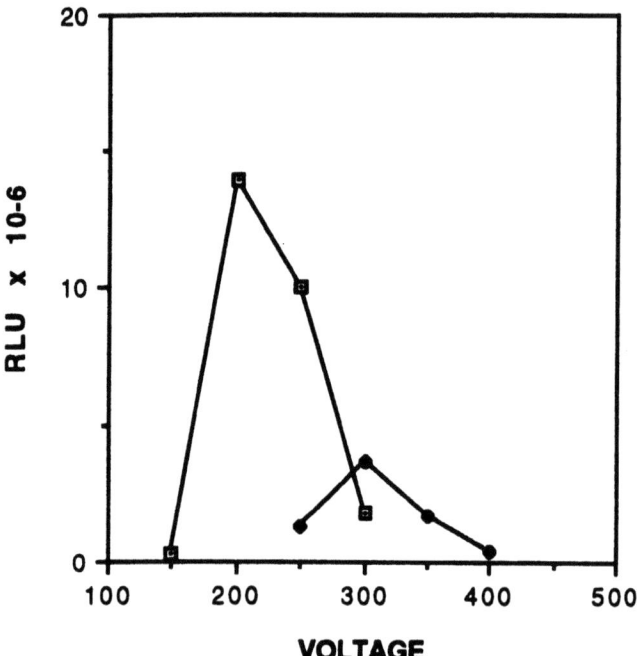

Fig. 1. Electroporation efficiency as a function of voltage and reaction volume: 5×10^6 Molt 13 cells were electroporated at the indicated voltages in either 0.3 mL (open symbols) or 0.8 mL (closed symbols) of complete medium. The time constants were 31–34 ms for the low volumes and 11.2–13.2 ms for high volumes. Cell survival measured by trypan blue exclusion at 24 h was 70% at the 0.3-mL activity peak and 30% at the 0.8-mL peak.

samples be treated as similarly as possible. Table 1 describes a typical protocol in which A, B, C, and D represent putative promoter fragments ranging in size from 0.18–4.8 kb cloned into the promoterless pXPI luciferase vector.

1. Use 5×10^6 cells/electroporation as a starting point in establishing electroporation conditions, and vary the cell number depending on the transfection efficiency of the particular cell line.
2. Prepare the cells as described above (Section 3.2.), and resuspend them in 290 µL of fresh medium for each of the samples to be electroporated.
3. Pipet 1.7 mL/well of culture medium into the appropriate number of wells of a 24-well culture plate, and place in a CO_2 incubator at 37°C.

Quantitation of Transient Gene Expression

Table 1
Protocol for Transient Assays of Reporter Constructs[a]

Plasmid	Size in kb	μg Plasmid, 0.5 pmol	μg RSV β-gal, 0.1 pmol	μg Carrier DNA	μg Total DNA
RSVluc	6.40	2.08	0.0	47.5	49.58
pXPI	5.90	1.92	0.5	47.5	49.92
pXPI + A	6.08	1.98	0.5	47.5	49.98
pXPI + B	6.18	2.00	0.5	47.5	50.00
pXPI + C	7.00	2.28	0.5	47.5	50.28
pXPI + D	10.70	3.48	0.5	47.5	51.48
RSV-β-Gal	8.00	0.0	0.5	47.5	48.00
X	—	0.0	0.0	47.5	47.50

[a]The sizes of A, B, C, and D are 0.18, 0.28, 1.1, and 4.8 kb, respectively.

4. Pipet 0.5 pmol (see Note 29) of each of the luciferase plasmids to be tested into the bottom of an electroporation cuvet in a fixed volume, usually 10 μL (see Note 30). Cap the cuvets to prevent evaporation if many samples are to be prepared.
5. Add carrier DNA to the resuspended cells to give a concentration of 47.5 μg/290 μL (see Note 31). Mix cells and DNA, and transfer a 290-μL aliquot to a cuvet containing 10 μL of TE without any plasmid (see Note 32).
6. Add 0.1 pmol/transfection of the RSV-β-gal control DNA to the remaining cell–carrier mixture. Mix, and then transfer 290 μL to each of the individual cuvets containing the luciferase reporter constructs.
7. Electroporate the samples at the voltage optimum determined by the protocol outlined in Section 3.3. Record the time constants, and follow the described procedure for transferring cells.
8. Harvest cells at 16–24 h, and process for enzyme assays (see Note 33).

4. Notes

1. The availability of a 1,2-dioxetane phenyl-β-D-galactoside substrate (Tropix, Bedford, MA), which gives rise to a chemiluminescent product, increases the sensitivity of the β-gal assay to the levels attainable with luciferase (20) if a luminometer or similar instrument is available. Differential heat inactivation of the endogenous eukaryotic β-gal (21) is required to take advantage of the increased sensitivity.
2. The luciferin stock solution is sufficient for 300 assays. It is stable for several months at 4°C or 1 yr frozen. Protect from light. Backgrounds may vary slightly by batch and source of luciferin, or as a function of the particular luminometer. One hundred fifty to 200 relative light units (RLU)

background is routine.
3. Plasmids for large-scale preparations are grown in *E. coli* JM109 in TB *(18)*, since this medium gives high yields of plasmid DNA and allows cultures to be grown for extended times.
4. Cultures that are grown for several days in TB yield very large bacterial pellets, and the volumes used for plasmid purification should be multiplied two- to fourfold. Alternative procedures to alkaline lysis may also be used, but we find this procedure the most convenient.
5. Carrier DNA should be made in large batches (2–3 L of culture) so that a single preparation can be used for many experiments. Test new carrier against the old batch before use, since DNA purity is critical for reproducible results.
6. Alternative procedures that omit the dialysis step give less consistent DNA preparations. DNA prepared using commercially available columns is perfectly good for making stable transfectants or for qualitative studies. However, it is frequently significantly contaminated with host DNA, and we have found that quantitation with these preparations is not reliable.
7. Allow sufficient time for DNA pellets to dissolve fully (1–2 h at 37°C or overnight at 4°C).
8. Plasmid carrier DNA yields can range from 5–10 mg/L. A 250-mL culture of a luciferase vector should yield approx 0.5 mg DNA.
9. All purified DNA is also checked by agarose gel electrophoresis before being used for transfections to determine whether there is significant contamination with bacterial DNA, which would contribute to the measured OD and distort calculations for molar equivalents of each construct.
10. To confirm quantitation based on OD of two plasmids, digest each with a restriction enzyme that gives a fragment of the same size, and compare the intensity of the resulting restriction fragment bands on an agarose gel. Undigested plasmid is also run on the gel to determine the percentage of nicked DNA.
11. For all the cells we have tested, complete medium is optimal.
12. The RSV-luc vector, which is highly active in most cells, is routinely used for determining optimum voltages. SV2-luc *(10)* and CMV-luc *(22)* vectors may also be used.
13. Care should be taken to be consistent in pipeting the DNA–cell mixture. When using 0.3 mL as the electroporation volume, small differences can make a significant difference in the column height in the cuvet and affect the efficiency of electroporation.
14. A pH gradient is established in the cuvet as evidenced by the pH indicator in the medium. Mixing diminishes this gradient.

15. Large fluctuations (20% or more) in the time constants from one sample to another or from one experiment to another indicate some problem with either a DNA preparation, a dirty or bad cuvet if being reused, or differences in sample volumes.
16. The 350-V sample may arc when electroporated, in which case the time constant may be high and the assay of this sample will be unreliable. A final volume of 300 µL in a 0.4-cm (800-µL) Bio-Rad cuvet maximizes DNA concentration and length of discharge (τ), both of which increase electroporation efficiency. However, this limits the maximum usable voltage to about 350 V. To test higher voltages, increase the volume electroporated (Fig. 1). Using smaller volumes per electroporation may increase the efficiency of transfection, but reproducibility for quantitation is diminished.
17. The 2.0-mL vol is sufficient for overnight culture of most cells tested. However, it may be necessary to resuspend some cells in a larger volume, in which case, use a six-well plate and 4.0 mL of medium.
18. For cells that attach during the incubation step, making quantitative recovery for assay difficult, use the following variation. After electroporation, pipet the cells into 12-mL culture tubes (instead of 12-well plates) containing 1.5–2 mL prewarmed medium, and incubate at 37°C in a CO_2 incubator overnight. To assay (Section 3.4.), centrifuge the cells at 1500 rpm for 10 min and remove the medium. Lyse the cells (Section 3.4., step 3) in the tubes, carefully washing down the sides to ensure complete recovery. Transfer to a 1.5-mL tube, and process as for suspension cells (Section 3.4., step 4).
19. We have tested transfected K562 cells for luciferase (and β-gal) activity at times from 1/2 to 24 h. Activity can be detected as early as 1 h, but enzyme increases are not linear with time until 4–5 h. Quantitative studies are best if assays are done at least 5 h after transfection. This time interval should be determined for the particular cells you are using. Longer incubations (e.g., 48 h) have no advantage, since activities are sometimes lower than at shorter times.
20. If you invert the tube as you aspirate when preparing all extracts and use a Pasteur pipet on a vacuum aspirator, you can process 20–30 samples in a few minutes. Run the pipet around the inside of the tube to remove residual medium. One pipet can be used for all samples, since the pellet should not be touched.
21. Most cells lyse within 30 s. Cells that resist lysis may be broken by quickly freezing and thawing.
22. Extracts are stable for at least a week at –20°C and longer at –70°C.
23. Do not add the cell extracts to all the tubes at the same time, since the enzyme is unstable in the luciferase assay buffer.
24. Because the luminometer readings are not linear above 5×10^6 RLU, any assay resulting in a value $>5 \times 10^6$ RLU is repeated using 5–10 µL of extract

and the result extrapolated to 50 µL.
25. Samples that have low enzyme concentrations may be incubated as long as 6 h, since the β-gal enzyme is very stable.
26. ONPG has a molar extinction coefficient at pH 7.3 of about 2500 absorbancy units/cm. To use only 10–30% of the substrate, the final OD should be below 2.0. This can be regulated for very active samples by the time of incubation or the amount of extract used for the assay.
27. Use of the usual sodium carbonate stop buffer *(23)* causes cloudiness with the luciferase lysis buffer. Substituting heat inactivation allows both luciferase and β-gal assays to be carried out using the same cell extract.
28. If it is necessary to increase the volume of the reaction to accommodate larger cuvets, the samples may be diluted with the phosphate-buffer solution. Typical background measurements for the 0.3-mL reaction are 0.06 absorbency units, and this increases somewhat with long incubations.
29. For most studies, we find that 0.5 pmol of luciferase plasmid is sufficient to give easily detectable levels of luciferase activity. For cells that are not efficiently transfected, this amount can be increased to as much as 2.0 pmol if higher signals are desired. Confirm that the response is linear in this range by also testing 1.0- and 3.0-pmol amounts.
30. This procedure is simplified if plasmids are at a common molar concentration rather than weight concentration (e.g., 1 pmol/20 µL).
31. Although the 5 pXPI constructs shown in Table 1 vary considerably in size, the total DNA concentration variation between the largest and smallest pXPI constructs is approx 3%. Total DNA concentration variations that are <5% are not significant.
32. The carrier only sample (x) is a control for contamination of carrier with either the luciferase or β-gal reporters, and provides the background values for both enzyme assays. The luciferase activity for the pXPI vector carrying fragment A expressed in RLU normalized by the β-gal control is calculated as follows:

$$\text{Luc}_{pXPI-A} = \text{luc}_{pXPI-A} - \text{luc}_x / \beta\text{gal}_{pXPI-A} - \beta\text{gal}_x \quad (1)$$

Quantitative results are usually expressed as the ratio of the activities of two constructs or the ratio of the activity of a construct to that of the luc vector with no additional insert. The RSV-luc and RSV-β-gal electroporations are included in each experiment to control for significant changes in transfection efficiencies in a given cell line from day to day.
33. Timing: For 10–20 samples, cell preparation, electroporation, and transfer to fresh medium require about 2 h. The following day, harvesting and lysing the cells require about 0.5 h, the luciferase assays about 0.5 h, and the β-gal assays 1–2 h.

References

1. Tatsuka, M., Orita, S., Yagi, T., and Kakunaga, T. (1988) An improved method of electroporation for introducing biologically active foreign genes into cultured mammalian cells. *Exp. Cell Res.* **178**, 154–162.
2. Takahashi, M., Furukawa, T., Nikkuni, K., Aoki, A., Nomoto, N., Koike, T., Moriyama, Y., Shinada, S., and Shibata, A. (1991) Efficient introduction of a gene into hematopoietic cells in S-phase by electroporation. *Exp. Hematol.* **19**, 343–346.
3. Klenchin, V. H., Sukharev, S. M., Chernomordik, L. V., and Chizmadzhev, Y. A. (1991) Electrically induced DNA uptake by cells is a fast process involving DNA electrophoresis. *Biophys. J.* **60**, 804–811.
4. Toneguzzo, F., Hayday, K. A. C., and Keating, A. (1986) Electric field-mediated DNA transfer: transient and stable gene expression in human and mouse lymphoid cells. *Mol. Cell. Biol.* **6**, 703–706.
5. Buschle, M., Brenner, M. K., Chen, I. S. Y., Drexler, H. G., Gignac, S. M., and Rooney, C. M. (1990) Transfection and gene expression in normal and malignant primary B lymphocytes. *J. Immunol. Methods* **133**, 77–85.
6. Anderson, M. L. M., Spandidos, D. A, and Coggins, J. R. (1991) Electroporation of lymphoid cells: factors affecting the efficiency of transfection. *J. Biochem. Biophys. Methods* **22**, 207–222.
7. Bronstein, I., Fortin, J., Stanley, P. E., Stewart, G., and Kricka, L. J. (1994) Chemiluminescent and bioluminescent reporter gene assays. *Anal. Biochem.* **219**, 169–181.
8. Hall, C. V., Jacob, P. E., Ringold, G. M., and Lee, F. (1983) Expression and regulation of *Escherichia coli* lacZ gene fusions in mammalian cells. *J. Mol. Appl. Gen.* **2**, 101–109.
9. Borrás, T., Peterson, C. A., and Piatigorsky, J. (1988) Evidence for positive and negative regulation in the promoter of the chicken δ1–crystallin gene. *Dev. Biol.* **127**, 209–219.
10. de Wet, J. R., Wood, K. V., DeLuca, M., Helinski, D. R., and Subramani, S. (1987) Firefly luciferase gene: structure and expression in mammalian cells. *Mol. Cell Biol.* **7**, 725–737.
11. Selden, R. F. and Rose, J. K. (1991) Optimization of transfection. *Curr. Protocols in Mol. Biol.* **1**, 9.9.1–9.9.3.
12. Gauss, G. H. and Lieber, M. R. (1992) DEAE-dextran enhances electroporation of mammalian cells. *Nucleic Acids Res.* **20**, 6739–7740.
13. Chu, G., Hayakawa, H., and Berg, P. (1987) Electroporation for the efficient transfection of mammalian cells with DNA. *Nucleic Acids Res.* **15**, 1311–1326.
14. McNally, M. A., Lebkowski, J. S., Okarma, T. B., and Lerch, L. B. (1989) Optimizing electroporation parameters for a variety of human hematopoietic cell lines. *BioTechniques* **6**, 282–286.
15. Nickoloff, J. A. and Reynolds, R. J. (1992) Electroporation-mediated gene transfer efficiency is reduced by linear plasmid DNAs. *Anal. Biochem.* **205**, 237–243.
16. Showe, M. K., Williams, D. W., and Showe, L. C. (1992) Quantitation of transient gene expression after electroporation. *Nucleic Acids Res.* **20**, 3153–3157.
17. Sukharev, S. I., Klenchin, V. A., Chernomordil, L. V., and Chizmadzhev, Y. A. (1992) Electroporation and electrophoretic DNA transfer into cells. The effect of DNA interaction with electropores. *Biophys. J.* **63**, 1320–1327.

18. Tartof, K. D. and Hobbs, C. A. (1987) Improved media for growing plasmid and cosmid clones. *Focus* **9,** 12.
19. Kingston, R. E. (1992) Introduction of DNA into mammalian cells. *Curr. Protocols in Mol. Biol.* **14(Suppl.),** 9.1.5–9.1.6.
20. Jain, V. K. and Magrath, I. T. (1991) A chemiluminescent assay for the quantitation of β-galactosidase in the femtogram range: application to quantitation of β-galactosidase in LacZ transfected cells. *Anal. Biochem.* **199,** 119–124.
21. Young, D. C., Kingsley, S. D., Ryan, K. A., and Dutko, F. J. (1993) Selective inactivation of eukaryotic β-galactosidase in assays for inhibitors of HIV-TAT using bacterial β-galactosidase as a reporter enzyme. *Anal. Biochem.* **215,** 24–30.
22. Pahl, H. L., Burn, T. C., and Tenen, D. G. (1991) Optimization of transfection into human myeloid cell lines using a luciferase reporter gene. *Exp. Hematol.* **19,** 1038–1041.
23. Rosenthal, N. (1987) Identification of regulatory elements of cloned genes with functional assays. *Methods Enzymol.* **152,** 704–720.

CHAPTER 18

Stable Integration of Vectors at High Copy Number for High-Level Expression in Animal Cells

James Barsoum

1. Introduction

The efficient production of proteins via expression of recombinant genes in cells other than those that naturally produce the protein has been critical to the study and the therapeutic utility of a wide variety of proteins. *E. coli* expression systems are useful for many proteins. However, large complex proteins are not expressed efficiently in bacteria. Also, bacterial expression cannot be used for eukaryotic proteins that require glycosylation for their activity. Some established animal cell lines can be used as production hosts since they are easily transfected, can grow to high cell densities, and can be propagated in low-serum or serum-free medium.

The level of expression from a single transfected gene in animal cell lines is usually too low to be useful. Transient transfections and viral vector infections can give high-level expression owing to the temporary presence of a high copy number of the introduced gene. However, stable transfectants are more convenient and are often required for large-scale production. Expression in stable lines can be elevated dramatically by increasing the number of integrated copies of the transfected DNA. The most commonly utilized gene amplification system *(1–5)* employs dihydrofolate reductase (DHFR) gene coamplification in Chinese hamster ovary cells having a deletion of the endogenous DHFR gene (DHFR–CHO cells; *6*). DHFR–CHO cells are transfected with the gene for the protein to be expressed along with the DHFR cDNA. These two genes can reside on one or two plasmids. DHFR expression is selected in medium lacking ribonucleotides and deoxyribonucleotides that are required for the growth of DHFR-deficient cells. Conventionally, the initial transfection results in cells having one or a low number of integrated plasmid copies. The DHFR and associated gene copy numbers are increased by selection in methotrexate (MTX), which is a competitive inhibitor of the DHFR enzyme. This amplification must take place in multiple steps, since the MTX concentration can only be increased in small increments. Unfortunately, since clones must be isolated and expanded after each amplification selection step, the total time required for the generation of a high-copy-number clone is 4–8 mo. The amount of effort is significant, and the time required can be prohibitive, especially when multiple proteins need to be expressed.

This chapter describes the use of electroporation to introduce into DHFR–CHO cells high-copy-number vector DNA in a single step. The electroporation and selection protocol is shown in Fig. 1. The gene to be expressed and the DHFR gene are present on a single plasmid separated by a spacer and a transcriptional termination signal. This plasmid is linearized prior to electroporation. A high concentration of plasmid DNA plus carrier DNA is transfected under optimum electroporation conditions. Selection in MTX leads to the isolation of clones having as many as 1000 copies of the transfected plasmid and high levels of protein production. DNA often integrates in long tandem repeats with the restriction site originally used for the linearization intact, indicating that the DNA may ligate in tandem repeats prior to integration. The time required from the start of the electroporation to the generation of an expanded population ($\sim 10^7$ cells) of a high-level expression clone is approx 1 mo. If expression is not sufficient following this electroporation and selection, a single round of gene amplification can significantly increase expression levels.

Insert gene to be expressed into pJOD-S or pMDR901
↓
Ethanol precipitate 200 µg <u>linearized</u> plasmid plus 200 µg carrier DNA
Resuspend DNA in 1X HeBS
Pass DHFR- CHO cells
↓
Electroporate 1-2 x 10^7 cells at 300 V, 960 µFD
↓
Seed in α+ medium
↓
2 days growth
↓
Seed 5 x 10^5 cells per 100 mm dish in α– medium
↓
Incubate 4-5 days
↓
Seed 5 x 10^5 cells per 100 mm dish in α– medium plus 0.5 µM MTX
↓
~14 days selection
↓
Pick clones
↓
Expand and assay protein production

Fig. 1. The protocol for high-copy-number electroporation and selection of CHO cells yielding stable high-level expression clones. *See text* for details.

2. Materials
2.1. Cells, Media, and Plates
1. Cells: DHFR-deficient cell line DUKX-CHO *(6)*.
2. α+ medium: MEM-α supplemented with 10 mg/L each of adenosine, cytidine, guanosine, uridine, 2'-deoxyadenosine, 2'-deoxyguanosine, and 2'-deoxythymidine and 11 mg/L 2'-deoxycytidine hydrochloride (MEM-α medium with nucleosides is sold by Gibco, Grand Island, NY), plus 10% fetal bovine serum (FBS; Hazleton, Lenexa, KS) and 4 mM glutamine (Mediatech, Washington, DC). All media are sterilized by filtration.
3. α– medium: MEM-α lacking ribonucleosides and deoxyribonucleosides (Gibco) plus 10% dialyzed FBS and 4 mM glutamine (*see* Note 1).
4. Phosphate-buffered saline (PBS): 138 mM NaCl, 2.7 mM KCl, 1.5 mM KH$_2$PO$_4$, 8.1 mM Na$_2$HPO$_4$.
5. Trypsin-EDTA: 0.05% trypsin, 0.53 mM EDTA · 4Na (1X liquid; Gibco).
6. Methotrexate (MTX; Sigma, St. Louis, MO) is dissolved in MEM-α at a 1-mM final concentration, filter-sterilized, and stored at –20°C.
7. Tissue-culture dishes: 100-mm tissue-culture dishes, 48-well and 12-well tissue culture plates (Corning, Corning, NY).

2.2. Plasmid Vectors
Either plasmid pJOD-S *(7)* or pMDR1 (Fig. 2; constructed by M. Rosa) is used in high-copy-number electroporation. Each contains all regulatory elements required for the expression of two genes (*see* legend to Fig. 2 and Note 2). The gene to be expressed is inserted into a *Sal*I site in the case of pJOD-S and either *Sal*I or *Not*I (*see* Note 3) in pMDR1. The expression of this gene is directed by the adenovirus 2 major late promoter (Ad MLP) augmented by the presence of an upstream SV40 enhancer. These plasmids also encode the murine DHFR cDNA. The transcriptional terminator from the human gastrin gene *(8)* lies between the two genes. They can be propagated in *E. coli* using ampicillin selection.

By way of example, the tissue plasminogen activator (tPA) gene was inserted into these vectors as described *(7)*. The other proteins cited in this chapter are all cell-surface molecules that were engineered into a soluble secreted form by deletion of the membrane-spanning domain (*see* refs. *9–11* for examples).

2.3. Reagents for Electroporation
1. Restriction enzyme: *Aat*II is used to digest the plasmid DNA prior to electroporation (*see* Note 6).

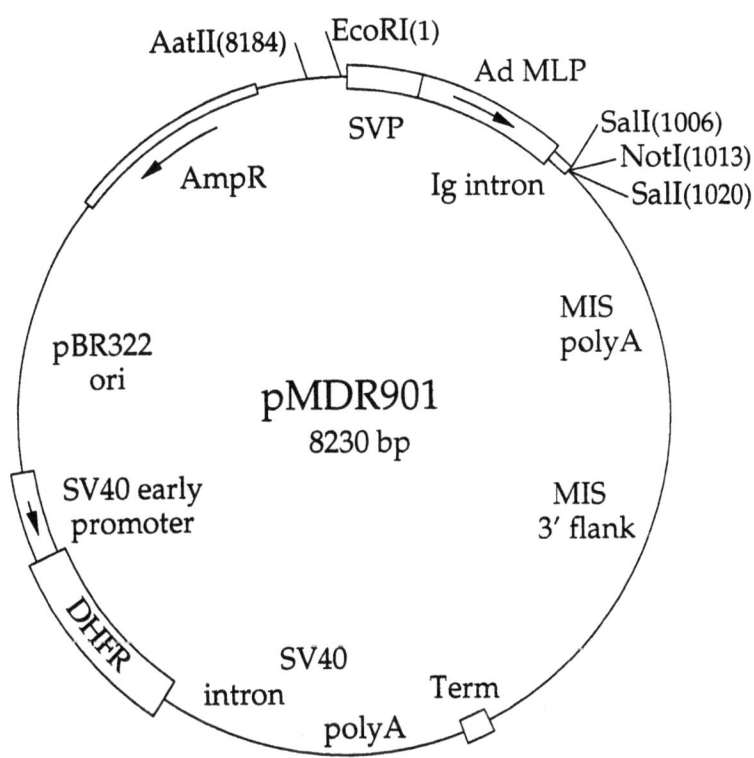

Fig. 2. Map of the high-copy-number expression plasmid pMDR901. pMDR901 encodes the murine DHFR cDNA driven by the SV40 early promoter, and followed by SV40 intron and polyadenylation (polyA) sequences. The gene to be expressed can be inserted into the *Sal*I or *Not*I sites downstream of the Ad MLP. Expression is augmented by the presence of an upstream SV40 enhancer. An immunoglobulin intron lies between the major late promoter and the *Sal*I cloning site. The polyA site and 3'-flanking DNA from the Müllerian inhibiting substance (MIS) gene are downstream of the cloning site. The DHFR gene and the inserted gene are separated by the human gastrin gene 3'-sequence (*see* Note 2). The plasmid backbone was derived from pBR322. SVP, SV40 early promoter; Ig, immunoglobulin; MIS, the Müllerian inhibiting substance gene; Term, the human gastrin gene 3' sequence; ori, origin of replication; AmpR, ampicillin resistance gene.

2. Carrier DNA: sonicated salmon sperm DNA (Boehringer Mannheim, Indianapolis, IN or Sigma). This carrier DNA must be of small size (300–2000 bp) and fairly clean. New shipments of carrier DNA are commonly

phenol-extracted, ethanol-precipitated, resuspended in H_2O, and analyzed by ethidium bromide staining of agarose gels prior to use. Additional sonication may be required to achieve the desired size range.
3. Electroporation buffer: 1X HEPES-buffered saline (HeBS): 20 mM HEPES-NaOH, pH 7.05, 137 mM NaCl, 5 mM KCl, 0.7 mM Na_2HPO_4, 6 mM dextrose. This buffer was originally used in calcium-phosphate transfections *(12)*.
4. Centrifuge tubes: 15-mL polypropylene centrifuge tubes (Corning).

3. Methods
3.1. Cell Propagation

Cells are propagated in 100-mm diameter tissue-culture dishes. To passage cells:

1. Rinse with PBS.
2. Add 2 mL of trypsin-EDTA.
3. Incubate 2–3 min at 37°C.
4. Dislodge cells by vigorously tapping the plate.
5. Add 8 mL of α+ medium to terminate the trypsin reaction.
6. Add 0.5 mL of this trypsin-treated cell suspension to a fresh 100-mm dish containing 10 mL of α+ medium (this is a 1:20 passage; CHO cells are usually passaged between 1:10 and 1:40).
7. Incubate dishes at 37°C in 5.5% CO_2 in a tissue-culture incubator.

3.2. Electroporation

1. Insert the gene to be expressed into a high-copy-number vector, such as pJOD-S or pMDR1.
2. Perform a large-scale plasmid preparation (*see* Note 4). One day prior to the electroporation, perform the following (steps 3–8).
3. Pass the DHFR-CHO cells, such that they are approx 50% confluent at the time of the electroporation (*see* Note 5).
4. Digest 200 µg of plasmid DNA with *Aat*II in order to linearize the vector by cleaving in the pBR322 sequence. Twenty units of *Aat*II are sufficient to linearize this amount of vector DNA in a few hours at 37°C (*see* Note 6).
5. Add 200 µg sonicated salmon sperm carrier DNA to the plasmid DNA in a 1.5-mL tube.
6. Add NaCl to a 0.1M final concentration and 2.5 vol of ethanol.
7. Pellet DNA at 11,000g for 10 min, aspirate the supernatant, and either dry the pellet in a Speed-Vac Concentrator or air-dry in a tissue-culture hood to prevent contamination.
8. Add 0.8 mL 1X HeBS to the DNA. Let the DNA resuspend by incubation overnight at room temperature (*see* Note 7).

9. Remove $1-2 \times 10^7$ cells from dishes by trypsin treatment (1 mL/100 mm dish). Perform this and all remaining steps at room temperature.
10. Add α+ medium (5 mL/100 mm dish), and transfer the cells to a 15-mL centrifuge tube.
11. Pellet at ~200g.
12. Completely aspirate the medium.
13. Resuspend the cell pellet by tapping the tube.
14. Add the 0.8 mL HeBS/DNA solution to the cells, and mix by pipeting.
15. Transfer cells to an electroporation cuvet having a 0.4-cm electrode gap.
16. Electroporate with one pulse at 300 V and 960 µFD, which produces a pulse of 10.0–11.5 ms duration (*see* Note 8).
17. Let the cells remain in the cuvet for 8–10 min.
18. Transfer the cells to two 100-mm dishes each having 10 mL α+ medium.
19. After the cells have firmly attached to these dishes (a few hours or the day after the electroporation), replace the medium with fresh α+ medium in order to remove dead cell debris and DNA (*see* Note 9).

3.3. Selection

1. Two days after the electroporation, remove cells from dishes by trypsin treatment and count cells.
2. Seed six 100-mm dishes each with 5×10^5 cells in α– medium (*see* Note 10).
3. Harvest cells by treatment with trypsin and count cells 4–5 d after α– medium selection is initiated.
4. Seed cells in multiple 100-mm dishes at 5×10^5 cells/plate in α– medium plus 0.5 µ*M* MTX. To maximize the chance for success, cells can be seeded at various MTX concentrations, such as 0.2, 0.5, and 0.8 µ*M*.
5. Feed cells with fresh medium every 3–5 d.
6. Pick MTX-resistant colonies 12–14 d after the initiation of MTX selection (*see* Note 11). Place clones into 48-well plates in the same medium in which they were selected.
7. Once dense, transfer the clones to 12-well plates.
8. When confluent, transfer to 100-mm dishes (*see* Note 12).

3.4. Protein Production Levels

Protein production levels are determined by a quantitative assay that measures specific proteolytic activity in the case of tPA *(7)* or by use of ELISA assays if a specific antibody is available. Protein secretion can be assayed at the 48-well stage as soon as the cells are dense. This allows an early elimination of clones that do not express well. For more precise quantitation, 100-mm plates are fed and 24 h later the medium is harvested for protein assay and cell number determination.

Table 1
Protein Secretion Levels of CHO Clones Generated
by High-Copy-Number Electroporation

Clones	mg/L/d in monolayer[a]	mg/L/d in suspension
tPA	24	ND[c]
tPA (amplified)[b]	45	ND
sCD4	19	ND
sCD4 (amplified)[b]	35	ND
Secreted protein #2	17	150
Secreted protein #3	18	80

[a] In monolayer cultures, approx 1×10^7 cells were present at the end of a 24-h period in a 100-mm tissue-culture dish in 10 mL of medium. Thus, the protein secretion levels in pg/cell/d was approx equal to the amount of mg/L/d.
[b] In the amplified clones, one round of amplification was performed by selection in sixfold (tPA) and eightfold (sCD4) higher MTX than the initial high-copy-number selection.
[c] Not determined.

For several genes, expression in the range of 20 pg/cell/d has been achieved (see Table 1 and Note 13). This is ≈20 mg/L/d when cells are grown in 100-mm dishes in 10 mL medium (~1×10^7 cells at confluence). When grown in suspension, CHO cells can produce three- to eightfold higher amounts of protein per volume of medium (Table 1 and Note 14). Most CHO clones adapt well to suspension. This can be done by seeding high numbers of cells in spinner flasks. If further information is required, refer to Section 4. for descriptions of selection in lower MTX (Note 15), gene copy numbers (Note 16), production of antibodies (Note 17), and the use of this protocol in cells other than DHFR–CHO (Note 18).

3.5. Amplification

If the expression level following the high-copy-number electroporation is not sufficient, one round of gene amplification can be performed. In order to determine the MTX concentrations to be used, MTX resistant clones should be tested in an MTX kill curve.

1. Seed 5×10^4 cells into multiple wells of six-well plates at various MTX concentrations. For a clone selected at 0.5 μM MTX, the kill curve could be 0.5–3.0 μM, increasing in small increments.

2. Five days of growth in this assay is sufficient to determine the highest MTX concentration at which a clone can survive without visible deleterious effects. Parental clones should be maintained at this MTX level (see Note 12).
3. For amplification selection, seed 1×10^4 and 1×10^5 cells/dish in multiple 100-mm dishes in α– medium at an MTX concentration that is threefold higher than the greatest MTX level that the parent can survive (additional higher MTX levels can also be employed, since this improves the chances of success with some clones; see Note 19).
4. Pick, expand, and assay clones as described previously (see Note 20).

4. Notes

1. It is critical that the serum in the selective medium be dialyzed since nondialyzed serum contains significant levels of nucleosides that will allow survival in the absence of DHFR activity.
2. Since transcriptional interference is sometimes observed with multiple genes expressed on the same plasmid (possibly because of transcripts extending beyond the polyA site and into the adjacent gene), the two genes are separated by approx 3000 bp, which includes the Müllerian inhibiting substance (MIS) 3'-flanking sequence, and the human gastrin gene 3' sequence (Term), which was reported to cause termination of RNA polymerase II migration (8).
3. pMDR1 is equivalent to pJOD-S, with the exception that the NotI site was removed from the MIS DNA and placed in a position immediately adjacent to SalI, so that a gene can be inserted into the SalI or NotI sites. Since NotI is a rare cutting enzyme, NotI linkers (New England Biolabs) can be added to most genes followed by digestion with NotI and insertion into the NotI site of pMDR1.
4. We have prepared DNA by both CsCl banding and with a Qiagen™ column (Qiagen) with no apparent differences in the results.
5. Actively growing cultures are preferable. One confluent 100-mm dish can be passed at 1:4 into two to four 100-mm dishes 1 d prior to the electroporation. This should provide $1-2 \times 10^7$ cells at the time of electroporation (the precise number of cells used is not critical).
6. It is critical that the input plasmid DNA be linear. Although the frequency of low-copy-number integrants in CHO cells is approximately equivalent between linear and supercoiled DNA, high-copy-number integration appears to require linearization of the input plasmid. Cleavage of DNA with restriction enzymes leaving 5' or 3' overhangs results in an approx 25-fold higher frequency of clones arising in 0.5 μM MTX than DNA cleaved with a restric-

tion enzyme leaving blunt ends *(7)*. Southern blot analysis indicates that in many cases the restriction site originally used to linearize the DNA is still intact. This suggests that plasmid copies may ligate in long tandem repeats via cohesive ends prior to integration *(7)*. If the inserted gene has an *Aat*II site, other restriction enzymes that cut once in the pBR322 backbone can be used. These include *Eco*RI and *Pvu*I (New England Biolabs).

7. This large amount of DNA should not be resuspended by pipeting or vortexing. It is better to let the DNA resuspend slowly. If the DNA is prepared on the day of the electroporation, rather than the day before, the DNA can be resuspended in a few hours at 37°C.

8. For different cell lines and/or electroporation devices, the optimal electroporation conditions must be determined. This can be done by transfection of a reporter plasmid that allows quantitation of protein production, such as CAT, or a colorimetric reporter, such as β-galactosidase, which will determine the percentage of cells in which the plasmid is expressed. The cell survival should be roughly 30%. The total amount of DNA in these test electroporations should be kept at 400 µg as the survival rate increases and the transfection efficiency decreases with lower DNA concentrations.

9. In an optimal electroporation, roughly 30% of cells will survive. If cell survival is >50%, a higher voltage should be used.

10. The α– medium selection allows the survival of cells that express at least low levels of DHFR. Four to 5 d after seeding in α– medium, healthy DHFR$^+$ CHO transfectants should be growing rapidly on a background of large irregular growth-arrested cells. In this protocol, 10–30% of cells should be DHFR$^+$. The initial α– selection is required, although the reason remains unclear. Clones should not be isolated after this brief α– selection, but clones must be picked after the MTX selection in order to achieve high-level protein production.

11. Although it is common to use cloning cylinders to pick colonies, a simpler procedure is to aspirate the medium, pipet a drop of medium onto the colony, and then lift the colony by scraping and pipeting with a pipetman.

12. MTX resistant clones produced by high-copy-number electroporation are often resistant to higher levels of MTX than the concentration in which they were selected. If they are propagated in lower concentrations of MTX than the highest concentration that they can survive, there is a possibility that they will lose DNA copies and decrease the protein production level over time. Therefore, it is advisable to perform an MTX kill curve as described in Section 3.4. and to maintain the cells in the highest MTX concentration in which they can survive. All clones generated to date have maintained a constant DNA copy number and protein produc-

tion level if maintained in an optimal MTX concentration. In the absence of selection, copy numbers and expression levels will decrease in many clones, although some have been shown to be stable when propagated in the absence of selection for up to 185 generations (7).

13. It is desirable to pick a large number of MTX-resistant colonies (≥20) to increase the odds of isolating a very high expression clone. In the case of tPA, the plasmid pJOD-tPA (7) was transfected by the above procedure, with MTX selection performed at 0.5 µM MTX. Twelve clones were isolated. The best clone secreted 21 pg/cell/d. Seven other clones produced between 3 and 13 pg/cell/d. In a separate electroporation, the plasmid pJODΔe-tPA (7; see Note 15) was used, and 24 clones were selected at 25 nM MTX. The best clone secreted 24 pg/cell/d whereas two others produced 16 and 19 pg/cell/d. Although this procedure has been successfully employed to express several proteins, the expression levels may differ significantly from protein to protein. Some of the factors involved may be posttranslational modifications, the efficiency of protein transport, and possible toxicity of the expressed protein. Expression levels of 20 mg/L/d have been routinely achieved, although a few proteins have been produced only at the 2–5 mg/L/d level.

14. CHO cells are a hardy cell line that adapts well to scale-up. They can be grown in low-serum and serum-free medium. In the latter case, careful media formulation is required. CHO cells quickly adapt to growth in suspension. This can be done by seeding a high number of cells in a spinner flask. Although there is variability beween different CHO clones, a three- to eightfold increase in protein production (mg/mL) is usually observed. This is owing primarily to an increase in cell density when cells are grown in suspension.

15. The plasmid pJODΔe (or pMDR902 if a *Not*I cloning site is desired) expresses lower levels of DHFR, so that selections an be performed in lower MTX concentrations. The original selections should be in 25 and 50 nM MTX. Other than decreasing the amount of MTX used, this vector has no significant advantages over pJOD-S and pMDR901.

16. The gene copy number achieved will vary. Since the position of chromosomal integration has a dramatic effect on transcription, there is no direct correlation between gene copy number and expression. High-level expression clones generated with this procedure may have anywhere from 50–1000 plasmid copies.

17. Antibodies have been successfully expressed using this system. In this case, the heavy and light chains can be inserted into the same plasmid with each gene driven by a separate adenovirus major late promoter, or the chains can be expressed by electroporation of equal amounts of two dis-

tinct pJOD-S or pMDR901 plasmids, with each plasmid expressing one of the chains. However, more efficient production of antibodies can be achieved by use of myeloma cell lines. One procedure uses glutamine synthetase selection in NS0 cells (13).

18. Unlike stepwise gene amplification, this procedure does not require the use of DHFR- cell lines (the procedure works well with the DHFR+ CHO parental line). This is because the MTX level of the initial selection is high enough that clones are unlikely to arise by amplification of an endogenous DHFR gene. The primary 5-d α– medium selection without cloning can be replaced with another dominant selectable marker, such as the *neo* gene (G418-resistance; 14). However, most cell lines do not take up and integrate as much DNA as CHO cells, and this procedure is far less efficient in these cells. In cell lines that are difficult to transfect, such as hepatocytes, high-copy-number electroporation has not been successful.

19. If a single round of amplification following the initial MTX selection is desired, at least three MTX-resistant clones should be put into amplification selection. Many clones do not appear to amplify their input DNA or do so at a very low frequency. Two to three different MTX concentrations should be used to increase the chances for success. As discussed in Section 3.4., the preferred MTX concentration for amplification selection is often roughly threefold greater than the highest MTX concentration in which the clone displays normal growth. Mechanisms other than DHFR amplification can result in resistance to high concentrations of MTX. Some cells arise that have decreased MTX uptake. However, these cells usually display abnormal growth and morphology. MTX resistant mutants of the DHFR gene may also arise at low frequency. These clones are not desirable, since resistance to >10 μM MTX can be achieved with only low copy numbers of MTX resistant DHFR.

20. The frequency of clones arising in amplification selection varies significantly, ranging from 10^{-2} to 5×10^{-5}. The tPA and sCD4 expression levels were increased twofold by amplification selection (Table 1), with a corresponding increase in gene copy numbers. However, a more dramatic increase in expression can be seen on amplification selection of clones having lower initial production levels. When clones were put into more than one round of amplification selection following a high-copy-number electroporation, no further increase in expression levels was achieved.

Acknowledgments

I thank Barbara Ehrenfels and Konrad Miatkowski for sharing information on the protein production levels of various cell lines.

References

1. Kaufman, R. J. and Sharp, P. A. (1982) Amplification and expression of sequences cotransfected with a modular dihydrofolate reductase complementary DNA gene. *J. Mol. Biol.* **159,** 601–621.
2. Haynes, J. and Weissmann, C. (1983) Constitutive long-term production of human interferons by hamster cells containing multiple copies of a cloned interferon gene. *Nucleic Acids Res.* **11,** 687–707.
3. Scahill, S. J., Devos, R., Van der Heyden, J., and Fiers, W. (1983) Expression and characterization of the product of a human immune interferon cDNA gene in Chinese hamster ovary cells. *Proc. Natl. Acad. Sci. USA* **80,** 4654–4658.
4. McCormick, F., Trahey, M., Innis, M., Dieckmann, B., and Ringold, G. (1984) Inducible expression of amplified human beta interferon genes in CHO cells. *Mol. Cell. Biol.* **4,** 166–172.
5. Kaufman, R. J., Wasley, L. C., Spiliotes, A. J., Gossels, S. D., Latt, S. A., Larsen, G. R., and Kay, R. M. (1985) Coamplification and expression of human tissue-type plasminogen activator and murine dihydrofolate reductase sequences in Chinese hamster ovary cells. *Mol. Cell. Biol.* **5,** 1750–1759.
6. Urlaub, G. and Chasin, L. A. (1980) Isolation of Chinese hamster cell mutants deficient in dihydrofolate reductase activity. *Proc. Natl. Acad. Sci. USA* **77,** 4216–4220.
7. Barsoum, J. (1990) Introduction of stable high-copy-number DNA into Chinese hamster ovary cells by electroporation. *DNA Cell Biol.* **9,** 293–300.
8. Sato, K., Ito, R., Baek, K.-H., and Agarwal, K. (1986) A specific DNA sequence controls termination of transcription in the gastrin gene. *Mol. Cell. Biol.* **6,** 1032–1043.
9. Fisher, R. A., et al. (1988) HIV infection is blocked *in vitro* by recombinant soluble CD4. *Nature* **331,** 76–78.
10. Miller, G. T., et al. (1993) Specific interaction of lymphocyte function-associated antigen 3 with CD2 can inhibit T cell responses. *J. Exp. Med.* **178,** 211–222.
11. Lobb, R., et al. (1991) Expression and functional characterization of a soluble form of vascular cell adhesion molecule 1. *Biochem. Biophys. Res. Commun.* **178,** 1498–1504.
12. Graham, F. L. and Van der Eb, A. J. (1973) A new technique for the assay of infectivity of human adenovirus 5. *Virology* **52,** 456–467.
13. Bebbington, C. R., Renner, G., Thomson, S., King, D., Abrams, D., and Yarranton, G. T. (1992) High-level expression of a recombinant antibody from myeloma cells using a glutamine synthetase gene as an amplifiable selectable marker. *Bio/Technology* **10,** 169–175.
14. Southern, P. J. and Berg, P. (1982) Transformation of mammalian cells to antibiotic resistance with a bacterial gene under control of the SV40 early region. *J. Mol. Appl. Genet.* **1,** 327–341.

CHAPTER 19

Electroporation of *Drosophila* Embryos

K. Puloma Kamdar, Thao N. Wagner, and Victoria Finnerty

1. Introduction

The ability to introduce DNA into embryos for the purpose of obtaining stable transformants is an important part of the technology that supports the exquisite genetic and molecular studies in *Drosophila*. As this ability to stably transform *Drosophila* embryos was developed, it became obvious that DNA introduced via microinjection could be transiently expressed in somatic cells of the embryo, larva, and even adult, regardless of whether it had also integrated into the germ-line nuclei (1). These observations prompted workers to examine the somatic expression of genes carried on microinjected λ phage to determine whether the clone carried a functional version of a particular gene (2). This provided an easier and much more rapid answer compared to making a stable transformant.

To increase the efficiency and ease of carrying out somatic expression studies, we have introduced DNA into *Drosophila* embryos by electroporation (3). The primary advantage of electroporation is that DNA can be introduced into a large number of embryos in a matter of minutes. In addition, if electroporation is performed using nondechorionated embryos, they require little if any special care as they develop. We have introduced both phage and plasmid DNA containing *Drosophila* genomic inserts into mutant embryos, and demonstrated wild-type or nearly wild-type levels of expression for aldehyde oxidase. In one case, the DNA carried the gene that encodes aldehyde oxidase (unpublished). In the other two cases, the genomic DNA carried one of the genes that provides a cofactor required for aldehyde oxidase activity *(3)*.

2. Materials

1. Electroporation buffer: 5 mM KCl, 0.1 mM sodium phosphate buffer, pH 7.8 *(4)*.
2. Dechorionation solution: household bleach (Clorox) diluted to 50% with distilled water.
3. Vinegar: 10% apple cider vinegar in water.
4. Yeast suspension: 10% baker's yeast dissolved in water.
5. Egg-laying chambers: These are made from disposable 50-mL polypropylene tubes with plug seal cap (catalog number 05-539-6, Fisher Scientific, Springfield, NJ). Cut the lower conical part of the tube off (at the 40-mL mark) and discard it. Tightly cover the threaded end of the tube with Nitex nylon mesh (500-µm pore size) held in place by a rubber band. Nylon cloth is available from Tetko, Inc. (Briarcliff Manor, NY).
6. Egg-collection baskets: Also made from the same 50-mL tubes. Cut a 1.5-cm^2 window in the center of the screw cap. Cover the threaded end of the tube with a 7-cm^2 piece of Nitex nylon mesh (100-µm pore size). Hold the mesh in place with the screw cap.
7. DNA: We have successfully used plasmid DNA from crude boiling preparations *(5)* and phage DNA prepared by sedimentation in CsCl *(6)*. After precipitation, the DNA is washed twice in 70% ethanol and resuspended in sterile distilled water.
8. Instant media (Carolina Biological, Burlington, NC): Autoclave and pour into Petri dishes.

3. Method

1. Use young embryos for electroporation (*see* Note 1). Have females void older embryos by allowing one or two 30-min egg-laying periods prior to egg collection (*see* Note 2).
2. Anesthetize flies with CO_2, and transfer 100–500 flies from a culture container to an egg-laying chamber (*see* Note 3). Plug the open side of the chamber with a large foam stopper pushed down far enough so that the flies have little if any room to move.
3. Moisten 5 cm^2 of blotter paper on a glass plate with 10% apple cider vinegar, and draw away excess vinegar from the blotter edges with a paper towel. Repeat with the yeast suspension. The blotter square should be moist, but not wet. Remove the blotter square to a clean, dry glass plate, and press the chamber firmly onto the paper. If necessary, place a small

glass plate or similar weighted object onto the chamber to hold it in place. After a 30-min egg-laying period, move the chamber to a freshly moistened square of blotter paper.
4. Wash the embryos from the blotter paper into an egg-collection basket with a stream of distilled water from a squeeze jar. If desired, dechorionate the embryos by immersing the basket in 50% bleach for 1 min (see Note 4). After dechorionation, thoroughly rinse the embryos in distilled water, and blot the excess water away by gently pressing a paper towel to the outside of the mesh screen. Using a 0.3-cm paintbrush, immediately place the embryos (as many as 500) into a cuvet with a 0.2-cm electrode gap containing 800 µL of the electroporation buffer and DNA at a concentration of 10 µg/mL.
5. Set up the electroporator using the capacitance extender (see Note 5). For dechorionated embryos in a 0.2-cm cuvet, pulse once with 350 V/cm at 960 µF. For nondechorionated embryos, pulse once with 1250 V/cm at 960 µF (see Note 6). After electroporation, incubate the embryos in the DNA solution in the cuvet for 2–3 min on ice.
6. Pour and rinse the dechorionated embryos out of the cuvet with a stream of sterile water (delivered from a pipet or squeeze jar) onto a Petri plate of barely moist instant media. Place the Petri plate in a humid chamber to allow eggs to develop. Nondechorionated embryos will develop on moist media without requiring a humid chamber.

4. Notes

1. Young embryos are used for electroporation. Prior to 90 min of development at 25°C, *Drosophila* embryos are an acellular syncthium of nuclei and cytoplasm, which allows introduced DNA the greatest opportunity to become incorporated into a variety of cell types when cellularization occurs.
2. The preharvest egg-laying period is important. Since adult females sometimes retain fertilized eggs, flies are encouraged to void these older embryos by allowing one or two 30-min egg-laying periods, and these embryos are discarded. Then, following the second 30-min egg-laying period, embryos are used for electroporation.
3. Laboratory strains of *D. melanogaster* deposit few if any eggs if their culture container is banged or jostled near to the time when they are expected to lay eggs. Because etherization also interferes with egg-laying, CO_2 is the best way to anesthetize these flies.
4. In general, we find higher expression levels in more tissues but with lower survival rates with dechorionated embryos *(3)*. The dechorionated embryos are fragile, subject to dessication, and necessarily require more care. With some strains, we found that we could give nondechorionated embryos two

pulses of 500 V/cm separated by a 30-s interval and obtain close to wild-type expression levels for our clones. It would be best to try non-dechorionated embryos first, and then dechorionate if sufficient expression levels are not obtained.
5. We use the Bio-Rad (Richmond, CA) Gene Pulser, model 165-2076 with capacitance extender model 165-2087. Cuvets are catalog number 165-2086, 0.2-cm electrode gap (Bio-Rad). For fly embryos, high capacitance and low voltage are required, which are the conditions used for mammalian cells. The pulse controller (intended for treating bacterial cells) is not used.
6. Voltage and genotype: We have found that various genotypes show a differential ability to survive electroporation. Some strains can be extremely sensitive, and conditions that give a significant survival rate will have to be established. We suggest conducting trials with your wild-type or control strain to find conditions that allow 70% survival with nondechorionated embryos.
7. Uses for electroporation in *Drosophila:* Most of our work has involved the rescue of aldehyde oxidase activity in mutant embryos. Aldehyde oxidase is a relatively stable enzyme in flies that facilitates efficient rescue of a null phenotype. However, we wondered whether we could observe other effects, such as interrupting a function by introducing a ribozyme by electroporation. In order to examine this as a potential use of electroporation for *Drosophila* embryos, we electroporated a plasmid-borne ribozyme directed against the *fushi tarazu* mRNA into wild-type embryos *(7)*. The ribozyme construct under a heat-shock promoter was provided by L. Pick and had been used to create stable transformants in wild-type flies *(7)*. They showed that when appropriately heat-shocked, about 40% of the embryos/larvae displayed the *fushi tarazu* phenotype to various degrees *(7)*. When electroporated into wild-type embryos, we found that about 40% showed an extreme *fushi tarazu* phenotype (unpublished). This suggests that electroporation of ribozymes could be used to create mutant phenocopies quickly. This strategy could be very useful for deciding whether a given construct should be used to make a stable transformant.

Since transient expression is controlled by adjacent promoter/enhancer elements, the nearly wild-type expression in most surviving larvae suggests that electroporation can be a powerful means for studying such regulatory regions without the necessity for germ-line transformation. Also, high levels of transient expression suggest that the rescue of developmental lethality may be possible.

DNA introduced via electroporation might be useful for creating stable transformants. In a preliminary experiment, we have obtained a few rosy[+]

transformants from a relatively small experiment, but we have not investigated transformation parameters further.

References

1. Rubin, G. M. and Spradling, A. C. (1982) Genetic transformation of *Drosophila* with transposable element vectors. *Science* **218,** 348–353.
2. Martin, P., Martin, A., Osmani, A., and Sofer, W. (1986) A transient expression assay for tissue-specific gene expression of alcohol dehydrogenase in *Drosophila. Dev. Biol.* **117,** 574–580.
3. Kamdar, P., VonAllmen, G., and Finnerty, V. (1992) Transient expression of DNA in *Drosophila* via electroporation. *Nucleic Acids Res.* **20,** 3226.
4. Spradling, A. (1986) P element-mediated transformation, in Drosophila, *A Practical Approach* (Roberts, D. B., ed.), IRL, New York, pp. 175–196.
5. Holmes, D. S. and Quigley, M. (1981) A rapid boiling method for the preparation of bacterial plasmids. *Anal. Biochem.* **114,** 193–197.
6. Yamamoto, K., Alberts, B., Benzinger, R., Lawhome, L., and Treiber, G. (1970) Rapid bacteriophage sedimentation in the presence of polyethylene glycol and its application to large scale virus purification. *Virology* **40,** 734–744.
7. Zhao, J. J. and Pick, L. (1993) Generating loss-of function phenotypes of the *fushi tarazu* gene with a targeted ribozyme in *Drosophila. Nature* **365,** 448–451.

CHAPTER 20

Transformation of Fish Cells and Embryos

Koji Inoue, Jun-ichiro Hata, and Shinya Yamashita

1. Introduction

Fish are the largest group of vertebrates. They have diverse features to adapt to a wide variety of environments, which make them excellent models for studies in various areas, including molecular and cellular biology *(1,2)*. Small freshwater species, such as medaka (*Oryzias latipes*) and zebrafish (*Brachydanio rerio*) are being studied in a growing number of laboratories because of their advantages as experimental animals *(2,3)*. The procedures for the preparation and maintenance of fish cell cultures do not differ from those of higher vertebrates. In general, fish cells grow in the same medium as mammalian cells and can be preserved by standard methods for freezing. A specific feature of fish cells is that they generally proliferate over a wide range of temperatures, and the optimal growth temperature reflects the environment that each species inhabits *(4,5)*. In addition, fish cells are thought to be naturally immortal. Even cells derived from nonneoplastic tissues can grow continuously *(6–8)*.

Transformation of fish cells and antibiotic selection of transformed cells are possible using standard procedures for mammalian cells. An interesting topic is the application of electroporation to the transformation of embryos, i.e., in the generation of transgenic fish, which provides useful models to study gene regulation and function *(9)*. Transgenic fish have been traditionally produced mainly by microinjection of DNA solu-

Table 1
Major Promoters Active in Cultured Fish Cells

Promoter	Fish cell line	Inducibility	Reference
MT-A (rainbow trout)	RTL-4	Inducible	18
MT-B (rainbow trout)	RTH	Inducible	13
MT-1 (mouse)	RTL-4	Inducible	18
MTIIA (human)	PSM, A2, EPC	Inducible	15
Hsp (*Drosophila*)	RTL-4	Inducible	14
X47 (*Xiphophorus*)	A2, EPC	Constitutive	15
Actin (carp)	EPC	Constitutive	16
RSV (chicken virus)	RTL-4	Constitutive	14
	PSM, A2, EPC		15
	EPC		16
CMV (human virus)	PSM, A2, EPC	Constitutive	15
SV40 (monkey virus)	RTH	Constitutive	13
	RTL-4		14
	EPC		16
Miw (chicken)	RTL-4	Constitutive	17

tion into fertilized eggs *(3)* as in other animals *(10)*. Microinjection is, however, a complicated operation that requires a great deal of skill. In contrast, electroporation is an easy and efficient method to produce transgenic fish strains. In this chapter, representative procedures for transforming a goldfish cell line, RBCF-1 *(6)*, and embryos of medaka by electroporation are described.

2. Materials
2.1. Plasmid Vectors

For selection of transformed cells, mammalian vectors for antibiotic resistance selections are applicable. For example, the use of pSV2neo has been described *(11)* as well as its derivative pSTneoB, constructed by Kato et al. *(12)* (*see* Note 1). For transgene expression, various regulatory elements are applicable. As shown in Table 1, many promoters and enhancers derived not only from fish, but also from other animals, were shown to be active in fish cell lines by transfection experiments using the chloramphenicol acetyltransferase gene as a reporter *(13–18)*. Some of these are constitutive elements, whereas others are inducible by heavy metals (metallothionein promoters) or heat shock (Hsp promoters).

Some of the promoters are also active in transgenic fish. For constitutive expression, long terminal repeat (LTR) sequences of Rous sarcoma virus (RSV) and human cytomegalovirus (CMV) function as powerful promoters. The pSV2 vector, which has been widely used in mammals, is also active. In addition to such well-known viral promoters, "all-fish" vectors containing the β-actin promoter *(17)* or antifreeze protein promoter *(19)* have been reported recently. To achieve inducible expression in transgenic fish, metallothionein promoters are available. Metallothionein promoters of rainbow trout have been shown to exhibit inducible expression in transgenic fish *(18)* in a dose-dependent manner *(20)*.

2.2. Solutions and Reagents for Transformation of Fish Cells and Embryos

1. Growth medium: Leibowits Medium L-15 supplemented with 15% fetal bovine serum (FBS) is used for culturing RBCF-1. Filter-sterilize through a 0.22-µm membrane.
2. Dulbecco's phosphate-buffered saline (PBS), Ca^{2+}-, Mg^{2+}-free: 8000 mg/L NaCl, 200 mg/L KCl, 1150 mg/L Na_2HPO_4 (anhydrous), 200 mg/L KH_2PO_4 (anhydrous). Sterilize by autoclaving.
3. Trypsin solution: Dissolve 50 mg trypsin (Gibco-BRL, Gaithersburg, MD) and 20 mg EDTA in 100 mL PBS. Filter-sterilize through a 0.22-µm membrane.
4. Mannitol buffer: $0.25M$ mannitol, 100 µM $CaCl_2$, 100 µM $MgCl_2$, 200 µM Tris-HCl, pH 7.5. Sterilize by autoclaving.
5. DNA stock solution: Dissolve plasmid DNA in sterilized mannitol buffer at a concentration of 1–1000 µg/µL. DNA solution can be stored for several months at 4°C if it is pure enough (*see* Note 2).
6. Selection medium: Dissolve appropriate amount of G418 (Geneticin, Gibco-BRL) in the growth medium (*see* Section 3.).
7. Crystal violet: 0.1% in PBS.

3. Methods
3.1. Transformation of Fish Cells

1. A growth curve should be determined for each cell line before transformation (*see* Note 3).
2. Determine the amount of drug required for selection before transformation, by inoculating cells into the growth medium containing various concentrations of the antibiotic and incubating until colonies grow to 1–3 mm in diameter. Colonies can be counted after staining with crystal violet. Generally, the minimal drug concentration resulting in 100% kill is the dose

required for selection. Some fish cell lines are highly resistant to G418. For RBCF-1, we use 500 µg/mL of G418.
3. Culture RBCF-1 cells to subconfluency in a 9-cm dish.
4. Remove the medium from the dish by aspiration. Add approx 0.5 mL of PBS and remove by aspiration.
5. Add 2 mL of trypsin solution, and remove immediately by aspiration. Incubate for 2 min at room temperature.
6. Add 2 mL of growth medium, and suspend cells gently. Centrifuge the cell suspension at 150g for 1 min at 4°C. Remove the supernatant.
7. Rinse the cells with mannitol buffer twice, and suspend them in fresh mannitol buffer by gentle pipeting.
8. Transfer the cell suspension to an electrode chamber (e.g., Shimadzu FTC-1; electrode gap, 2 mm). Add an appropriate volume of the DNA stock solution (final DNA concentration, 100 µg/mL; final cell density, 2×10^7 cells/mL) (see Note 4). Mix by gentle pipeting.
9. Apply one to three electric pulses of 3 kV/cm; pulse interval, 1 s; pulse length, 50 µs (see Notes 5 and 6).
10. Transfer the cell-DNA suspension into a 9-cm dish containing 10 mL of growth medium. Incubate under usual culture conditions for 1–2 d.
11. Replace the growth medium with selection medium, and incubate until colonies form (7–10 d).

3.2. Transformation of Medaka Embryos

1. Maintain a school of medaka (4 males and 8 females in a 40-L aquarium) under controlled photoperiod (14 h light and 10 h dark) at 26°C. They spawn at the beginning of light periods.
2. Separate males from females 1 d before the experiment, and keep them separate during the dark period.
3. Mate females and males at the onset of the light period. They usually spawn eggs within a few minutes after mating (see Note 7).
4. Take egg clusters from the abdomen of the females 20 min after mating. Cut off the attaching filaments with a fine scissors (see Note 8).
5. Mix eggs and 800 µL of DNA, and transfer to the electrode chamber, (e.g., Shimadzu, FTC–3, electrode distance, 2 mm).
6. Apply 5 electric pulses of 750 V/cm; pulse interval, 1 s; pulse length, 50 µs (see Notes 5 and 9).
7. Rinse treated eggs with distilled water several times, and incubate separately at 26°C in 96-well microplates filled with distilled water until they hatch (8–15 d). Hatchlings should be transferred to an aquarium (see Note 10).

4. Notes

1. The plasmid pSTneoB *(12)* is a selection vector that has a herpes simplex virus thymidine kinase promoter in addition to an SV40 promoter. Electroporation conditions for cultured cells can be determined by introducing this or other appropriate vectors and counting colonies formed in the selection medium. The ranges of pulse conditions we usually test for fish cell lines are 2.5–5 kV; pulse number, 1–5.
2. Since contamination by nicked plasmid DNA or fragments of genomic DNA reduce survival rate, plasmid DNA should be as pure as possible. Purification by ultracentrifugation in a CsCl gradient is highly recommended.
3. Since transformation efficiencies are influenced by the growth phase of cells used, it is important to determine the growth curve of each cells under the conditions used for each experiment.
4. We usually use DNA solutions at a concentration of 50–200 µg/mL. The use of higher DNA concentrations does not always increase the transformation efficiency, because DNA also affects viability of cells.
5. The instrument we use for applying pulses is the Shimadzu GTE-1, which generates square-wave pulses.
6. When transforming cultured cells, the temperatures of the electrode chamber and the cell-DNA suspension are important factors in obtaining reproducible results. Some researchers place the electrode chamber on ice.
7. For further information about maintenance of medaka, *see* ref. *21*. Since survival after pulse treatment and transformation efficiencies significantly varies among lots of eggs, it is important to control the quality of fertilized eggs to obtain high transformation efficiencies. An index of the egg quality is the survival rate of untreated eggs. Eggs that exhibit low survival rates without pulse treatment are generally more damaged by pulse treatment than those of high survival, and thus yield poor results. The parental fish that spawn eggs of high quality should be selected and maintained carefully.
8. For transformation of fish embryos, removal of the chorion may increase the transformation efficiency. The chorion can be removed manually *(22)* or by enzyme treatment *(23,24)*.
9. Electroporation of fish embryos or gametes using other instruments has also been reported, for example, Bio-Rad (Richmond, CA) Gene Pulser *(25)*, Beakon model 6000 *(26,27)*, and homemade systems *(28,29)*. Various electrode chambers are also available. Specific pulse conditions should be determined for each type of equipment.

10. Transformed cells are selected with antibiotics, but selection of transgenic embryos using antibiotics has never been successful *(30)*. Since no convenient marker is available to select transformants at present, transgenic embryos are identified by analyzing DNA of all surviving embryos. To determine optimal pulse conditions, a rough estimate is possible by determining conditions that cause death of a small proportion (about 10–20% in the case of medaka) of embryos after pulse application.

References

1. Powers, D. A. (1989) Fish as model systems. *Science* **246**, 352–358.
2. Kimmel, C. B. (1989) Genetics and early development of zebrafish. *Trends Genet.* **5**, 283–288.
3. Ozato, K., Inoue, K., and Wakamatsu, Y. (1989) Transgenic fish: biological and technical problems. *Zool. Sci.* **6**, 445–457.
4. Sato, M., Mitani, H., and Shima, A. (1990) Eurythermic growth and synthesis of heat shock proteins in primary cultured goldfish cells. *Zool. Sci.* **7**, 395–399.
5. Nicholson, B. L. (1988) Fish cell cultures: an overview, in *Invertebrate and Fish Tissue Culture* (Kuroda, K., Kurstak, E., and Maramorosch, K., eds.), Japanese Scientific Society Press/ Springer-Verlag, pp. 191–194.
6. Shima, A., Nikaido, O., Shinohara, S., and Egami, N. (1980) Continued *in vitro* growth of fibroblast-like cells (RBCF-1) derived from the caudal fin of the fish *Carassius auratus. Exp. Gerontol.* **15**, 305–314.
7. Shima, A. and Setlow, R. B. (1985) Establishment of a cell line (PF line) from a gynogenetic teleost, *Poecilia formosa* (Girard) and characterization of its repair ability of UV-induced DNA damage. *Zool. Sci.* **2**, 477–483.
8. Komura, J., Mitani, H., and Shima, A. (1988) Fish cell culture: establishment of two fibroblast-like cell lines (OL-17 and OL-32) from fin of the medaka, *Oryzias latipes. In Vitro* **24**, 294–298.
9. Inoue, K., Yamashita, S., Hata, J., Kabeno, S., Asada, S., Nagahisa, E., and Fujita, T. (1990) Electroporation as a new technique for producing transgenic fish. *Cell. Differ. Dev.* **29**, 123–128.
10. Palmiter, R. D. and Brinster, R. L. (1986) Germ-line transformation of mice. *Ann. Rev. Genet.* **20**, 465–499.
11. Southern, P. J. and Berg, P. (1982) Transformation of mammalian cells to antibiotic resistance with a bacterial gene under control of the SV40 early region promoter. *J. Mol. Appl. Genet.* **1**, 327–341.
12. Kato, K., Takahashi, Y., Hayashi, S., and Kondoh, H. (1987) Improved mammalian vectors for high expression of G418 resistance. *Cell Struct. Funct.* **12**, 575–580.
13. Zafarullah, M., Bonham, K., and Gedamu, L. (1988) Structure of the rainbow trout metallothionein B gene and characterization of its metal-responsive region. *Mol. Cell. Biol.* **8**, 4469–4476.
14. Inoue, K., Akita, N., Yamashita, S., Shiba, T., and Fujita, T. (1990) Constitutive and inducible expression of a transgene directed by heterologous promoters in a trout liver cell line. *Biochem. Biophys. Res. Commun.* **173**, 1311–1316.

15. Friedenreich, H. and Schartl, M. (1990) Transient expression directed by homologous and heterologous promoter and enhancer sequences in fish cells. *Nucleic Acids Res.* **18,** 3299–3305.
16. Liu, Z., Moav, B., Faras, A. J., Guise, K. S., Kapuscinski, A. R., and Hackett, P. B. (1990) Development of expression vectors for transgenic fish. *Bio/technology* **8,** 1268–1272.
17. Inoue, K., Yamashita, S., Akita, N., Mitsuboshi, T., Nagahisa, E., Shiba, T., and Fujita, T. (1991) Histochemical detection of foreign gene expression in rainbow trout. *Nippon Suisan Gakkaishi* **57,** 1511–1517.
18. Inoue, K., Akita, N., Shiba, T., Satake, M., and Yamashita, S. (1992) Metal-inducible activities of metallothionein promoters in fish cells and fry. *Biochem. Biophys. Res. Commun.* **185,** 1108–1114.
19. Du, S. J., Gong, Z., Flether, G. L., Shears, M. A., King, M. J., Idler, D. R., and Hew, C. L. (1992) Growth enhancement in transgenic Atlantic salmon by the use of an "all fish" chimeric growth hormone gene construct. *Bio/technology* **10,** 176–181.
20. Kinoshita, M., Toyohara, H., Sakaguchi, M., Kioka, N., Komano, T., Inoue, K., Yamashita, S., Satake, M., Wakamatsu, Y., and Ozato, K. (1994) Zinc-induced activation of rainbow trout metallothionein-A promoter in transgenic medaka. *Fisheries Science* **60,** 307–309.
21. Ozato, K., Inoue, K., and Wakamatsu, Y. (1992) Gene transfer and expression in medaka embryos, in *Transgenic Fish* (Hew, C. L. and Fletcher, G. L., eds.), World Scientific Publishing, Singapore, pp. 27–43.
22. Iwamatsu, T. (1983) A new technique for dechorionization and observations on the development of the naked egg in *Oryzias latipes*. *J. Exp. Zool.* **228,** 83–89.
23. Sakai, Y. (1961) Method for removal of chorion and fertilization of the naked egg in *Oryzias latipes*. *Embryologia* **5,** 357–368.
24. Stuart, G. W., McMurray, J. V., and Westerfield, M. (1988) Replication, integration and stable germ-line transmission of foreign sequences injected into early zebrafish embryos. *Development* **103,** 403–412.
25. Buono, R. J. and Linser, P. J. (1992) Transient expression of RSVCAT in transgenic zebrafish made by electroporation. *Mar. Mol. Biol. Biotechnol.* **1,** 271–275.
26. Powers, D. A., Hereford, L., Cole, T., Creech, K., Chen, T. T., Lin, C. M., Kight, K., and Dunham, R. (1992) Electroporation: a method for transferring genes into the gametes of zebrafish (*Brachydanio rerio*), channel catfish (*Ictalurus punctatus*), and common carp (*Cyprinus carpio*). *Mar. Mol. Biol. Biotechnol.* **1,** 301–308.
27. Zhao, X., Zhang, P. J., and Wong, T. K. (1993) Application of beakonization: a new approach to produce transgenic fish. *Mol. Mar. Biol. Biotechnol.* **2,** 63–69.
28. Müller, F., Ivics, Z., Erdélyi, F., Papp, T., Varadi, L., Horváth, L., Maclean, N., and Orbán, L. (1992) Introducing foreign genes into fish eggs with electroporated sperm as a carrier. *Mol. Mar. Biol. Biotechnol.* **1,** 276–281.
29. Xie, Y., Liu, D., Zou, J., Li, G., and Zhu, Z. (1993) Gene transfer via electroporation in fish. *Aquaculture* **111,** 207–213.
30. Yoon, S. J., Hellerman, E. M., Gross, M. L., Liu, Z., Schneider, J. F., Faras, A. J., Hackett, P. B., Kapuscinski, A. R., and Guise, K. S. (1990) Transfer of the gene for neomycin resistance into goldfish, *Carassius auratus*. *Aquaculture* **85,** 21–33.

CHAPTER 21

Electroporation of Cardiac Cells

Leslie Tung

1. Introduction

High-intensity, pulsed electrical shock is deliberately applied to the heart in the clinical setting for the treatment of cardiac arrhythmia, e.g., during defibrillation. However, overdoses of electrical shock can result in toxic instead of therapeutic effects *(1)*. Successful defibrillation is thought to occur when the bulk of the heart is subjected to a minimum level of potential gradient *(2)*. Unfortunately, it is difficult to achieve a uniform intensity throughout the heart *(3)* or even underneath the shock electrode *(4)*. In practice, the intensity will be much higher at the electrodes, by perhaps as much as 25 times the minimum level of gradient *(5,6)*. At high levels of shock, a multitude of pathological effects can arise in these regions of high potential gradients. For example, morphological and ultrastructural changes in tissue are well documented *(7)*. Another major consequence of the resulting myocardial injury in the heart is production of arrhythmias *(8,9)*, which may result in unsuccessful defibrillation. It is well known that in defibrillation success curves, there is an optimal level of shock above which the success for defibrillation decreases *(10)*. Undesirable side effects include conduction block *(11)*, loss of pacemaker activity *(9)*, decreased level of excitability *(12)*, idioventricular beats *(13)*, and runs of ventricular tachyarrhythmia *(6)*.

There is substantial experimental and theoretical evidence to support the idea that electroporation may be the primary mechanism underlying the toxic effects of high-level shocks *(14)*. Electroporation can occur in a wide variety of cell types when the transmembrane potential exceeds 0.2–1 V, and these levels of potential might be expected to occur in myocardial cells near the shock electrodes *(8,15)*, where maximum gradients can reach levels exceeding well over 100 V/cm *(6,11)*. Direct evidence for electroporation comes from tracer experiments using fluorescent FITC-labeled dextran *(16)*. In aggregates of chick cells, the formation of pores depends on shock intensity. Pores of 45–60 Å in diameter result from the application of a series of six, 10-ms duration, 200 V/cm pulses. More recently, it has been shown that the electrical conductance of the cardiac cell membrane undergoes a significant increase when ramp or rectangular, 5- or 10-ms pulses with amplitude in the range of 300–800 mV are applied across the membrane *(17–19)*.

With pore formation in the cell membrane, the transmembrane potential will be shunted to zero, resulting in a period of standstill *(20)*. A reduction in resting potential may also account for the changes in excitability, conduction velocity, electrocardiogram, and contractility observed experimentally in whole hearts *(14)*. If the pore lifetime is long compared with the diffusion time for a given ion or molecule, a significant loss or gain of the ion or molecule could occur in the cell. Resting potential recovers with a time-course that depends on the shock level *(20)*.

The greatest injury to the cell may be the result of an increase in calcium influx through the electrically induced pores, producing a "calcium overload" that results in contractile oscillations and arrhythmogenic currents. The rise in intracellular calcium may also account for the formation of granules in the mitochondria, loss of mitochondrial function, and hypercontractive appearance of contraction bands, activation of membrane phospholipases, and, ultimately, development of tissue necrosis *(14)*. Uncoupling of cardiac cells can occur, reducing conduction velocity or even preventing the propagation of electrical activity. With permeabilization of the cell membrane, osmotically driven water flow can occur, producing intracellular edema and swelling of intracellular organelles. Large pores can lead to the loss of essential enzymes, electrolytes, and metabolites to the extracellular medium.

Electroporation-mediated injury to the myocardium is determined by such factors as shock amplitude and duration; waveform shape; size and placement of the electrodes; number of shocks applied; and time between shock pulses *(14)*. Recovery from electroporation must involve, first, the resealing of the field-induced membrane pores and, second, restoration of normal levels of intracellular electrolytes and metabolites. Ongoing research is directed at a basic understanding of the parameters that govern injury at the tissue and cellular levels, and may lead to improvements in the therapeutic use of electrical shock in cardiac muscle. One promising direction of potential benefit is the influence of waveform shape on the recovery process of the cell membrane following electrical shock. For example, it has been suggested that asymmetrically weighted biphasic waveforms can diminish the time that pores are open because the opposite polarity of the second phase may reorient the phospholipids in the cell membrane *(9)*. Recent experiments suggest that such waveform shapes may ameliorate the severity of the shock-induced conductance change to the membrane and accelerate the postshock recovery time of the membrane back to its normal low conductance state *(19)*.

This chapter reviews methods used to characterize the electroporation event in cardiac cell membranes in the context of toxic effects associated with defibrillation. As yet, there have been no attempts to utilize electroporation in a therapeutic manner.

2. Materials

1. High-voltage, high-current amplifier: Linear amplifiers are desirable, since waveforms of arbitrary shape can be applied. Such amplifiers need to be capable of delivering at least several hundred volts at several amps of current (e.g., model PA04, Apex Microtechnology Corp, Tucson, AZ). Commercial electroporation or defibrillation devices can be used, but with some limitations in terms of waveform shape and calibration (*see* Note 1).
2. Microelectrodes (*see* Note 2 for compensation of electrical artifact): These electrodes are used to impale cardiac cells and record the transmembrane potential. Microelectrodes are formed by any of a number of commercial instruments (e.g., model P-87, Sutter Instruments Co., Novato, CA) by heating glass capillary tubing with a hot filament while tension is applied along the long axis. The microelectrode tip is typically <1 μm in diameter, and is filled by applying suction to the pipet while the tip is immersed in the filling solution. After a small amount of liquid is pulled into the pipet tip, the pipet can be back filled using a syringe and fine needle. *See* Note 3 for variations on this method.

3. Suction electrodes: These electrodes predate microelectrodes as a means by which monophasic action potentials resembling the transmembrane potential could be recorded. The electrode consists of glass or polyethylene capillary tubing, which contains a fine silver-silver chloride wire. The wire is exposed at its tip and protrudes slightly from the mouth of the tubing. When the tubing is pressed against the surface of cardiac tissue and suction is applied, a monophasic action potential can be recorded *(21)* *(see* Note 4).
4. "Optrode": Electro-optical signals resembling the transmembrane potential can be recorded with the use of fluorescent potentiometric dyes. The newer class of dyes used today are the styryl-pyridinium dyes *(22)*. Compared with older dyes, these dyes have a high signal, on the order of 10% change in fluorescence with a 100-mV change in transmembrane potential, and have lower toxicity *(see* Note 5), greater photostability, and more rapid spectral response times. Di-4-ANEPPS is currently the dye of choice and is commercially available (Molecular Probes, Eugene, OR). Its molecular formula is $C_{28}H_{36}N_2O_3S$ and has a mol wt of 480.7. It is a red dye and is not readily soluble in water (a nonpolar solvent, such as ethanol or DMSO, must be used). In solution, the peak absorption wavelength is 496 nm, and fluorescence emission is 705 nm. However, the emission wavelengths shift to lower values when the dye is bound to membrane. In practice, an argon (514 nm) or green HeNe laser (543 nm) can be used for excitation of the dye, and fluorescence collected at wavelengths >570 nm *(23)*. Compared with microelectrode recordings, the electro-optical signal has advantages and disadvantages *(see* Note 5). An "optrode" utilizing either single or multiple optical fibers can be used to deliver excitation light to a point on the tissue and to collect the fluorescence emitted by the dye *(23,24)*. In this way, it is possible to study changes in cellular transmembrane potential directly under a shock electrode *(25)*. By adding suction at the optrode tip, the fiber–tissue interface can be mechanically stabilized, thus minimizing the motion artifact (Fig. 1) *(23)*.
5. Fluorescently labeled tracers: FITC-dextran (fluorescein isothiocyanate-labeled dextran, Sigma, St. Louis, MO) can be obtained in varying molecular weights ranging 30–120 Å in diameter. By tracking the flux of tracer across the cell membrane, the presence of electrically induced pores can be confirmed *(16)*. Fluorescence can be measured by illuminating the specimen cell with 488-nm wavelength light (argon laser) and detecting at wavelengths >515 nm.
6. Platinizing solution: A solution of $0.025N$ hydrochloric acid, 3% platinum chloride, and 0.025% lead acetate (available premixed from Fisher Scientific, Pittsburgh, PA).

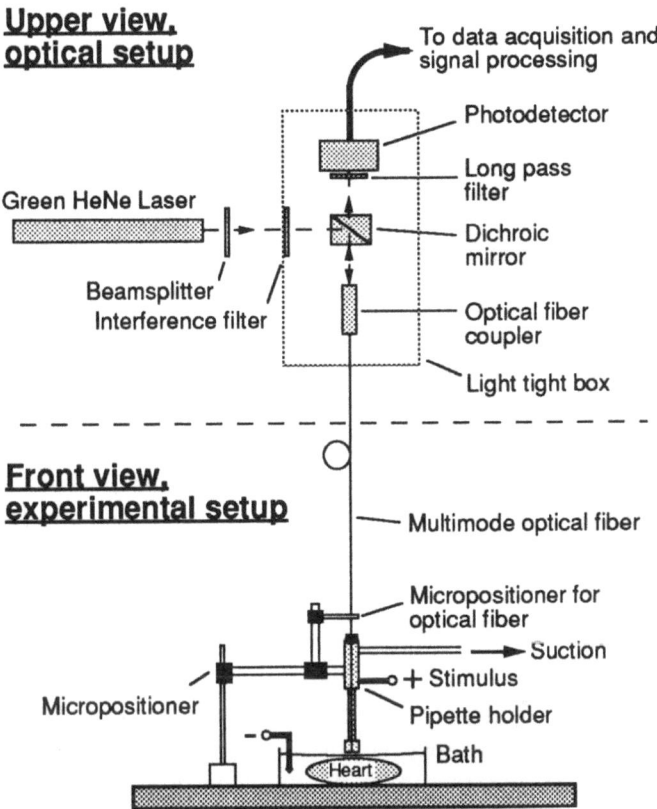

Fig. 1. "Optrode" used for electro-optical recordings of cardiac cellular transmembrane potentials. A portable fluorimeter is interfaced to a multimode optical fiber that carries both excitation to and fluorescence from the experimental preparation. The end of the optical fiber forms an "optrode" and can be positioned on the surface of cardiac muscle preparations stained with a fluorescent voltage-sensitive dye. The "optrode" also can be configured to contain a unipolar stimulus electrode, so that electro-optical potentials can be recorded directly under the stimulus electrode in response to currents with known density. Adapted from ref. *23*.

3. Methods

Electrical field intensities have been measured in the region surrounding defibrillating electrodes and shown to exceed strengths of 100 V/cm *(11)*. For isolated, in vitro studies of electroporation, graded levels of potential gradient can be applied at different structural levels of cardiac muscle, including tissue, cell, and membrane patch. Among the methods described below are three approaches used in our laboratory to investigate the electrical injury of cardiac cell membranes resulting from high electric fields: voltage-sensitive dyes, a "micropaddle" system, and the patch-clamp method.

3.1. Tissue Level

Although electroporation has not yet been demonstrated conclusively with defibrillation pulses applied to tissue, it is well established that ultrastructural and electrophysiological changes are correlated to regions of high current density, as described earlier. There is a reduction of resting potential, shortening of action potential duration, and reduction of action potential amplitude *(26–28)*. An increase in pacing threshold also will be observed as the tissue becomes less excitable. Preliminary data have been obtained that suggest that electroporation occurs at the anodal electrode at lower potential gradients than at the cathodal electrode *(30)*. The following procedure is based on experiments performed on isolated muscle strands *(26–29)*, in which the potential gradient applied to the muscle can be well controlled.

1. Construct a rectangular chamber with a volume of 2–4 mL. Place two parallel plate electrodes constructed from chlorided silver, stainless-steel, or platinum foil (0.001- to 0.004-in. thick) on opposite ends of the chamber, and connect them to a high-voltage, high-current amplifier (*see* Note 6). The applied electric field can be estimated by dividing the voltage across the electrodes by the distance between the electrodes (typically 1–2 cm) or measured directly by an exploring electrode *(27,28)* (*see* Note 6, Eq. [9]).
2. Mount an isolated trabeculae or dissected strand of muscle in the chamber between the two electrodes. Fill the chamber with normal physiological saline solution.
3. Using a microelectrode, suction electrode, or optrode, monitor the transmembrane potential with increasing levels of shock. At the onset of electroporation, the transmembrane potential will be shunted toward 0 mV as the cell membrane becomes short-circuited (Fig. 2). It is important to place the reference (indifferent) electrode properly if electrodes are used (*see* Note 2).

3.2. Cell Cultures

The method described here was developed by Jones and coworkers to characterize the effects of strong electric fields on cultured chick cells *(16,20)*.

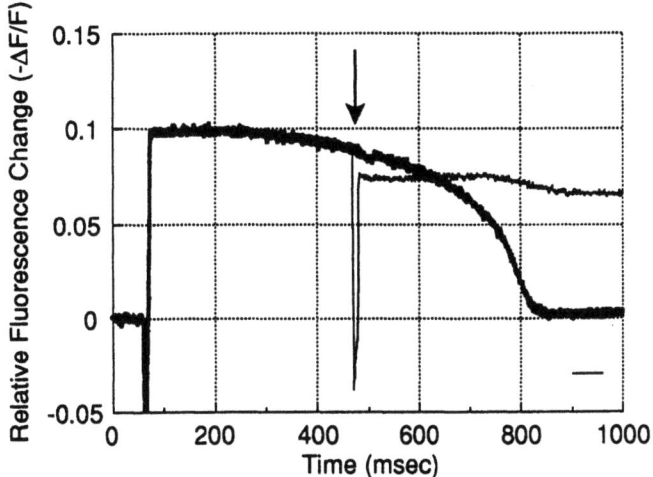

Fig. 2. Effect of high-intensity shock on tissue transmembrane potential. Bold trace: optical action potential measured under an anode during electrical stimulation at low intensity (21 mA/cm^2). The fluorescence change has been normalized to the background fluorescence level. Light trace: similar action potential, but with a second, high-intensity shock (345 mA/cm^2) applied during the plateau phase (arrow). Both traces are single sweep, with no signal averaging. Reproduced from ref. 8. Calibration of the electro-optical signal with the known electrical values of the action potential suggests that the transmembrane potential is driven toward 0 mV by the shock pulse.

1. Place a Plexiglas™ insert containing a 1-in. square hole into a 60-mm Falcon culture dish, on which cells can be cultured.
2. Mount platinum-platinum black electrodes on opposite sides of the chamber, and connect them to a stimulator capable of voltages up to 500 V. The electric field in the chamber can be calculated as the voltage across the electrodes divided by the distance between them (*see* Note 6, Eq. [9]).
3. Fill the chamber with culture medium, and (optionally) cover with a thin layer of Klearol (white mineral oil) (Witco Chemical Co., Greenwich, CT) to prevent evaporation.
4. Place a 1-Ω resistor in series with the electrodes so that the voltage across it is a measure of the current through the electrodes. In this way, the quality of the wave shape can be assessed (*see* Note 6, Eq. [3]).
5. To measure cellular transmembrane potentials, an exploring glass micropipet *electrode mounted in a micromanipulator* can be inserted into the bath. It is critical to place the reference (indifferent) electrode properly (*see* Note 2).

6. Replace the growth medium of the cells by medium containing FITC-dextran of a given mol wt for a period of 5 min. Then apply a series of six 5-ms rectangular wave shocks, 10 s apart, with intensity ranging up to 200 V/cm.
7. Since the FITC-dextran in the bath will swamp any change in signal from within the cell, it is necessary to wash out the dextrans in the bath solution following the shock pulse (e.g., at 10 min postshock). The necessity for washout therefore places a limitation on the time resolution available with this approach. By detecting the presence or absence of fluorescence with a fluorescence microscope, it can be determined which mol-wt dextrans enter the cell. Consequently, an estimate can be obtained of the pore size at some time interval following the shock.

3.3. Single Cells: "Micropaddle" System

The method described here was developed to study the effects of electric shock on single, isolated frog and guinea pig heart cells *(31,32)*. A miniaturized, parallel plate electrode system (Fig. 3) was developed to take advantage of the small cell size, which is typically $10 \times 3 \times 300$ μm (frog) or $30 \times 10 \times 100$ μm (guinea pig). Typical dimensions for the paddles are 800–1200 μm in length, 20–40 μm in width, 100–200 μm in height, and 400–600 μm in separation. The small size of the paddles provides several significant advantages. First, the large volume of the bath surrounding the paddles serves as a heat sink and minimizes Joule heating effects. Second, the geometric leverage gained in the small size permits high-intensity electric fields to be produced with relatively small currents and voltages *(see* Note 6, Eq. [1]). Potential gradients up to 800 V/cm can be obtained with a waveform generator constructed from conventional + 15 V analog circuits *(31)*. Using this system, electroporation (as indicated by hypercontracture of the cell) has been shown to follow an inverse hyperbolic strength-duration relation similar to that for excitation *(32)*.

The micropaddle system can be constructed by the following steps.

1. Strip approx 3 mm of the insulation off the ends of 0.005-in. diameter Teflon™-coated platinum wire (A-M Systems, Everett, WA) using a razor blade. Sandwich the bare wire tips between two machined flat metal bars, and carefully compress the entire assembly in a vise so that the bare wire end is pressed into a rectangular cross-section.
2. Cement two such wires together using epoxy or dental wax so that the ends form a parallel set of electrodes approx 0.5 mm apart. The far ends of the wires can be soldered into a board mount connector (2400 series, 3M Corp., Austin, TX) so that the electrode assembly can be plugged on or off an electrode holder (Fig. 3).

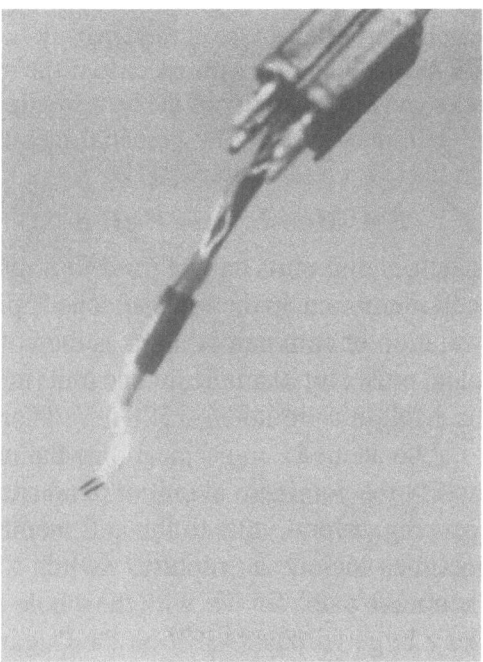

Fig. 3. Photomicrograph of micropaddles. The bare ends of Teflon™-coated platinum wires are pressed into rectangular cross-sections and fixed with epoxy to form parallel plate electrodes. The other ends of the wire are soldered to a circuit board mount connector. For scale, the square width of the connector is 5 mm.

3. To lower the electrode impedance and reduce the current density at the electrode surface (and therefore decrease the extent of electrolysis), it is helpful to coat the surface with platinum black. Place the electrodes in platinizing solution. Connect the electrodes to a resistor and battery as a cathode, with the anode consisting of a large-area platinum electrode. Apply a dc current at a density of approx 10–30 mA/cm^2 for a duration of approx 5–15 min until the coating can be visualized under a microscope.
4. To generate a rectangular pulse of potential gradient, apply a rectangular constant current pulse through the electrodes (*see* Notes 6 and 7, Eq. [3]).
5. Using this system, a current of 50 mA can produce a current density of approx 10 A/cm^2, which in physiological saline solution corresponds to a potential gradient of approx 800 V/cm. Because of the relatively large electrode impedance and polarization potentials, the micropaddle electrode voltage divided by the intraelectrode distance cannot be used as an accu-

rate measure of the electric field (*see* Note 6, Eq. [8]). In principle, the electric field can be derived from the current density (*see* Note 6, Eq. [5]), but two factors must be known with precision: the surface area of the paddle electrodes and the resistivity of the bath solution. Instead, the system should be calibrated directly by potential measurements using an exploring electrode *(32)* (*see* Note 6, Eq. [1]).

3.4. Membrane Patch

Conventional patch-clamp units can be used to impose potential gradients across the cell membrane in the "cell-attached" patch-clamp mode (Fig. 4A). One limitation of commercial units is the voltage range available for the command pulse; for example, in one unit (model 3900, Dagan Corp, Minneapolis, MN), it is limited to ±500 mV. Therefore, these voltages may or may not be adequate to permeabilize the membrane. It may be necessary to modify the commercial unit or to construct a customized unit capable of applying several volts to the cell membrane *(18)*. Commercial units sometimes include a "rupture" switch that facilitates the breakage of the membrane patch for use with the whole-cell clamp mode. This feature applies a large voltage (+13V on the Dagan 3900 unit) for a user-selected time to the membrane patch, and it is likely that electroporation of the patch is the result.

1. Ramp or rectangular voltage steps can be applied to the membrane during voltage clamp. Ramp wave shapes have no discontinuity in voltage (at the onset of the ramp) and therefore do not generate current spikes that would arise from intrinsic capacitance (located primarily at the tip of the patch electrode). Ramp wave shapes also avoid the possibility of membrane conditioning that would result from successive, subthreshold shocks. A single ramp pulse is sufficient to reveal the breakdown threshold and kinetics of membrane conductance change *(18)* (Fig. 4B1 and B2). On the other hand, rectangular pulses are advantageous in observing the latency kinetics of membrane breakdown following the onset of the pulse *(19)* (Fig. 4C1–3).
2. It is useful to superpose a low-amplitude, square-wave pulse train on the applied voltage. In this way, the membrane conductance prior to, during, and following the test pulse can be monitored (points a, c, and d, Fig. 4C3).
3. Whole-cell recordings also can be performed under voltage-clamp conditions *(33)*. With hyperpolarizations of several hundred millivolts, a substantial increase in fluctuations will be observed in the current trace *(17)*, consistent with electroporation.

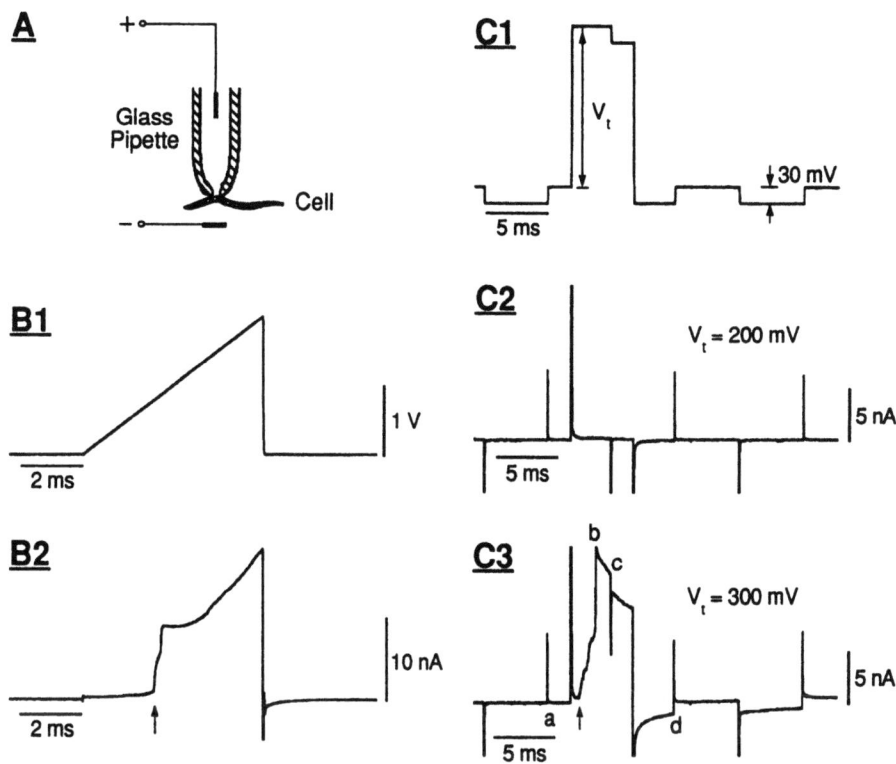

Fig. 4. Electroporation of cardiac membrane patches. (A) Configuration of the "cell-attached" patch-clamp method. (B1) Voltage ramp waveform applied to the pipet. (B2) Current waveform resulting from the ramp of (B1). Initially, the current is near zero, indicating a very low membrane conductance. After a delay of about 2 ms in this experiment, there is a sudden rise in conductance, as indicated by the arrow. (C1) Rectangular waveform applied to the pipet. A 5-ms test waveform with variable amplitude V_t is applied to the pipet. V_t is increased with successive trials. A low-amplitude (30 mV) train of square-wave pulses with a 10-ms period also is imposed on the membrane patch. In this way, membrane conductance can be monitored prior to, during, and following the test pulse. (C2) Membrane current for the voltage waveform of panel (C1) with $V_t = 200$ mV. Aside from rapid capacitive transients, the current is near zero, indicating a very low membrane conductance. (C3) Membrane current for the voltage waveform of (C1) with $V_t = 300$ mV. In this case, electroporation occurred during the test waveform as indicated by the increase in current (arrow) terminating in a very rapid jump (point b). Membrane conductance, indicated by the magnitude of the current steps for the low amplitude pulse train, is initially near zero prior to the test pulse (point a), substantially increased during the test pulse (point c), and still increased, but somewhat lower following the test pulse (point d).

4. Notes

1. Commercial electroporation units (e.g., Gene Pulser, Bio-Rad, Richmond, CA) are high-voltage, high-current devices capable of delivering up to 2500 V at tens of amps of current. When used with low-resistance media (e.g., physiological saline solution), the output waveform decays exponentially with a time constant determined by the internal (variable) capacitance of the device and the resistance of the external load. Direct monitoring of the voltage output or current will be essential for quantitative measurements of shock intensity. Similarly, external defibrillators (e.g., model M1724A, Hewlett-Packard, Cupertino, CA) are capable of delivering several thousand volts at tens of amps of current. However, there are several complications in their use as a pulse generator. First, the unit is generally calibrated in terms of energy delivered into a nominal 50-Ω load *(34)* (energy = voltage × current, integrated over the duration of the pulse), and has a range from 2–360 J. For a nominal 50-Ω load, the waveform is a slightly underdamped sinusoid (or can be a truncated exponential) with duration on the order of 10 ms. The internal capacitor and inductor of the device are fixed, so that the waveform duration is not variable for a given load. If the actual load differs from the nominal 50-Ω valve, the shape, amplitude, and duration of the waveform will be affected. An additional complication is that the shock electrodes have an inherent impedance that results in a voltage drop across the electrode during the flow of shock current. Thus, the voltage delivered by the device to the external load will be less than the voltage stored in the unit prior to the shock (on which the energy is calculated).

2. Shock artifacts occur on microelectrode recordings during the field pulse. The amplitude and direction of the artifact will depend on the location of the microelectrode relative to the reference (indifferent) electrode (typically a silver-silver chloride electrode inserted into a capillary tube filled with physiological saline-agar solution). Nulling of the shock artifact is achieved by placing the indifferent electrode on the same equipotential line as the microelectrode (for a rectangular chamber, the equipotential lines will lie parallel to the shock electrodes). Low-level stimuli can be applied to the electrodes, and fine adjustment of the reference electrode can be made to null out the shock artifact. Alternatively, a virtual ground technique can be used with a potentiometer to zero out the shock artifact *(27)*. However, even if the procedure just described is followed, perfect compensation cannot be achieved if the two electrodes have differing frequency responses. For example, glass microelectrodes have a low pass frequency characteristic, whereas metal electrodes can have a high

pass frequency characteristic *(35)*. In this case, it would be impossible to subtract out the shock artifact signal at all frequencies. Another approach is to circumvent the artifact by attenuating or disconnecting the electrode signal from the recording amplifier during the instant of the shock *(11,28)*.

3. Variations include the use of capillary tubing containing a fine glass fiber (e.g., Kwik-Fil capillary tubing, World Precision Instruments, New Haven, CT) that facilitates filling up to the tip, and a flexibly mounted electrode that reduces the tendency of the microelectrode to dislodge during tissue contraction, thus improving overall mechanical stability *(36)*.

4. The monophasic action potential (MAP) recorded by suction electrodes is always smaller in amplitude than the true transmembrane potential, and absolute values cannot be obtained. Furthermore, the MAP is relatively unstable, and decays in amplitude over a period of tens of minutes.

5. Compared with microelectrode recordings, electro-optical recordings are simpler to maintain over long periods of time, and are signal averaged over many cells. The signal is free of electrical artifacts and electrical interference. On the other hand, one of the greatest impediments toward the use of voltage-sensitive dyes in cardiac muscle is the presence of motion artifact. Strategies that have been used to solve this problem include *(22)*:
    a. Bathing the tissue in solutions containing very low concentrations of calcium ion (the activator for muscle contraction) or calcium channel blockers;
    b. Washing the tissue with uncouplers of excitation-contraction coupling;
    c. Dual-wavelength measurement;
    d. Mechanical immobilization of the tissue.

    One other drawback with the use of potentiometric dyes is that there is some toxicity associated with the dye and possibly its solvent, making it unsuitable for clinical use.

6. Measurement of applied potential gradients (electric field): It is assumed here that two plate electrodes are mounted on opposite ends of a rectangular bath, separated by a distance L and having cross-sectional area A equal to that of the bath (Fig. 5A). The electrodes are connected to an amplifier that delivers an output voltage V_o and current I. The equivalent circuit model for the system is shown in Fig. 5B. Each electrode is represented by a generalized impedance (Z) in a series with a half-cell potential (E_{hc}). Electric field (E) is defined as the potential gradient or voltage drop per unit distance. Therefore, for the one-dimensional configuration of Fig. 5A, the potential gradient is linear across the bath, and we have:

$$E = (V_1 - V_4) / L = (V_2 - V_3) / \ell \qquad (1)$$

where V_1, V_2, V_3, and V_4 are potentials at various points along the bath.

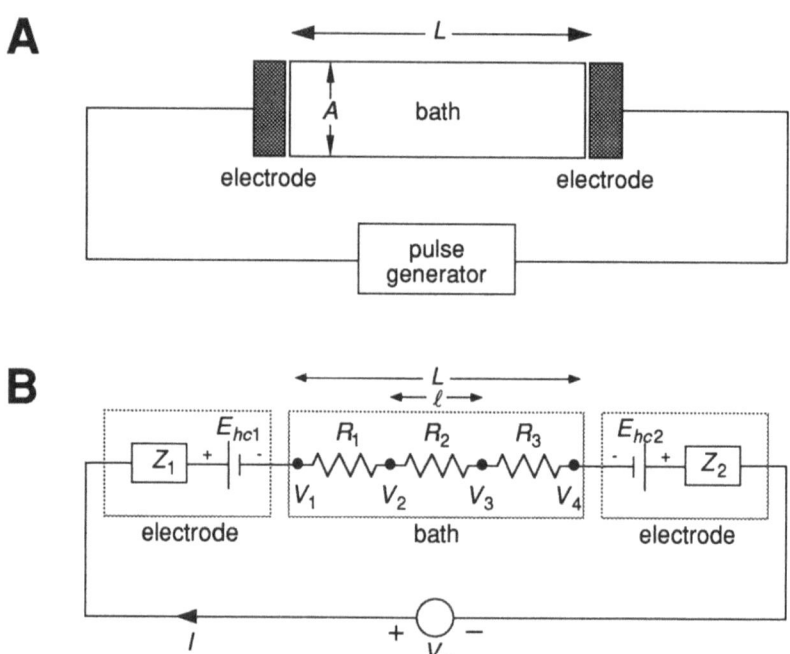

Fig. 5. Electrical model for voltage pulses applied via electrodes to a conductive bath solution. (**A**) Configuration of electrodes and bath. The electrodes are placed at opposite ends of the bath, which has length L, and connected to a pulse generator. The cross-section of the electrodes and bath are both assumed to have area A. (**B**) Equivalent electrical circuit of (A). Each electrode is represented as a complex impedance Z in series with a dc polarization (half-cell) potential E_{hc}. The bath is purely resistive at frequencies less than about 100 kHz, and is represented by a series of resistances representing the lumped resistance between various points in the bath. V_1 and V_4 are the bath potentials adjacent to the electrodes, and V_2 and V_3 potentials at two intermediate points in the bath (separated by a distance ℓ). The voltage output of the pulse generator is V_o, and current flow across the electrodes is I.

Because the field is constant throughout the bath, the potential gradient is equal to the voltage drop between any two points, divided by the distance between the points. Equation (1) shows that large fields can be obtained for a given voltage difference if the distance between the two electrodes is reduced (the basis of the "micropaddle" system).

At frequencies <100 kHz, physiological saline solution behaves as a resistive medium, so that we have:

$$V_1 - V_4 = (R_1 + R_2 + R_3)I \equiv R_b I \qquad (2)$$

where R_b is the total bath resistance. Combining (1) and (2):

$$E = (R_b / L) I \qquad (3)$$

Sometimes in the literature, current density is used interchangeably with potential gradient as we will see below. By definition, current density J is the current per cross-sectional area:

$$J \equiv (I / A) \qquad (4)$$

and combining Eqs. (3) and (4):

$$E = [(R_b A / L)] J \equiv \rho J \qquad (5)$$

We see that E is proportional to J, where $\rho = (R_b A / L)$ is defined to be the resistivity of the bath. Equation (5) in fact is the general expression for Ohm's law in a volume conductor.

Returning to Eq. (3), often it is not the quantity I that is measured, but rather, V_o. From the equivalent circuit of Fig. 5B, we have:

$$V_o = (Z_1 + R_1 + R_2 + R_3 + Z_2) I + (E_{hc1} - E_{hc2}) \qquad (6)$$

so that:

$$I = [V_o - (E_{hc1} - E_{hc2})] / [Z_1 + R_b + Z_2]. \qquad (7)$$

Finally, Eqs. (3) and (7) yield:

$$E = [R_b / (Z_1 + R_b + Z_2)] [V_o - (E_{hc1} - E_{hc2})] / L \qquad (8)$$

If $Z_1 + Z_2 \ll R_b$ and $E_{hc1} - E_{hc2} \ll V_o$, then Eq. (8) is approximately

$$E \approx (V_o / L) \qquad (9)$$

Only then can the potential gradient be approximated as the electrode voltage divided by the interelectrode distance.

7. The difference in using current-controlled waveforms rather than voltage-controlled waveforms is shown in Fig. 6, which shows the interelectrode current and voltage measured simultaneously from the micropaddles under both controlled conditions. Panel A was obtained for a rectangular voltage pulse. In this case, the current response is not proportional to the voltage trace, indicating that the total impedance of the electrical pathway (electrodes plus bath) is not resistive, owing to the impedance and polarization

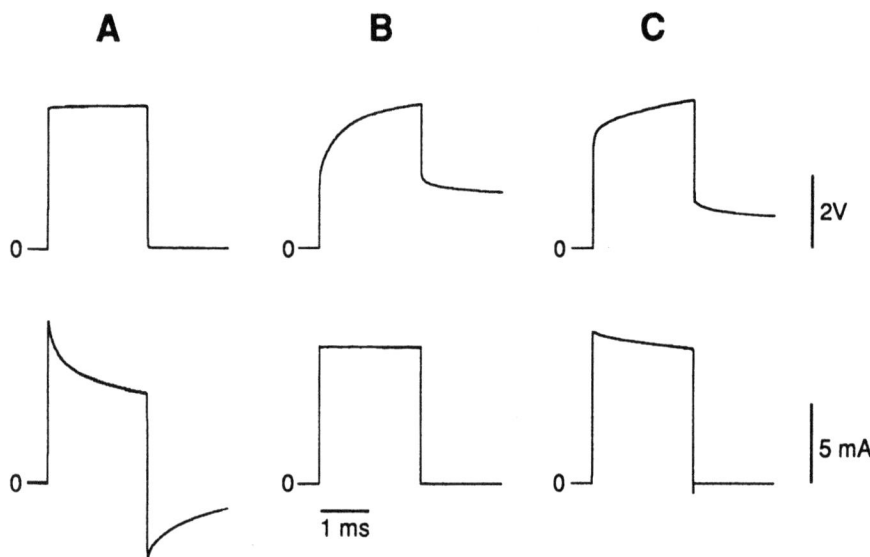

Fig. 6. Voltage and current waveforms measured from micropaddle, platinum/platinum black electrodes (as shown in Fig. 3) immersed in physiological saline solution. The upper row is voltage measured across the electrodes; the lower row is total current injected between the electrodes. The left-hand panels (**A**) were obtained under voltage-controlled conditions; the center panels (**B**) under current-controlled conditions; the right-hand panels (**C**) with a commercial stimulator (model SD9, Grass Instruments, Quincy, MA) with its output terminals isolated from ground and output voltage range at the maximum setting (10–100 V). Reproduced from ref. *(14)*.

of the electrodes (*see* Note 6, Eq. [7]). Therefore, the voltage drop across the electrodes is not a good measure of the voltage drop within the bath (although it is a good approximation when large voltages are involved, as described for the muscle strand experiments). Since the electric field is proportional to current density (*see* Note 6, Eq. [5]), the electric field is not rectangular as desired and surges at the beginning of the pulse primarily from electrode capacitance. A substantial current continues to flow after the cessation of the pulse. Panel B was obtained for current-controlled conditions. The voltage response is nonlinear and shows a very slow decay (90% decay time is 2.3 s) following the cessation of the current pulse. However, in this case, the current (and hence electric field) is rectangular. Panel C was obtained using a commercial stimulator (model SD9, Grass

Instruments, Quincy, MA), with its output terminals isolated from ground. Under these conditions with these electrodes, the stimulator is a close approximation to a constant current source. The time-dependent effects seen in all three panels can be attributed to the time-varying impedance and polarization of the platinum electrodes used, which contribute significant artifacts at the low voltages, and high current densities associated with the micropaddle system. Nonpolarizable electrodes, such as silver-silver chloride, cannot be used since free silver ion can be toxic to single cardiac cells. Therefore, current-controlled waveforms are essential for these types of studies. Current-based shock delivery has been proposed for clinical use, in part for similar reasons to minimize electrical instability and nonlinearity owing to the electrode/tissue interface, and in part to eliminate the dependence of the shock pulse on the geometry and impedance of the medium *(31)*.

Acknowledgment

This work was supported by a grant from the Maryland affiliate of the American Heart Association and NIH grant HL48266.

References

1. Ideker, R. E., Hillsley, R. E., and Wharton, J. M. (1992) Shock strength for the implantable defibrillator: can you have too much of a good thing? *Pacing Clin. Electrophysiol.* **15,** 841–844.
2. Ideker, R. E., Wolf, P. D., and Tang, A. S. L. (1994) Mechanisms of defibrillation, in *Defibrillation of the Heart* (Tacker, W. A., ed.), Mosby, Baltimore, MD, pp. 15–45.
3. Ideker, R. E., Wolf, P. D., Alferness, C., Krassowska, W., and Smith, W. M. (1991) Current concepts for selecting the location, size and shape of defibrillation electrodes. *Pacing Clin. Electrophysiol.* **14(Part I),** 227–240.
4. Kim, Y., Zieber, H. G., and Wang, F. E. (1990) Uniformity of current density under stimulating electrodes. *Crit. Rev. Biomed. Eng.* **17,** 585–619.
5. Tang, A. S. L., Wolf, P. D., Afework, Y., Smith, W. M., and Ideker, R. E. (1992) Three-dimensional potential gradient fields generated by intracardiac catheter and cutaneous patch electrodes. *Circulation* **85,** 1857–1864.
6. Wharton, J. et al. (1992) Cardiac potential and potential gradient fields generated by single, combined, and sequential shocks during ventricular fibrillation. *Circulation* **85,** 1510–1523.
7. Van Fleet, J. F. and Tacker, W. A. (1994) Cardiac damage from transchest and ICD defibrillator shocks, in *Defibrillation of the Heart* (Tacker, W. A., ed.), Mosby, Baltimore, MD, pp. 259–298.
8. Tung, L., Tovar, O., Neunlist, M., Jain, S. K., and O'Neill, R. J. (1994) Effects of strong electrical shock on cardiac muscle tissue. *Ann. NY Acad. Sci.* **720,** 160–175.

9. Jones, J. L. (1994) Waveforms for implantable cardioverter defibrillators (ICDs) and transchest defibrillation, in *Defibrillation of the Heart* (Tacker, W. A., ed.), Mosby, Baltimore, MD, pp. 46–81.
10. Schuder, J. C., Gold, J. H., Stoeckle, H., McDaniel, W. C., and Cheung, K. N. (1983) Transthoracic ventricular defibrillation in the 100 kg calf with symmetrical one-cycle bidirectional wave stimuli. *I.E.E.E. Trans. Biomed. Eng.* **30,** 415–422.
11. Yabe, S., Smith, W. M., Daubert, J. P., Wolf, P. D., Rollins, D. L., and R. E. Ideker, R. E. (1990) Conduction disturbances caused by high current density electric fields. *Circ. Res.* **66,** 1190–1203.
12. Guarnieri, T., et al. (1988) Increased pacing threshold after an automatic defibrillator shock in dogs: effects of Class I and Class II antiarrhythmic drugs. *Pacing Clin. Electrophysiol.* **11,** 1324–1330.
13. Witkowski, F. X., Penkoske, P. A., and Plonsey, R. (1990) Mechanism of cardiac defibrillation in open-chest dogs with unipolar DC-coupled simultaneous activation and shock potential recordings. *Circ.* **82,** 244–260.
14. Tung, L. (1992) Electrical injury to heart muscle cells, in *Electrical Trauma: The Pathophysiology, Manifestations, and Clinical Management* (Lee, R. C., Cravalho, E. G., and Burke, J. F., eds.), University of Cambridge Press, Cambridge, pp. 361–400.
15. Tung, L. and Borderies, J.-R. (1992) Analysis of electrical excitation of cardiac muscle cells. *Biophys. J.* **63,** 371–386.
16. Jones, J. L., Jones, R. E., and Balasky, G. (1987) Microlesion formation in myocardial cells by high-intensity electric field stimulation. *Am. J. Physiol.* **253,** H480–H486.
17. Bonvallet, R. and Christé, G. (1988) Membrane responses to large hyperpolarizations in trabecles and single cells of frog atrium. *Gen. Physiol. Biophys.* **7,** 433–477.
18. O'Neill, R. J. and Tung, L. (1991) A cell-attached patch clamp study of the electropermeabilization of amphibian cardiac cells. *Biophys. J.* **59,** 1028–1039.
19. Tovar, O. and Tung, L. (1992) Electroporation and recovery of the cardiac cell membrane with rectangular voltage pulses. *Am. J. Physiol.* **263,** H1128–H1136.
20. Jones, J., Lepeschkin, E., Jones, R. E., and Rush, S. (1978) Response of cultured myocardial cells to countershock-type electric field stimulation. *Am. J. Physiol.* **235,** H214–H222.
21. Hoffman, B. F., Cranefield, P. F., Lepeschkin, E., Surawicz, B., and Herrlich, H. C. (1959) Comparison of cardiac monophasic action potentials recorded by intracellular and suction electrodes. *Am. J. Physiol.* **111,** 177–186.
22. Salama, G. (1988) Optical measurement of transmembrane potential in heart, in *Spectroscopic Membrane Probes*, vol. III (Loew, L., ed.), CRC, Boca Raton, FL, pp. 137–199.
23. Neunlist, M., Zou, S., and Tung, L. (1992) Design and use of an "optrode" for optical recordings of cardiac action potentials. *Pflugers Arch.* **420,** 611–617.
24. Dillon, S. (1991) Optical recordings in the rabbit heart show that defibrillation strength shocks prolong the duration of depolarization and the refractory period. *Circ. Res.* **69,** 842–856.
25. Neunlist, M. and Tung, L. (1994) Optical recordings of ventricular excitability of frog heart by an extracellular stimulating point electrode. *Pac. Clin. Electrophysiol.* **17,** 1641–1654.

26. Fogelson, L. J., Tung, L., and Thakor, N. V. (1988) Electrophysiologic depression in myocardium by defibrillation-level shocks. *Proc. Annu. Int. Conf. I.E.E.E. Eng. Med. Biol. Soc.* **10,** 963–964.
27. Knisley, S. B., Smith, W. M., and Ideker, R. E. (1994) Prolongation and shortening of action potentials by electrical shocks in frog ventricular muscle. *Am. J. Physiol.* **266,** H2348–H2358.
28. Kodama, I., Shibata, N., Sakuma, I., Mitsui, K., Iida, M., Suzuki, R., Fukui, Y., Hosoda, S., and Toyama, J. (1994) After effects of high-intensity DC stimulation on the electromechanical performance of ventricular muscle. *Am. J. Physiol.* **267,** H248–H258.
29. Li, H. G., Jones, D. L., Yee, R., and Klein, G. J. (1991) Defibrillation shocks increase myocardial pacing threshold: an intracellular microelectrode study. *Am. J. Physiol.* **260,** H1973–H1979.
30. Tung, L. and Neunlist, M. (1994) Asymmetrical injury effects of high intensity anodal and cathodal shock on cardiac transmembrane potentials (abstr.). *Am. Heart J.* **128,** 634.
31. Mulligan, M. R., O'Neill, R. J., Zei, P., and Tung, L. (1988) Graded effects of pulsed electric fields on contractility of single heart cells. *Proc. Annu. Int. Conf. I.E.E.E. Eng. Med. Biol. Soc.* **10,** 902–903.
32. Tung, L., Sliz, N., and Mulligan, M. R. (1991) Influence of electrical axis of stimulation on excitation of cardiac muscle cells. *Circ. Res.* **69,** 722–730.
33. Hamill, O. P., Marty, A., Neher, E., Sakmann, B., and Sigworth, F. J. (1981) Improved patch-clamp techniques for high-resolution current recording from cells and cell-free membrane patches. *Pflugers Arch.* **391,** 85–100.
34. Lerman, B. B., Halperin, H. R., Tsitlik, J. E., Brin, K., Clark, C. W., and Deale, O. C. (1987) Relationship between canine transthoracic impedance and defibrillation threshold. Evidence for current-based defibrillation. *J. Clin. Invest.* **80,** 797–803.
35. Geddes, L. A. and Baker, L. E. (1989) *Principles of Applied Biomedical Instrumentation*, 3rd ed., Wiley, New York, pp. 315–452.
36. Fedida, D., Sethi, S., Mulder, B. J., and ter Keurs, H. E. (1990) An ultracompliant glass microelectrode for intracellular recording. *Am. J. Physiol.* **258,** C164–C170.

CHAPTER 22

Electroporation for Gene Therapy

Kathryn E. Matthews, Sukhendu B. Dev, Frances Toneguzzo, and Armand Keating

1. Introduction

Interest continues to grow rapidly in clinical gene therapy. Although many protocols approved by the US NIH Recombinant Advisory Committee involve the genetic marking of cells to determine, for example, the source of relapse after autologous bone marrow transplantation for certain malignancies (1), increasing numbers are concerned with the correction of inherited single-gene disorders.

Despite extensive clinical activity in this field, major limitations remain regarding such basic issues as efficiency of gene transfer, longterm expression of the transgene, and identification of the most appropriate target cells (2). After a decade of prominence, retrovirus-mediated transfer is undergoing re-evaluation because of a variety of drawbacks, and increased attention is currently directed toward other approaches that involve adenoviruses, adeno-associated viruses, and the herpes simplex virus, especially for in vivo transfer (3–5). In this context, it is timely to consider a physical method of gene transfer, such as electroporation, as an alternative strategy for the development of gene therapy protocols.

1.1. Why Choose Electroporation?

Of the physical techniques of gene transfer currently available, electroporation has numerous advantages. It has been extensively studied, and is simple, rapid, and relatively nontoxic to target cells (6). Moreover, instruments that generate either an exponential decay-type or square-wave pulse are readily available commercially and are easy to operate.

Other advantages include the ability to transfect populations, such as nonreplicating cells, that are resistant to other gene-transfer methods; the minimal need for manipulation of the transgene; the complete avoidance, if desired, of any viral sequences; and the lack of size constraints on the transgene *(7)*. In addition, we and others have shown that progenitor populations as well as fully differentiated cells can be stably transfected. Moreover, data from electroporation of permanent hematopoietic cell lines indicate that transgenes are integrated in single- or low-copy number *(8)*.

Despite these favorable features, important limitations of electroporation remain and include low gene-transfer efficiency compared to retrovirus-mediated transfer, the possibility of integration of the transgene in concatameric form, and lack of data regarding long-term in vivo expression of transgenes. Further studies are in progress to improve transfer frequency, reduce target cell toxicity, and explore approaches that enhance homologous recombination.

1.2. Electrotransfection of Human Bone Marrow Progenitor Cells

Bone marrow is a particularly suitable organ for gene therapy, because it is easily accessible, can be manipulated in vitro, and contains pluripotent stem cells capable of reconstituting hematopoiesis. Transfected bone marrow stem cells have the potential to remain throughout the lifetime of the individual and thus may correct a single gene defect permanently. In the development of gene therapy protocols, the transfer and expression of the exogenous gene in human hematopoietic stem cells are assessed indirectly by analyzing the cells comprising hematopoietic colonies using semisolid methylcellulose assays for hematopoietic progenitors *(9)*.

Electroporation as a method of gene delivery to human hematopoietic progenitor cells offers several advantages in addition to those described above. Plasmid DNA containing selectable marker genes, such neo^r, or genes of clinical interest, can be introduced into a wide variety of human hematopoietic cells with comparative ease. The conditions for efficient transfection of a variety of human myeloid and lymphoid cell lines are well established *(10–12)*. Toneguzzo and Keating *(13)* showed that a selectable gene (in this instance, neo^r) introduced into human marrow cells was stably expressed in the progeny of committed granulopoietic progenitor cells (CFU-GM) at a frequency of 2.7%. This expression frequency can be increased to 4.5% with the use of plasmids containing more active promoters from which the neo^r gene is transcribed (K. E. M. and A. K., unpublished data).

In order to determine the actual frequency of gene transfer into hematopoietic progenitors, individual granulopoietic and early erythroid colonies can be analyzed by polymerase chain reaction amplification of the marker gene. In contrast to the relatively low levels of transgene expression, gene-transfer frequency into CFU-GM and erythroid progenitor cells (BFU-E) was about 27% (K. E. M. and A. K., unpublished data).

1.3. Efficiency of Electroporation

As is the case with retroviral gene transfer into mammalian cells (14), transfection efficiency with electroporation can be improved by increasing the proportion of cycling target cells. Takahashi et al. (12) were able to increase the transient transfection efficiency of K562 (a permanent leukemic cell line with myeloid features) from 4 to 8% with cell-cycle synchronization. Also, by increasing the proportion of marrow progenitor cells in S-phase, the frequency of transiently transfected granulopoietic progenitors could be nearly doubled to 2% (15). We showed that incubation of bone marrow progenitor cells with the cytokine IL-3 prior to electroporation increased the proportion of progenitor cells in S-phase, and also increased the frequency of successfully transfected granulopoietic and erythroid progenitors from 27 to 40% (K. E. M. and A. K., unpublished data). The mechanism by which the proliferative status of the progenitor cell and its susceptibility to electrotransfection are related has yet to be elucidated.

2. Materials
2.1. Plasmid DNA

Prepare plasmids using two cycles of cesium chloride density gradient centrifugation. Linearize plasmid DNA using an appropriate restriction enzyme (see Note 1).

2.2. Solutions

1. HeBS buffer: 20 mM HEPES (4-2-hydroxyethyl-1-piperazineethane sulfonic acid), 137 mM NaCl, 5 mM KCl, 0.7 mM Na$_2$HPO$_4$, 6 mM dextrose.
2. Iscove's modified Dulbecco's medium (IMDM) (Life Technologies, Burlington, ON).

3. Long-term culture medium (LTCM): McCoy's 5A medium (Life Technologies) supplemented with 12.5% fetal calf serum, 12.5% horse serum, 1% each of glutamine, sodium pyruvate, sodium carbonate, and vitamins (100X stock solution obtained from Life Technologies), 0.8% essential amino acids, 0.4% nonessential amino acids (all obtained from Life Technologies), and $10^{-7}M$ hydrocortisone (Sigma, St. Louis, MO).
4. Trypsin buffer: A solution of 0.25% is made by dissolving bactotrypsin (Difco, Detroit, MI) in a buffer consisting of $0.075M$ KCl and $0.015M$ sodium citrate.
5. PBS: $Mg^{2+}Ca^{2+}$-free phosphate-buffered saline, pH 7.23.
6. Ficoll-Paque (density 1.077 g/mL, Pharmacia, Uppsala, Sweden).
7. PCR lysis buffer: 50 mM KCl, 10 mM Tris-HCl, pH 8.3, 1.5 mM MgCl$_2$, 2 µg/µL proteinase K (Sigma).

3. Methods

3.1. Preparation of Bone Marrow Mononuclear Cells

1. Dilute human bone marrow cells with an equal volume of IMDM.
2. Pass cells over a Ficoll-Paque gradient as follows: Layer 10 mL of diluted cells over 4 mL of Ficoll-Paque and centrifuge at 400g for 30 min.
3. Collect the mononuclear fraction at the interface and wash three times with a volume of 10–15 mL of 1X PBS. Approximately 20×10^6 cells are used for each electroporation experiment. Resuspend cells in PBS to a maximum volume of 0.4 mL.

3.2. Electroporation of Bone Marrow Mononuclear Cells

1. Add linearized plasmid DNA to the cells to a concentration of 100 µg/mL.
2. Incubate this mixture on ice for 10 min prior to the transfection and then transfer to an electroporation cuvet with a gap of 2 mm between the electrode plates.
3. Pulse once with an exponentially decaying pulse of 2.0 kV (peak discharge of 1.7 kV and pulse length of 0.2 ms) (*see* Note 2). Remove cells from the electroporation chamber, and incubate on ice for 10 min (*see* Note 3).
4. Dilute cells with an equal volume of IMDM supplemented with 20% fetal bovine serum, and incubate in a 37°C incubator for 1 h to recover. Viable cells are then counted using the trypan blue exclusion method *(9)*.

3.3. Determining the Frequency of Gene Transfer into Hematopoietic Progenitors

1. Plate transfected and sham electroporated (i.e., in the absence of DNA) mononuclear marrow cells in methylcellulose cultures, and place in a 37°C humidified incubator containing 5% CO_2. These culture conditions, which

have been described previously (9), allow for the detection of colonies derived from progenitors committed to granulopoietic (CFU-GM), erythroid (BFU-E), and multipotential pathways (CFU-GEMM).

2. After 14 d of growth in methylcellulose, pick individual colonies into 30 µL of PCR lysis buffer, and assay for the presence of transferred sequences using the polymerase chain reaction (16).

3.4. Preparation of Human Stromal Cells

Passaged human stromal cells are a highly proliferative population, sharing the phenotypic characteristics of the predominant cell type in the adherent layers of human long-term marrow cultures (17). Evidence from our work and that of others suggests that the stromal cells derived from the bone marrow can be removed, manipulated in vitro, and re-engrafted in a murine model (18,19). The transfected stromal cell may therefore provide an attractive means of gene delivery. Stromal cell populations can be readily transfected by electroporation. Using optimal conditions derived from transient transfection assays, Keating et al. (20) were able to obtain long-term expression of several exogenous genes, including neo^r and SV40 large T-antigen. A gene of clinical interest, the human Factor IX cDNA, has also been electrotransfected into and expressed by passaged human stromal cells (L. Fouillard and A. K., unpublished data).

1. Generate marrow stromal cells by placing the bone marrow mononuclear cells obtained from density gradient separation (see Section 3.1.) in long-term culture as previously described (17,21). Serially passage the adherent layer from these cells from 4–8 times over a 3–5-wk period.
2. In preparation for electroporation, trypsinize stromal cell layers, resuspend in 5 mL of LTCM, and then wash twice with 10 mL of PBS. Resuspend cells in PBS to a maximum final volume of 400 µL at a concentration of $1–2 \times 10^7$/mL.

3.5. Electroporation of Stromal Cells

1. Add linearized plasmid DNA to the cells at a final concentration of 100 µg/mL. Incubate cells and DNA on ice for 10 min prior to electroporation. Transfer cells to an electroporation cuvet with a 2-mm gap between electrode plates.
2. Pulse once with an exponentially decaying pulse of 1.0 kV. Following electroporation, incubate on ice for 10 min, and then add an equal volume of medium. Place cells at 37°C for a recovery period of 1 h. A cell viability of approx 30% can be expected. Restore viable stromal cells to long-term culture medium (LTCM).

4. Notes

1. We find linearized plasmid DNA to be more effective in creating long-term stably transfected cells than supercoiled plasmids *(10)*. However, when using electroporation devices that discharge square-wave pulses, others have found supercoiled plasmids to be more effective *(22)*.
2. We have optimized the voltage and capacitance settings for maximum transient transfection efficiency using the BTX 600 instrument (Biotechnologies and Experimental Research, Inc., San Diego, CA) and BTX electroporation cuvets with a gap width of 2 mm. Maximal transient activity in a variety of human hematopoietic cell lines was obtained using an exponentially decaying pulse at a generated electric field strength of approx 6 kV/cm. Electroporation chambers with different gap widths can be used provided the voltage is adjusted to render the same electric field strength.
3. In general, optimal transient and stable transfection efficiencies are achieved using the same parameters *(23)*.
4. Although electroporators that produce exponentially decaying pulses are usually preferred for the transfection of human hematopoietic cells *(11,13,23)*, Takahashi et al. *(12)* have demonstrated that, in some instances, electroporation using a square-wave pulse is more efficient.
5. Some controversy exists as to whether cells should be incubated on ice prior to and after electroporation *(24)*. Chu et al. *(25)* have found, for example, that some mammalian cell lines are more efficiently transfected at room temperature, whereas others find that myeloid leukemia cell lines are more efficiently transfected at 4°C. When using an exponentially decaying pulse for electroporation, both mouse *(26)* and human *(13)* bone marrow mononuclear cells are more efficiently transfected at 4°C. These data suggest that the optimal temperature is cell-dependent and should be empirically determined.

References

1. Brenner, M. K., Rill, D. R., Moen, R. C., Krance, R. A., Mirro, J., Jr., Anderson, W. F., and Ihle, J. N. (1993) Gene-marking to trace origin of relapse after autologous bone-marrow transplantation. *Lancet* **341**, 85,86.
2. Keating, A. and Leslie, K. (1993) Gene therapy and bone marrow transplantation, in *Clinical Bone Marrow Transplantation* (Atkinson, E., ed.), Cambridge University Press, Cambridge, UK, pp. 715–724.
3. Larrick, J. W. and Burck, K. L. (1991) *Gene Therapy.* Elsevier, New York.
4. Yang, N.-S. (1992) Gene transfer into mammalian somatic cells in vivo. *Crit. Rev. Biotechnol.* **12(4)**, 335–356.
5. Wolff, J. A., Jr. (ed.) (1994) *Gene Therapeutics.* Birkhauser, Boston.
6. Chang, D. C., Chassy, B. M., Saunders, J. A., and Sowers, A. E. (eds.) (1992) *Guide to Electroporation and Electrofusion.* Academic, San Diego.

7. Keating, A. and Toneguzzo, F. (1988) Gene transfer by electroporation, in *Experimental Hematology Today* (Baum, S. J., Dicke, K. A., Lotzova, E., and Pluznik, D. H., eds.), Springer-Verlag, New York, pp. 71–74.
8. Toneguzzo, F., Keating, A., Glynn, S., and McDonald, K. (1988) Electric field-mediated gene transfer: characterization of DNA transfer and patterns of integration in lymphoid cells. *Nucleic Acids Res.* **16**, 5515–5533.
9. Keating, A. and Toor, P. (1990) In vitro clonal culture of human hematopoietic progenitor cells, in *Methods in Molecular Biology, vol. 5. Animal Cell Culture* (Pollard, J. W. and Walker, J. M., eds.), Humana, Clifton, NJ, pp. 339–346.
10. Toneguzzo, F., Hayday, A., and Keating, A. (1986) Electric-field mediated gene transfer: transient and stable gene expression human and mouse lymphoid cells. *Mol. Cell. Biol.* **6**, 703–706.
11. Pahl, H. L., Burn, T. C., and Tenen, D. G. (1991) Optimization of transient transfection into human myeloid cell lines using a luciferase reporter gene. *Exp. Hematol.* **19**, 1038–1041.
12. Takahashi, M., Furukawa, T., Nikkuni, K., Aoki, A., Nomoto, N., Koike, T., Moriyama, Y., Shinada, S., and Shibata, A. (1991) Efficient introduction of a gene into hematopoietic cells in S-phase by electroporation. *Exp. Hematol.* **19**, 343–346.
13. Toneguzzo, F. and Keating, A. (1986) Stable expression of selectable genes introduced into human hematopoietic stem cells by electric field meditated DNA transfer. *Proc. Natl. Acad. Sci. USA* **83**, 3496–3499.
14. Miller, D. G., Adam, M. A., and Miller, A. D. (1990) Gene transfer by retrovirus vectors occurs only in cells that are actively replicating at the time of infection. *Mol. Cell. Biol.* **10**, 4239–4242.
15. Takahashi, M., Furukawa, T., Tanaka, I., Nikkuni, K., Aoki, A., Kishi, K., Koike, T., Moriyama, Y., and Shibata, A. (1992) Gene introduction into granulocyte-macrophage progenitor cells by electroporation: the relationship between introduction efficiency and the proportion of cells in S-phase. *Leuk. Res.* **16**, 761–767.
16. Mullis, K. B. and Faloona, F. A. (1987) Specific synthesis of DNA in vitro via a polymerase-catalyzed chain reaction. *Methods Enzymol.* **155**, 335–351.
17. Keating, A., Just-Mitchell, K., Toor, P., Klein, M., and Sodek, J. (1986) Passaged human marrow stromal cells: a unique cell population. *Exp. Hematol.* **14**, 426.
18. Wu, D.-D. and Keating, A. (1991) Engraftment of donor-derived bone marrow stromal cells. *Exp. Hematol.* **19(6)**, 485.
19. Nolta, J. A., Hanley, M. B., and Kohn, D. B. (1994) Sustained human haematopoiesis in immunodeficient mice by cotransplantation of marrow stroma expressing human interleukin-3: analysis of gene transduction of long-lived progenitors. *Blood* **83**, 3041–3051.
20. Keating, A., Horsfall, W., Hawley, R., and Toneguzzo, F. (1990) Effect of different promoters on the expression of genes introduced into haematopoietic and marrow stromal cells by electroporation. *Exp. Hematol.* **18**, 99–102.
21. Keating, A., Powell, J., Takahashi, M., and Singer, J. W. (1984) The generation of human long-term cultures from marrow depleted of Ia (HLA-DR) positive cells. *Blood* **64**, 1159–1162.

22. Xie, T.-D. and Tsong, T.-Y. (1993) Study of mechanisms of electric-field-induced DNA transfection. Effects of DNA topology on surface binding, cell uptake, expression and integration into host chromosomes of DNA in the mammalian cell. *Biophys. J.* **65,** 1684–1689.
23. Potter, H. (1988) Electroporation in biology: methods, application and instrumentation. *Anal. Biochem.* **174,** 361–373.
24. Rols, M.-P., Delteil, C., Serin, G., and Teissie, J. (1993) Temperature effects of electrotransfection of mammalian cells. *Nucleic Acids Res.* **22,** 540.
25. Chu, G., Hayakawa, H., and Berg, P. (1987) Electroporation for the efficient transfection of mammalian cells with DNA. *Nucleic Acids Res.* **15,** 1311–1324.
26. Narayanan, R., Tare, N. S., Benjamin, W. R., and Grubler, U. (1989) A sensitive technique to monitor gene transfer and expression in bone marrow stem cells. *Exp. Hematol.* **17,** 832–835.

PART III

ELECTROFUSION PROTOCOLS

CHAPTER 23

Electrofusion of Mammalian Cells

Kenneth L. White

1. Introduction

Electrofusion has developed into an extremely efficient method for the fusion of mammalian cells. Cell–cell fusion has become an important tool for the study of cell biology, molecular biology, and bioproduction of important biological substances. Although other fusion methodologies have been successfully used, most have inherent disadvantages. Chemical fusogens, such as polyethylene glycol (PEG), are relatively efficient inducers of fusion. However, viability of the resulting cells is reduced owing to the toxicity of the fusogenic chemical *(1,2)*. Alternatively, use of attenuated fusogenic virus (i.e., Sendai) has also provided efficient fusion rates. However, batch-to-batch variability in fusion rates is a significant disadvantage of this technique. Electrofusion provides both extremely efficient fusion rates using a wide variety of mammalian cell types and high postfusion survival. Although the parameters for each new cell type used must be empirically determined, electrofusion can be applied to a wide range of cells because of dielectric properties that can be induced in most cells. Indeed, there are no known electrofusion resistant cells *(3,4)*.

The efficiency of electrofusion is higher than any of the other approaches currently used. If necessary, the entire fusion process can be observed by use of a microscope. This provides the opportunity for immediate modification of parameters used if problems are observed (i.e., alignment, lysis, or fusion rates). The fusion parameters can be well defined, controlled, and highly repeatable using the electronics associated with the electrofusion apparatus. Not only is there a wide variety of cell types that can be electrofused, but electrofusion also enables the fusion of artificial vesicles (i.e., liposomes) to each other or to cells. The time required for cells to be exposed to fusion conditions is greatly reduced using electrofusion as compared to other fusion techniques. Most electrofusion pulse durations are <1 ms. Therefore, cells are only exposed to fusion conditions for a limited time and immediately returned to the controlled environment of the incubator. Cells induced to fuse by chemical or biological (virus) means often require exposure to fusion conditions for 30–90 min.

Various stages of embryonic blastomeres from porcine *(5)*, bovine *(6)*, leporine *(7)*, ovine *(8)*, and murine *(9)* embryos have been successfully fused to enucleated oocytes. These types of experiments provide important information about developmental competence and differentiation events, as well as the importance of nuclear–cytoplasmic interaction relative to subsequent development. In addition, electrofusion has allowed efficient production of human and mouse monoclonal antibody (MAb)-producing hybridoma lines *(10,11)*.

During the electrofusion process, cells are placed into a fusion chamber and initially exposed to an ac alignment pulse (3–250 V) for 5–30 s, which induces a dielectrophoresis of cells *(6,12,13)*; this is followed by a high-voltage (1.21–12 kV/cm) dc fusion pulse. Cells are placed into a low conducting fusion medium to facilitate the dielectric potential within the cells. Cells are then induced to align based on their attraction to the appropriate dipole of the electrodes *(14)*. In addition, the ac alignment pulse facilitates increased association between adjacent cells. It is the enhanced interaction between adjacent cell membranes that is important to successful fusion following delivery of the dc fusion pulse *(15)*.

2. Materials
2.1. Apparatus

To maximize fusion efficiency, the minimum specifications for the apparatus are that they have both low-voltage ac alignment capabilities and the ability to generate a high-voltage dc fusion pulse. The ac alignment is important for facilitating cell–cell contact, which is paramount for success of fusion. In addition, the duration of both the ac and dc pulses must be controllable, and the number (frequency) of the dc pulse must be adjustable.

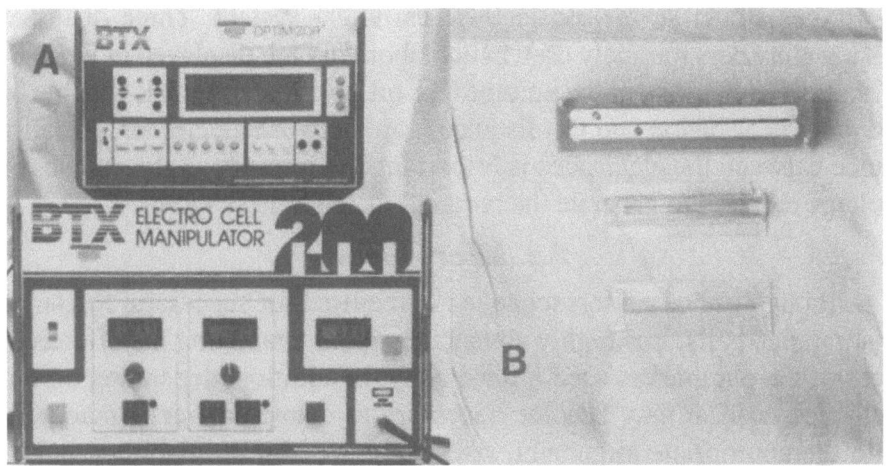

Fig. 1. (**A**) The BTX, Inc. (San Diego, CA), BTX 200 Cell Manipulator™ (bottom) used in conjunction with the BTX Optimizer™ (top) cell-fusion apparatus. The Cell Manipulator has digital controls for the ac pulse amplitude (V) and duration (S), as well as the dc pulse amplitude (V), duration (µs), and frequency (number of pulses) for precise setting of these parameters. The BTX Optimizer™ provides the ability to confirm the delivered chamber current across electrodes. This capability is critical to ensure the accuracy and repeatability of fusion conditions. (**B**) Examples of three different commonly used fusion chambers in mammalian cell fusion. Top—This chamber has a 1.0-mm gap between electrodes and a volume of 500 µL, often used for fusion of large numbers of cells in suspension. Middle—This chamber has a 3.2-mm electrode gap and a volume of 800 µL, also often used for fusion of large numbers of cells in suspension as well as embryonic cells. Bottom—This chamber has a 400–500 µm electrode gap and a volume of 100 µL; it is used exclusively for embryonic cell fusion.

An appropriate apparatus is the BTX 200 Cell Manipulator™ used in conjunction with the BTX Optimizer™ (Fig. 1A). The function of the Optimizer™ is primarily that of an oscilloscope to evaluate the actual voltage generated in the chamber following delivery of both the ac alignment and dc fusion pulses. Often there is a significant disparity between the voltage indicated on the fusion apparatus and the pulse actually generated in the chamber.

There are several fusion chamber options available for use. In most cases, the manufacturer of the fusion apparatus will have various chamber types available for use in their particular system. Three different fusion chambers routinely used in our laboratory are displayed in Fig. 1B. These differ in terms of the volume (0.8 mL to 100 µL) each is capable of containing at fusion and the distance (gap) between electrodes. The distance between the electrodes is important and impacts on the amount of voltage required to generate the required fusion field (kV/cm).

2.2. Microscope

Although use of a microscope is not required for successful fusion of mammalian cells, it is highly advantageous for evaluating the effectiveness of the parameters used immediately. In addition, depending on the cell type used, it may become necessary to manipulate cells manually into the appropriate alignment, rather than use the ac alignment pulse. This is particularly true when attempting to fuse two cells that exhibit a significant disparity in volume or cells that may have a nonspherical shape. The type of microscope used will often be determined by what is available in the laboratory and the size of the particular cells to be fused. The microscopes used in our laboratory are a Wild stereomicroscope with magnifications of 40–100× and a Nikon inverted microscope with magnifications of 100–400×. When manual manipulation of cells is required, the Wild stereoscope is the one of choice owing to its greater field of view at lower magnification, which often facilitates manipulation of the cells. Compound or upright microscopes are more difficult to use primarily because of the short working distances of the objectives available, which prevent acceptable visualization of the cells in the chamber.

2.3. Fusion Media

Several fusion media are effective in supporting fusion of mammalian cells *(16–19)*. Primarily, fusion media are of low conductance to promote the establishment of cellular dielectric potentials within cells, which facilitates intercellular contact and reduces the current-generated temperature increase in media immediately following delivery of alignment and fusion pulses (*see* Note 1). Two extremely common fusion media are Zimmermann's fusion medium *(20)* and a 0.3*M* mannitol-based fusion medium *(16)*. Currently, in our laboratory, a modified cell-fusion medium (CFM) is used to induce successful fusion with various cell types.

CFM: 0.1 mM Ca(C$_2$H$_3$O$_2$)$_2$, 0.5 mM Mg(C$_2$H$_3$O$_2$)$_2$·4HO, 1.0 mM K$_2$HPO$_4$ (dibasic), 0.1 mM glutathione, 0.01 mg/mL BSA, 0.28 mM

sucrose. Prepare by placing calcium acetate in 100 mL of purified (i.e., Milli-Q) water and stirring using a magnetic stir plate until dissolved. Add remaining compounds to 600 mL of purified water (sucrose will dissolve more slowly than other components). Slowly add calcium acetate solution to 600 mL of solution while stirring; this prevents calcium from precipitating. Allow all compounds to dissolve and mix thoroughly. Bring solution to 1 L with additional purified water, and adjust pH to 7.0 using 1N acetic acid, if necessary. After verifying pH, filter-sterilize fusion medium with a 0.2-µm filter. Aliquot into 100-mL sterilized bottles, and store at 40°C for up to 2 mo (see Note 2).

3. Methods

The methods described in this section are general methodologies for fusion of both multiple cells (i.e., hybridoma cells) and cells of limited number (i.e., embryonic cells). These methods will provide a background for further development of specific parameters as applicable to the specific cells utilized. Other chapters within this book provide more specific information regarding specific species of cells and cell types.

3.1. ac Alignment Pulse

Optimum alignment parameters, as with fusion parameters, must be empirically determined for each cell type. Therefore, the methods described in this section, as well as subsequent sections, will provide general methodologies that are subject to modification. An initial evaluation of alignment parameters can be easily made by use of a microscope. If effective, cells should immediately begin to orient as indicated in Fig. 2. If cells fail to move into this position, modification of the voltage and duration is required (see Note 3).

1. Remove cells from culture vessel, wash three times in culture medium without fetal bovine serum (FBS), and place cells back in a 37°C, 5% CO_2 incubator until needed (up to 3 h).
2. Turn on fusion apparatus, and allow 15–20 min for warm-up. During warm-up time, place fusion chamber on heated (37°C) stage and tape it to the stage to prevent movement (see Note 4). Attach electrodes to the fusion apparatus. Wash the fusion chamber channel twice using a Pasteur pipet with fresh warm (37°C) fusion medium, and replace the channel with fusion medium, such that the meniscus of the medium is directly below the top of the electrodes (see Note 5). Replace fusion medium following each fusion replicate (see Note 6).

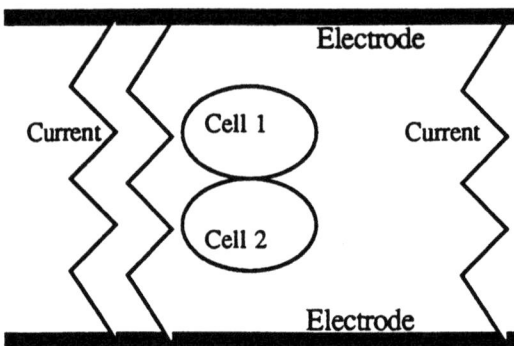

Fig. 2. Alignment of cells within fusion chamber. Chamber electrodes are represented by bold parallel lines and generated current as angular thin lines. The adjacent membranes of cells to be fused must be parallel to the electrodes, which are also perpendicular to the current generated. Only adjacent, closely associated membranes in this orientation will be induced to fuse.

3. Immediately prior to fusion, cells are resuspended, centrifuged, and washed twice with and placed in fusion medium. If using fibroblasts, melanomas, lymphocytes, stem cells, or hybridoma-type (HT) cells, place 2.5×10^7 cells/mL in fusion chamber (volume depends on the type of chamber used; see Fig. 1B). If using embryonic cells, place cells in warm fusion medium, and hold for 1–2 min before fusion. Cells will initially float and then sink to the bottom of the dish. After equilibration in the fusion medium, transfer a maximum of 10 pairs of cells into the chamber.
4. Nonembryonic (HT) cells are exposed to a 100 V/cm ac alignment pulse for 30 s and immediately followed by the dc fusion pulse. Embryonic cells are exposed to a 6-V, ac pulse for 10 s immediately followed by the dc fusion pulse.

3.2. dc Fusion Pulse

The dc fusion parameters will depend on the type, volume, and volume disparity between precursor hybrid cells. Table 1 lists various cell types and their fusion parameters.

1. If using a BTX or similar fusion apparatus, the dc fusion pulse will automatically follow the ac alignment pulse. The technician should hear two clicks following automatic delivery of pulses. If using manual pulse mode, deliver dc pulse immediately following the alignment pulse before cells have the opportunity to move out of alignment (see Note 7).

Table 1
Electrofusion Parameters of Various Cell Types

Cell type	dc Voltage, kV/cm	Duration, μs	No. of pulses	Reference
B-lymphocytes	7.69	20	2	21
Bovine blastomeres	1.21	70	1	6
	0.50	100	2	22
Lymphocyte—SP2	3.00	15	3	23
Lymphocyte—NS1	12.0	600	1	11
Porcine blastomeres	1.20	30	1	5
Murine blastomeres	1.56	99	1	24
	3.33	50	1	25
Ovine blastomeres	1.25	80	1	8
Sperm—oocyte	4.50	30	1	26
	0.19	30	1	27
Leydig—adrenocortical	3.75	15	9	28

2. Allow cells to remain undisturbed in the chamber for at least 60 s following the last dc pulse (*see* Note 8).
3. Transfer potentially fused cells into final culture medium, and place in incubator (*see* Note 9).
4. If fusing embryonic cells, remove cells from culture after 1 h, and evaluate fusion rate. Those pairs of embryonic cells that do not fuse within 30 min can be placed back in the fusion chamber and fusion procedure repeated.
5. If fusion parameters used are successful, fusion should be completed within 30 min postpulse delivery (Fig. 3) (*see* Note 10).
6. If selection schemes are to be used (i.e., with hybridoma cells), allow fused cells to be cultured in control (nonselective medium) for 48–72 h prior to transferring to selective medium. This allows cells to "recover" from the fusion process to ensure subsequent cell death is not related to the stress of the fusion process.

4. Notes

1. With an ac amplitude of 25 V for 1 min, the temperature of the cell-fusion medium within the chamber will increase to 33–34°C and stabilize. If electrolyte contamination of fusion medium occurs, this temperature will increase. Therefore, it is critical to be meticulous in preparing the fusion medium, wash the fusion chamber following fusion with "pure" (Milli-Q) water eight times, and wash the fusion chamber prior to use with fusion

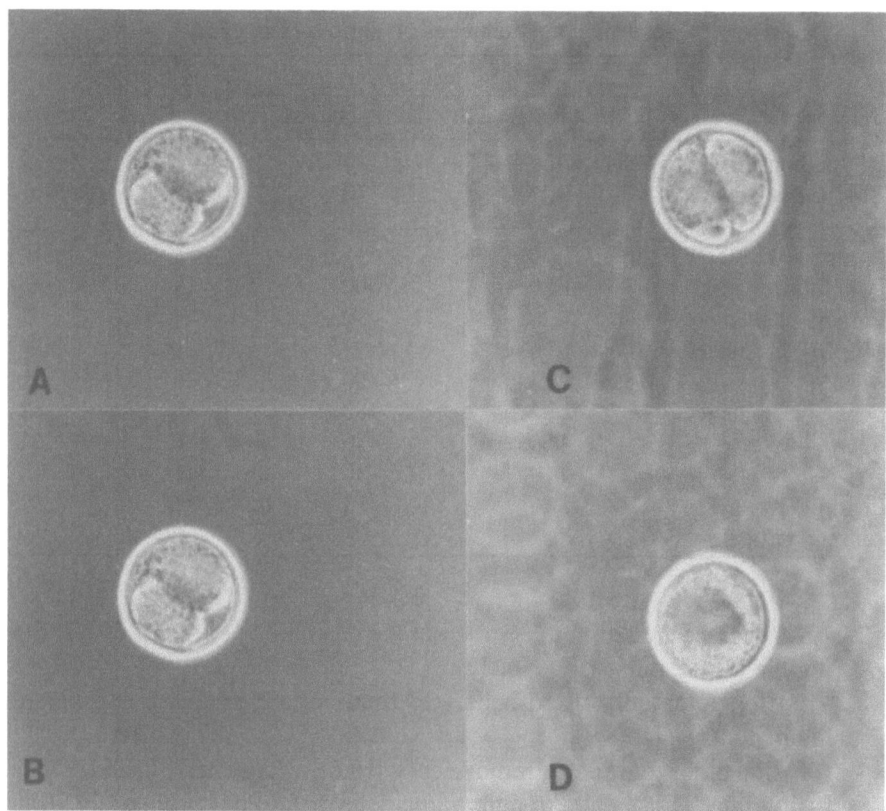

Fig. 3. Cell fusion of blastomeres (embryonic cells) in a two-cell bovine embryo. Cells (surrounded by an acellular zona pellucida) are placed in fusion medium (as previously described) within a 3.2-mm fusion chamber. (**A**) Embryonic cells immediately prior to initiation of fusion cycle (ac/dc pulses). (**B**) Embryonic cells approx 5 min postfusion cycle; note slight alteration in cell–cell interface and cell shape of top blastomere. (**C**) Embryonic cells approx 15 min postfusion cycle; note the dramatic alteration in the shape of the cells and the less distinct cell–cell interface. (**D**) Embryonic cell approx 25 min postfusion cycle; induced cell–cell fusion is complete with the formation of a single hybrid cell.

medium, both of which will help prevent electrolyte contamination. After conclusion of fusion cycles, wash the chamber eight times with 70% ethanol and then eight times with "pure" water, dry, and store in a closed container.

2. Fresh CFM analyzed with the chamber conductance button of the BTX Optimizer™ should indicate 19–26 mho. As electrolyte contamination increases, the conductance value will also increase. Conductance of the chamber with fresh fusion medium must be measured without cells immediately prior to initiating a set of fusion experiments and as often during the experiments as warranted.
3. If cells fail to align, check the conductance of the fusion medium. The fusion medium may have been "contaminated" with electrolytes (either in the process of making the medium or from an improperly cleaned chamber). If the conductance of the medium is not the problem, increase the ac voltage in 1.5-fold steps and re-evaluate the alignment after each voltage increase. Repeat until alignment occurs. If a fivefold increase in voltage fails to align cells, cells may need to be prewashed in medium that contains 2 mg/mL of bovine serum albumin (BSA) prior to the fusion medium wash. The BSA will help charge membranes and facilitate development of cellular dipoles.
4. Temperature will affect fusion results. Keep fusion conditions relative to temperature as consistent as possible. Remember to control the temperature of the fusion medium, the cells, and the chamber by placing on the chamber heated microscope stage. Fusion medium must be maintained in the incubator until use, and cells should be added to the chamber immediately prior to initiating the fusion cycle.
5. Allow the fusion apparatus to make a complete cycle (ac/dc pulse) with fusion medium in the chamber prior to adding cells. This is to check the fusion parameters prior to initiating a set of fusion experiments.
6. It is important to change the fusion medium after each fusion cycle to ensure uniformity of results. Medium will begin to evaporate very rapidly, which will result in alterations in osmolarity and concomitant changes in conductance.
7. Remember that the electrical field delivered to the cells (kV/cm) is dependent on the distance between the fusion chamber electrodes. Therefore, this calculation must be made to determine the voltage setting of the apparatus. In addition, the voltage setting will change as different chambers are used unless chambers have the same electrode distance.
8. We have found that membrane pores can exist for a minimum of 5 min postfusion *(24)*, which could result in the intracellular influx of ions across their concentration gradient. Therefore, if decreased viability results 12–24 h following fusion, maintenance of cells within fusion medium for an increased time (7–10 min) may increase cell survival postfusion.
9. Immediately after delivery of a fusion pulse, the adjacent membranes are disorganized and the cells are susceptible to lysis. In addition, the subsequent reorganization (fusion) of the adjacent membranes is occurring and

the cell-to-cell association is fragile. Therefore, gentle transfer of cells following delivery of a fusion cycle is important to facilitate cell fusion. Using the protocol described here, we commonly achieve fusion frequencies of ≥70%. If assessment of postfusion viability is required, allow cells to "recover" for 1 h after the fusion pulse, then sample cells, and stain with a viability stain (i.e., trypan blue, fluorescein diacetate, acridine orange, and so forth).

10. The two most common problems experienced are either elevated cell lysis (>25%; usually resulting from excessive voltage or pulse duration) or low fusion rates (<50%; usually resulting from insufficient voltage or too long a pulse duration). Bracketing of fusion parameters (i.e., high, moderate, low) will facilitate the identification of optimum dc pulse parameters.

References

1. Finaz, C., Lefevre, A., and Teissie, J. (1984) Electrofusion: a new, highly efficient technique for generating somatic cell hybrids. *Exp. Cell Res.* **150**, 477–482.
2. Haydu, Z., Lazar, G., and Dufits, D. (1977) Increased frequency of polyethylene glycol induced protoplast fusion by dimethylsulfoxide. *Plant Sci. Lett.* **10**, 357–360.
3. Stenger, D., Kaler, K. V. I. S., and Hui, S. W. (1991) Dipole interactions in electrofusion: contributions of membrane potential and effective dipole interaction pressures. *Biophys. J.* **59**, 1074–1084.
4. Kaler, K. V. I. S. and Jones, T. B. (1990) Dielectrophoretic spectra of single cells determined by feedback-controlled levitation. *Biophys. J.* **57**, 173–182.
5. White, K. L., Bunch, T. D., Reed, W. A., Wang, S., and Yue, C. (1993) Bovine oocyte activation is mediated by IP3-sensitive intracellular calcium pools. *Biol. Reprod. Suppl.* **1**, 346.
6. Collas, P., Blaise, J. J., and Robl, J. M. (1992) Influence of cell cycle stage of the donor nucleus on development of nuclear transplant rabbit embryos. *Biol. Reprod.* **46**, 492–500.
7. Smith, L. C. and Wilmut, I. (1989) Influence of nuclear and cytoplasmic activity on the development in vivo of sheep embryos after nuclear transplantation. *Biol. Reprod.* **40**, 1027–1035.
8. Rickords, L. F. and White, K. L. (1992) Effect of electrofusion pulse in either electrolyte or nonelectrolyte fusion medium on subsequent murine embryonic dlvelopment. *Mol. Reprod. Dev.* **32**, 259–264.
9. Ohnishi, K., Chiba, J., Goto, Y., and Tokunaga, T. (1987) Improvement in the basic technology of electrofusion for generation of antibody-producing hybridomas. *J. Immunol. Methods* **100**, 181–189.
10. Takahashi, Y., Suzuki, K., Niimura, T., Kano, T., and Takashima, S. (1991) A production of monoclonal antibodies by a simple electrofusion technique induced by AC pulses. *Biotech. Bioeng.* **37**, 790–794.
11. Schmitt, J. J., Zimmermann, U., and Neil, G. A. (1988) Efficient generation of stable antibody forming hybridoma cells by electrofusion. *Hybridoma* **8**, 107–115.

12. Zimmermann, U., Gessner, P., Wander, M., and Foung, S. K. H. (1990) Electroinjection and electrofusion in hypo-osmolar solution, in *Electromanipulation in Hybridoma* (Borrebaeck, C. and Hagen, I., eds.), Stockton, New York, pp. 1–30.
13. Glassy, M. (1988) Creating hybridomas by electrofusion. *Nature* **333,** 579–580.
14. Burt, P. H., Pethig, R., Gascoyne, P. R. C., and Becker, F. F. (1990) Dielectrophoretic characterization of friend murine erythroleukaemic cells as a measure of induced differentiation. *Biochim. Biophys. Acta* **1034,** 93–101.
15. Willadsen, S. M. (1986) Nuclear transplantation in sheep embryos. *Nature* **320,** 63–65.
16. Sukharev, S. I., Bandrina, I. N., Barbul, A. I., Fedorova, L. I., Abidor, I. G., and Zelenin, A. V (1990) Electrofusion of fibroblasts on the porous membrane. *Biochim. Biophys. Acta* **1034,** 125–131.
17. Dimitrov, D. S. and Sowers, A. E. (1990) Membrane electroporation—fast molecular exchange by electroosmosis. *Biochim. Biophys. Acta* **1022,** 381–392.
18. Zimmermann, U. (1986) Electrical breakdown, electropermeabilization and electrofusion. *Rev. Physiol. Biochem. Pharmacol.* **105,** 75–256.
19. Zimmermann, U. (1987) Electrofusion of cells, in *Methods of Hybridoma Formation* (Bartal, A. H. and Hirshaut, Y., eds.), Humana, Clifton, NJ, pp. 97–147.
20. Pratt, M., Mikhalev, A., and Glassy, M. C. (1987) The generation of Ig-secreting UC 729-6 derived human hybridomas by electrofusion. *Hybridoma* **6,** 469–477.
21. Iwasaki, S., Kono, T., Fukatsu, H., and Nakahara, T. (1989) Production of bovine tetraploid embryos by electrofusion and their developmental capability in vitro. *Gamete Res.* **24,** 261–267.
22. Kowalski, M., Hannig, K., Klock, G., Gessner, P., Zimmermann, U., Neil, G. A., and Sammons, D. W. (1990) Electrofused mammalian cells analyzed by free-flow electrophoresis. *Biotechniques* **9,** 332–341.
23. Prather, R. S., Sims, M. M., and First, N. L. (1989) Nuclear transplantation in early pig embryos. *Biol. Reprod.* **41,** 414–418.
24. Rickords, L. F. and White, K. L. (1992) Electrofusion-induced intracellular Ca^{+2} flux and its effect on murine oocyte activation. *Mol. Reprod. Dev.* **31,** 152–159.
25. Kaufman, M. H. and Webb, S. (1990) Postimplantation development of tetraploid mouse embryos produced by electrofusion. *Development* **110,** 1121–1132.
26. Rickords, L. F., White, K. L., and Wiltbank, J. N. (1990) Effect of microinjection and two types of electrical stimuli on bovine sperm-hamster egg penetration. *Mol. Reprod. Dev.* **27,** 163–167.
27. Minhas, B. S. and Kim, H. (1994) Electroporation of microsurgically fertilized murine oocytes. *Biomed. Prod.* 22–23.
28. Podesta, E. J., Solano, A. R., Molina, Y., Vedia, L., Paladini, A., Jr., Sanchez, M. L., and Torres, H. N. (1984) Production of steroid hormone and cyclic AMP in hybrids of adrenal and Leydig cells generated by electrofusion. *Eur. J. Biochem.* **145,** 329–332.

CHAPTER 24

Stabilizing Antibody Secretion of Human Epstein Barr Virus-Activated B-Lymphocytes with Hybridoma Formation by Electrofusion

Susan Perkins and Steven K. H. Foung

1. Introduction

Epstein Barr virus (EBV) can be used to transform human B-lymphocytes to derive populations of cells secreting specific antibodies of interest. Isolating monoclonal or stable populations of these cells, however, has proven very difficult (1). In our laboratory, we have developed methods to immortalize specific antibody-producing cells by fusing secreting EBV-activated lymphocytes to mouse–human heteromyeloma cell lines with electrofusion, followed by cloning (2). This methodology has allowed us to produce human hybridomas secreting 1–200 µg/mL of IgG specific for HCMV (3), HTLV-I (4), and HCV (unpublished) using several different mouse–human heteromyeloma fusion partners. Because as few as 5×10^4–10^6 EBV-activated B-cells can be successfully fused with a high degree of efficiency and consistency (up to one hybrid for each 100–1000 input EBV-activated cells), they can be fused as soon as antibody can be detected in a microtiter well, before the cells lose secretion or are overgrown by nonsecreters (5). High efficiency is achieved by varying the electrical parameters depending on the specific cells, the cell number, and the medium in which the cells are fused (see Table 1).

Table 1
Examples of Fusion Voltage Used with Different Fusion Partners in Different Fusion Media to Immortalize Antigen-Specific Antibody Secreted by EBV-Activated Lymphocytes from Peripheral Blood

Hybrid name	Antibody secreted to	IgG, µg/mL	# EBV fused	Heteromyeloma fusion partner	Fusion medium	dc Fusion voltage	Range in hybridoma formation efficiency (5)[a]
Z10 (18)	CMV	5	2×10^6	SBC-H2O	Iso-osmolar (14)	3.0 kV/cm 3 pulses/15 µs	34–68
X2-16 (19)	CMV	100	1.4×10^6	$K_6H_6/B5$	Iso-osmolar (14)	3.0 kV/cm 3 pulses/15 µs	6–18
IH-9 (20)	HTLV-I	3	10^5	$K_6H_6/B5$	Hypo-osmolar 75L3	1.0 kV/cm 3 pulses/15 µs	141–258
WA-04 2B10 (4)	HTLV-I	40	1.5×10^5	H73C11	Hypo-osmolar 100L3	1.25 kV/cm 1 pulse/10 µs	57–283
WA-11 1F5 (4)	HTLV-I	70	1.5×10^6	$K_6H_6/B5$	Iso-osmolar 300L3	1.75 kV/cm 3 pulses/15 µs	40–96
JB-04 1D7	HCV	90	10^6	$K_6H_6/B5$	Iso-osmolar 300L3	1.75 kV/cm 3 pulses/15 µs	11–66
JB-16 2D6	HCV	14	1.5×10^5	H73C11	Hypo-osmolar 100L3	1.25 kV/cm 1 pulse/10 µs	194–388
JB-17 1D2	HCV	60	1.5×10^5	$K_6H_6/B5$	Hypo-osmolar 100L3	1.25 kV/cm 3 pulses/15 µs	80–200

[a] (The number of wells with growth × The range in colony#/well in over 1/2 the wells) / 10^5 input EBV-activated B-cells

During electrofusion, cells are fused by first aligning them in an alternating current (ac) and then fusing with a direct current (dc) of high intensity and short duration *(6)*. Highest efficiency has been achieved using hypo-osmolar fusion media, which cause the cells to swell and their membranes to become more permeable. Many of the EBV-activated B-cells, however, do not tolerate the hypo-osmolar media well, so the parameters producing the most hybrids do not necessarily produce the most secreting hybrids. Fusing a larger cell number in iso-osmolar medium is often necessary to produce immortal secreting cells.

Since fusion parameters can vary depending on such factors as the state of activation of the cells and the growth medium, it is critical that a consistent system of cell isolation, care, and feeding be developed. In our system, B-lymphocytes are isolated from the peripheral blood lymphocytes of a seropositive individual by negative selection, and activated using a range of EBV concentrations to maximize stimulating as wide a range of B-cells as possible. Selection is made by assays for antibody secretion, rather than by prior cell selection. The EBV-activated B-cells tend to be very heterogeneous, and a small number will fuse over a wide range of dc voltages. Since the individual populations of cells must be fused early when they are few in number and cannot be used to optimize fusion, the fusion parameters most frequently used are those that work best to fuse the heteromyeloma fusion partner cells to themselves. The number of hybrids that will grow out from a fusion, however, can also be affected by such factors as the type and number of washes before fusion, the ratio of EBV-activated B-cells to heteromyeloma cells, the fusion medium, the plating medium, the plating density, and the use of feeders *(7)*.

The specific procedure presented is for the fusion of a small number of EBV-activated B-lymphocytes to the mouse–human heteromyeloma cell line K_6H_6/B5 *(8)* in the hypo-osmolar medium 100L3. Because cell-isolation and cell-culture techniques are so intertwined with the fusion procedure, the steps leading up to cell fusion and those after it to immortalize the antibody secreted are relevant, and important details of each step are discussed in Section 4. (*see* Notes 1–7, Section 4.1., and Notes 8–11, Section 4.2.). A scientist with tissue-culture experience, but unfamiliar with electrofusion, can be successful in producing hybrids by the following procedure, but maximizing the yield of secreting hybrids takes time and experience to develop a consistent system (*see* Notes 12–15, Section 4.3.).

2. Materials

2.1. Electrofusion Equipment

1. Power supply: Biojet CF (Biomed, Theres, Germany). Source of power must be capable of delivering a range of alternating and direct currents.
2. Helical fusion chambers: Each chamber consists of platinum electrodes of 200-μm diameter wound in a helix with 200-μm spacing that screws into a receptacle to hold the cells (provided by U. Zimmermann [6]).

2.2. Cell Lines

1. K_6H_6/B5: A heteromyeloma fusion partner derived from human lymphoid cells from a lymphoma patient fused to the mouse myeloma line NS-1-Ag 4 (courtesy of R. Levy [8]). K_6H_6/B5 fuses well and produces many high immunoglobulin-secreting hybrids.
2. EBV-activated B-lymphocytes secreting specific antibody: To produce your own lines, isolate lymphocytes from a seropositive donor and activate them as outlined in Section 3.1. and Notes 1–4.

2.3. Cell-Culture Media

Medium composition will affect cell fusability and hybrid growth.

2.3.1. Component Reagents

1. Iscove's Modified Dulbecco's Medium (IMDM) (1X): Gibco (Grand Island, NY).
2. Fetal calf serum (FCS): Heat-inactivate for 1 h at 56°C, and store at –20°C.
3. L-Glutamine (100X stock): 200 mM, store at –20°C (Gibco).
4. FM (100X stock): 0.45 g sodium pyruvate, 0.1 g bovine insulin, 1.32 g cis-oxalacetic acid (Sigma, St. Louis, MO). Dissolve in 100 mL of double-distilled water (ddH_2O) with a stir bar for 1 h, and freeze in aliquots at –20°C. Freeze/thaw only once.
5. 2-Mercaptoethanol (2-ME)(1000X stock): In a fume hood, dilute 0.5 mL of 2-ME (Bio-Rad, Richmond, CA) into 6.6 mL of ddH_2O, add 5 mL of the dilution to 95 mL of ddH_2O, and 0.20 μ filter-sterilize. Store aliquots at –20°C, thaw only once, and then store 4°C.
6. HT (100X stock): 0.776 g thymidine and 0.2772 g hypoxanthine (Sigma). Dissolve in 200 mL of ddH_2O at 70°C. Sterile filter, and store in aliquots at –20°C.
7. Aminopterin (1000X stock): Wearing gloves (toxic), dissolve 0.0176 g of aminopterin (Sigma) in 5 mL of NaOH, add 40 mL of ddH_2O, adjust pH to 7.0–7.3, and bring to 50-mL vol. Store in aliquots at –20°C.
8. Ouabain (1000X stock): 0.584 g ouabain (Sigma). Dissolve in 100 mL of IMDM, and freeze in aliquots at –20°C. Ouabain is light-sensitive.
9. Penicillin and streptomycin (P/S) (100X stock): 10,000 U/mL penicillin G sodium and 10,000 μg/mL streptomycin sulfate (Gibco). Store at –20°C.

2.3.2. Media

0.20 μ Filter-sterilize all media, and store at 4°C. After 1 mo, add fresh L-glutaminine and refilter. Discard after 2 mo.

1. Complete growth medium: IMDM with 1X FM, 1X 2-ME, 2 mM L-glutamine, and 10–30% FCS.
2. Hybridoma selection medium (HAT/ouabain): Complete growth medium containing 15% FCS, 1X HT, and 1X aminopterin (HAT). Add 1X ouabain just prior to the first feeding. Cover the medium bottle with aluminum foil to protect it from the light.
3. Hybridoma pre- and postselection medium (HT/IMDM): Complete growth medium containing 15% FCS and 1X HT. For preselection medium, on the day of fusion only, the medium is made with IMDM containing no pH indicator.
4. IMDM containing P/S: IMDM with 1X P/S. Place in outer wells of 96-well tissue-culture trays to prevent evaporation.

2.4. Fusion Media

Use fusion medium for up to 1 yr. Refilter-sterilize enough for use the day of each fusion.

1. L3 fusion medium: 0.0536 g magnesium acetate, 0.0079 g calcium acetate (E. M. Science, Cherry Hill, NJ), 0.5 g bovine serum albumin (Serva Biochemicals, Paramus, NJ), and sorbitol (E. M. Science). Dissolve in 500 mL of ddH$_2$O, 0.20 μ filter-sterilize, and store at 4°C. Fusion medium osmolarity is adjusted by varying the amount of sorbitol.
2. 300L3 (300 mosM, iso-osmolar medium): 500 mL of L3 fusion medium containing 25.5 g of sorbitol.
3. 100L3 (100 mosM, hypo-osmolar medium): 500 mL of L3 fusion medium containing 8.5 g of sorbitol.

2.5. Other Reagents

1. Trypan blue 0.02% (Gibco): for counting cell lines. Store at room temperature.

3. Methods

For electrofusion of a small number of EBV-activated B-cells to K_6H_6/B5 in hypo-osmolar medium to immortalize antibody secretion, all work is performed using sterile technique under a laminar flow hood.

3.1. Cell-Growth Conditions

Grow cells in complete IMDM growth media at 37°C with 6–6.5% CO_2.

1. K_6H_6/B5: Grow in 10% FCS/IMDM in a horizontal flask at relatively high density, and feed three times a week.
2. EBV-activated B-lymphocytes: Feed cells secreting the antibody of interest 30% FCS/IMDM two times a week. To create the EBV-activated B-cell lines to fuse:
    a. Isolate peripheral blood lymphocytes (PBL) (*see* Note 1).
    b. Separate B-cells out of the PBL population (*see* Note 2).
    c. Activate B-cells with EBV (*see* Note 3).
    d. Assay for antigen-specific activity immediately before fusion (*see* Note 4).

3.2. The Day Before Fusion

1. Expand EBV-activated B-cells to fuse by transferring cells from 96-well round-bottom plates to 96-well flat-bottom plates in 30% FCS/IMDM using a P200 pipetman (*see* Note 5). Also expand cells from wells with no antigen-specific activity to use as cell-count controls and negative controls.
2. Feed K_6H_6/B5 with 10% FCS/IMDM so that cultures will be in good health and at approx 5×10^5 cells/mL when fused.
3. If feeders are to be used, prepare one tray for each fusion (*see* Note 9).

3.3. Fusing

Wash 10^5 EBV-activated B-cells and 2×10^5 K_6H_6/B5 cells once in 300L3, fuse in 100L3 with 3 dc pulses of 1.25 kV/cm, and plate in the 60 inner wells of one 96-well microtiter tray (*see* Note 11).

1. Sterilize the helical fusion chambers in 70% alcohol for 10–15 min, and let them air-dry under the hood for 2–3 h before fusion.
2. Program the power supply to deliver an alignment current of 1 MHz 6 V ac and a fusion current of 25 V, 3 pulses of 15 µs dc with 1 s between pulses (1 MHz 300 V/cm ac and 1.25 kV/cm dc). The ac current should run for 30 s before the dc fusion pulses, and then for 30 s following them, dropping gradually from 6 V ac to zero over the 30 s.
3. Count cells available to fuse in a counting chamber using 0.02% trypan blue to discriminate live from dead cells by mixing 50 µL cells with 450 µL trypan blue. Count K_6H_6/B5 by transferring some cells to a 15-mL conical tube. Estimate the number of EBV-activated B-cells indirectly. Since it is undesirable to lose a significant number for a cell count, harvest 50 µL of cells from a well with no antigen-specific activity that was activated at the same time with the same concentration of EBV, and that visually has approximately the same cell density, cell size distribution, and activation level as the well to be fused.

4. Pool 2×10^5 K_6H_6/B5 and 10^5 EBV-activated B-cells to fuse in a 15-mL conical tube. Add the fusion partner to the tube first; then mix the well of EBV-activated B-cells by pipeting gently with a P200 pipetman, and add the calculated volume into the fusion partner.
5. Centrifuge the conical tube at 200–450g for 8 min. Aspirate.
6. Wash cells once in iso-osmolar medium (300L3) at 5×10^5 cells/mL. Resuspend the cells gently, but thoroughly with a P1000 pipetman containing 300 µL of wash medium, and then underlay the suspension with the remaining 300 µL of medium, using the same tip to minimize cell loss and maximize the thoroughness of the wash. Centrifuge as above. Aspirate to a dry pellet.
7. Resuspend the pellet in 250 µL of 100L3 gently, but thoroughly with a P1000, and add all of the cells to the receptacle of the helical chamber. Minimize the number of air bubbles created, and avoid transferring them into the chamber. Screw in the helix, and place the assembly in a rack with the receptacle end up to hold the cells in place and prevent them from pooling in the receptacle. Timing with this step is critical, since the cells must be fused exactly 10 min after suspension in the hypo-osmolar medium (100L3). If more than one fusion is performed at the same time, leave 2 min between the addition of the hypo-osmolar medium to each pellet to facilitate correct timing.
8. Fuse 10 min after the cells are placed in 100L3 by attaching the chamber to the power supply with a cable and running the preset fusion program (*see* step 2, this section).
9. After fusing, replace the chamber in the rack with the receptacle down. Cells are extremely fragile at this point, and any unnecessary movements of the chamber should be avoided.
10. Twelve minutes after fusion, wash cells out of the helical chamber into preselection medium (HT/IMDM with no pH indicator), which is premeasured in a 15-mL conical tube for plating at 5×10^3 cells/well (60 wells/fusion) (*see* Note 10).
    a. Unscrew the helix.
    b. Pipet 1 mL of preselection medium out of the conical tube with a P1000, and rinse cells out of the helix into the receptacle by holding the helix inverted at an angle and guiding the medium with the pipet to rinse all surfaces.
    c. Repeat rinse of the helix.

    The cells are very fragile, and fusion is still taking place, so it is very important to pipet gently and avoid air bubbles. Since the hybrids will not tolerate the hypo-osmolar medium for long, it is critical to wash them into the plating medium within 12 min after fusion.

11. Use a 5-mL plastic pipet to mix the cells very gently in the receptacle and return them to the conical tube containing the preselection medium. Rinse the receptacle once with medium from the conical tube.
12. Plate the cells in the 60 inner wells of a 96-well flat-bottom plate using one or two drops each of approx 60-µL vol/well with the same 5-mL pipet. Place IMDM containing P/S in the outer wells if it is not already there.
13. Place trays of newly plated hybrids in the incubator as quickly as possible.
14. A successful fusion checked microscopically the day it is plated should contain:
    a. At least some obvious EBV-activated B-cells to indicate that they made it through the fusion process intact.
    b. Many cells larger than either the EBV-activated B-cells or the fusion partner to indicate that fusion occurred (*see* Note 8).

3.4. Hybridoma Feeding Schedule

D 1 — Day of fusion: Plate in HT/IMDM with no pH indicator.
D 2 — Feed cells an equal volume of HAT/ouabain: 2 drops/well (each of approx 60 µL) with a 5-mL serologic pipet.
D 6 — Feed cells HAT/ouabain: Aspirate one-third of the medium in each well and feed each well one drop of approx 75 µL of new medium with a 10-mL pipet.
D 10— Feed as for d 6.
D 14— Feed cells as for d 6 with HT/IMDM, and twice/wk thereafter in the same manner.

By d 14, some hybrids will be visible macroscopically and the control wells of unfused cells should be dead (*see* Note 8). Assay hybrids when the medium is yellow enough to detect antibody, but not so yellow that it will kill the cells. They will be ready for the initial screening assay 2.5–7 wk after fusion. Assay the supernatant over the hybrids for antigen-specific antibody, and clone reactive wells of interest (*see* Note 6).

4. Notes
4.1. Procedural Overview

1. Using sterile technique, isolate PBL from a donor with serum antibody to the antigen of interest. Draw blood into acid citrate dextrose anticoagulant, maintain at room temperature, and separate as soon as possible after drawing by Hypaque/Ficoll gradient centrifugation *(9,10)*. Count PBL yield in 0.02% trypan blue to discriminate live from dead cells and 7% acetic acid to distinguish lymphocytes from other cells with similar specific gravity.
2. Isolate B-lymphocytes by rosetting out the T-lymphocytes with sheep red blood cells that have been treated with 2-amino-ethylisothiouronium bro-

mide (AET) to enhance the T-cell receptors *(11,12)*. Count nonrosetting cells in 7% acetic acid. Good EBV activation requires as pure a B-cell population as possible.

3. Activate B-lymphocytes by plating 10^4 nonrosetting lymphocytes/well in 96-well U-bottom microtiter plates in 30% FCS/IMDM containing 5–40% (v/v) supernatant from the marmoset line B95-8 as the source of EBV *(13)*. Fewer cells/well do not activate consistently, whereas more cells/well often result in overgrowth by nonsecreters. A range of virus concentrations is used to maximize the chance of activating the widest variety of B-cells. Feed cells 30% FCS/IMDM twice a week until medium is yellow enough from cell growth to harvest for assay (10–25 d), at which point the wells will contain $1–3 \times 10^5$ cells/well.

4. Harvest supernatant and assay for antigen-specific activity. In our experience, it is critical that an antigen-specific screening assay be developed with which the donor's serum is reactive before EBV stimulation has been set up. The assay should be specific, sensitive, be able to test a large number of samples, be performed with only approx 100 µL of supernatant, and take 1–2 d maximum. Because all of the cell manipulations thus far have been designed to maximize the number of B-cells activated, and many populations of EBV-activated B-cells will not maintain secretion for long, timely screening for specific activity is imperative.

5. Expand reactive wells of EBV-activated B-cells to 96-well flat-bottom trays overnight before fusing to maximize cell health. Individual wells of cells may be fused the day after activity is detected in hypo-osmolar medium, as detailed in this procedure, and/or several wells pooled and fused in iso-osmolar medium. In addition, a well of cells may be enriched (i.e., cloned at approx 10^3 cells/well in 96-well flat-bottom trays over human fibroblast feeders, irradiated with 100 Gy), and then reactive wells pooled and fused in iso-osmolar medium *(14)*.

6. Screen hybrids with the same antigen-specific assay used to select the EBV-activated B-cells. Clone reactive wells by plating limiting dilutions in 96-well plates over human fibroblast feeders shortly after detection, and/or up to several weeks later by using soft agar or limiting dilutions *(15,16)*. Perform more assays to characterize the antibodies further in the interim to maximize the chances of isolating hybrids secreting different antibodies with different activities.

7. Hybrid yield is determined by counting macroscopic growth in the wells when the supernatant is yellow enough for assay. Because of the uncertainty and inaccuracy involved in counting a large number of colonies in a large number of wells, we calculate the efficiency of hybrid yield as a range of likely values, rather than as a single number. It is an estimate of

the number of colonies formed in a fusion for each 10^5 input EBV-activated B-cells. The hybridoma formation efficiency range is defined as the number of wells with growth multiplied by the range in colony number in over half of the wells divided by 10^5 input EBV-activated B-cells *(5)*.

4.2. Further Technical Details of the Electrofusion Procedure

8. Plate medium controls of each cell line fused, and feed them with each set of fusions. Place 1 mL of K_6H_6/B5 and 100 µL of EBV-activated B-cells from a nonreactive well each into a 12 × 75 tube, centrifuge, aspirate, resuspend in 0.5 mL of HT/IMDM with no pH indicator, and plate one drop of each suspension in three wells each. If fibroblast feeders are used, also plate one drop of the same medium without cells in three wells over the feeders. After 2 wk in selection medium, both the EBV-activated B-cells and K_6H_6/B5 will be dead. Ouabain kills many of the fibroblast feeders, and causes others to lift off and appear like single viable hybrid cells, so the feeder control is useful in the first week or two of the life of the fusion to discriminate slow-growing hybrid cells from the feeders. The day fusions are plated, many cells can be observed with bulbs and strange projections that look like they might be fused, but these can also just be the heteromyeloma fusion partner reacting to the stress of fusion, since it is a hybrid itself.

9. Many hybrids will grow without feeders, but most of our secreting hybrids have been developed by plating over human fibroblasts. Grow the 60 inner wells of 96-well flat-bottom tissue-culture plates to near confluence in 10% FCS/IMDM, with IMDM containing P/S in the outer wells. Using Hff-B human foreskin fibroblasts from a 4-d-old infant (courtesy of L. Rasmussen, Stanford University), one T75 flask can be harvested into 20–50 mL of 10% FCS/IMDM to make 4–10 feeder trays the day before fusion. Wash cells on the flask twice with phosphate-buffered saline (PBS) containing no Ca^{2+} or Mg^{2+}, followed by 7–15 min in the incubator with 1–2 mL Versene 1:5000 (Gibco) and 0.2% trypsin (Sigma) at 1:1 ratio just prior to harvest. On the day of fusion, aspirate medium in the 96-well plates, irradiate cells with 60 gy, replate the 60 inner wells with one drop HT/IMDM with no pH indicator using a 10-mL serologic pipet (approx 75 µL/drop), and fill the outer wells with IMDM with P/S.

10. Cells that have just been fused are washed into medium without pH indicator, because the indicator is toxic to the cells still swollen with open pores. The volume of preselection medium needed for each fusion is calculated for plating at 2 drops/well with a 5-mL serologic pipet (approx 60 µL/drop). Newly fused hybrid cells are fragile and should be manipulated as

Electrofusion of EBV Activated B-Cells 305

little as possible, and plated as gently but quickly as possible. The goal is to plate the maximum number of viable hybrid cells. If feeders are used and each well already contains one drop of that medium, only 1 drop/well is available to wash cells out of the chamber. In our hands, that is approx 3.5 mL for 60 wells.

11. Because fusions are performed using a very small number of cells and the EBV-activated B-cells are sensitive to the hypo-osmolar medium, attention to detail is important to minimize cell loss. Employ plastic pipets to prevent cells from sticking to glass; aspirate medium from over cell pellets as quickly as possible after centrifugation to prevent cells from eluting; manipulate cells as gently as possible to avoid air bubbles that might kill them. Cell viability is critical since dead cells will prevent cell fusion. Since timing is also critical, all supplies, equipment, and reagents must be prepared ahead of time and available for immediate use. Only a limited number of fusions can be performed at one time. In our hands, timing can be kept accurately with a maximum of four fusions.

4.3. Optimizing the Procedure for Different Cell Types and Cell Numbers

12. Approximate fusion parameters are developed empirically by observing cell fusion under the microscope using open-chamber slides with two platinum wires of 200-μm diameter and 200 μm apart (courtesy of U. Zimmermann *[6]*). Fusion parameters are then altered based on hybrid yield and percent secreting hybrids produced in helical chamber fusions. Antigen-specific secreting hybrids have been produced with EBV-activated B-cells and other mouse–human heteromyeloma fusion partners in this manner using alternative fusion parameters, including SBC-H2O *(3,14)*, and H73C11 *(4,13)*.

13. Fusion parameters vary with the osmolarity of the fusion medium, as well as with the cells being fused. In general, the mouse–human heteromyeloma cell lines swell more, fuse more rapidly, and many more will fuse in hypo-osmolar media (75–100L3) than in iso-osmolar medium (300L3) if a lower dc voltage is employed *(17)*. For example, K_6H_6/B5 fuses well to itself in 300L3 with 35 V dc, but only 20 V dc are required to fuse cells in 75L3. Small numbers of EBV-activated B-cells are currently fused to K_6H_6/B5 using 100L3, rather than 75L3, because many EBV-activated B-cells do not tolerate the hypo-osmolarity well and will not live through the fusion process, or will not grow out as hybrids whose secretions can be stabilized. For efficient fusion and hybrid growth in 100L3, however, 25 V dc are required, rather than the 20 V used with 75L3. In another example, when fusing H73C11 growing in HB104™ with 1% FCS/IMDM with small

numbers of EBV-activated B-cells, we found that it was necessary to raise the osmolarity of the fusion medium to 125L3 and the fusion voltage by 5 V dc to promote significant hybrid growth.

14. When a larger number of EBV-activated B-cells are available for fusion (2×10^5–10^6), iso-osmolar medium (300L3) is employed. Because more cells are involved and they tend to be less fragile in this medium, they are centrifuged for 10 min at 450g. Cells are still fused at the ratio of 1 EBV:2 fusion partner, but the voltage must be increased compared to that used in hypo-osmolar media (3 pulses of 15 μs at 35 V dc with K_6H_6/B5). Cells take longer to fuse in this medium, so they are left in the helical chamber for 30 min after fusion before washing out and plating. Since a smaller percentage of the cells will fuse, it is necessary to plate at a higher density (2–3 × 10^4 cells/well with K_6H_6/B5 hybrids).

15. Because of cell heterogeneity, B-lymphocyte populations are treated in several different ways to optimize the chances of immortalizing the desired antibodies. Nonrosetting B-lymphocytes are EBV-activated at a range of marmoset supernatant concentrations. A single well of the EBV-activated cells with activity of interest may be divided and part expanded to check stability of secretion, part fused in hypo-osmolar medium, and part enriched so that reactive wells can be pooled and fused in iso-osmolar medium. Electrofusion is a flexible tool that can be adapted to all these populations, but there are no set fusion parameters.

References

1. James, K. and Bell, G. T. (1987) Human monoclonal antibody production, current status and future prospects. *J. Immunol. Meth.* **100,** 5–40.
2. Foung, S. K. H. and Perkins, S. (1989) Electric field-induced cell fusion and human monoclonal antibodies. *J. Immunol. Meth.* **116,** 117–122.
3. Foung, S. K. H., Perkins, S., Bradshaw, P., Rowe, J., Rabin, L., Reyes, G. R., and Lennette, E. T. (1989) Human monoclonal antibodies to human cytomegalovirus. *J. Infect. Dis.* **159,** 436–443.
4. Perkins, S., Rehman, M., Rowe, J., and Foung, S. K. H. (1992) Generation of human monoclonal antibodies to HTLV-I by microfusion techniques. Second International Conference on Human Antibodies and Hybridomas. Cambridge, England.
5. Foung, S., Perkins, S., Kadafer, K., Gessner, P., and Zimmermann, U. (1990) Development of microfusion techniques to generate human hybridomas. *J. Immunol. Meth.* **134,** 35–42.
6. Zimmermann, U. (1986) Electrical breakdown, electropermeabilization and electrofusion. *Rev. Physiol. Biochem. Pharmacol.* **105,** 175–256.
7. Perkins, S., Zimmermann, U., and Foung, S. K. H. (1991) Parameters to enhance human hybridoma formation with hypo-osmolar electrofusion. *Hum. Antibod. Hybridomas* **2,** 155–159.

8. Carroll, W. L., Thielemans, K., Dilley, J., and Levy, R. (1986) Mouse × human heterohybridomas as fusion partners with human B cell tumors. *J. Immunol. Meth.* **89,** 61–72.
9. Boyum, A. (1968) Isolation of monoclonal cells and granulocytes from human blood. *Scand. J. Clin. Lab. Invest.* **97,** 77–87.
10. Foung, S. K. H., Perkins, S., and Engleman, E. G. (1985) Peripheral blood lymphocyte separation from whole blood or buffy coats, in *Human Hybridomas and Monoclonal Antibodies* (Engleman, E. G., Foung, S. K. H., Larrick, J., and Raubitschek, A., eds.), Plenum, New York, pp. 435–436.
11. Saxon, A., Feldhaus, J., and Robbins, R. A. (1976) Single step separation of human T and B cells using AET treated rosettes. *J. Immunol. Meth.* **12,** 285–288.
12. Foung, S. K. H., Coutre, S., and Engleman, E. G. (1985) Separation of human T and non-T lymphocytes from peripheral blood, in *Human Hybridomas and Monoclonal Antibodies* (Engleman, E. G., Foung, S. K. H., Larrick, J., and Raubitschek, A., eds.), Plenum, New York, pp. 437–440.
13. Rehman, S. M. M., Perkins, S., Zimmermann, U., and Foung, S. K. H. (1992) Human hybridoma formation by hypo-osmolar electrofusion, in *Guide to Electroporation and Electrofusion* (Chan, D. L., Chassy, B. M., Saunders, J. A., and Sowers, A. E., eds.), Academic, San Diego, CA, pp. 523–533.
14. Perkins, S., Zimmermann, U., Gessner, P., and Foung, S. K. H. (1989) Formation of hybridomas secreting human monoclonal antibodies with mouse-human fusion partners, in *Electromanipulation in Hybridoma Technology, a Laboratory Manual* (Borrebaeck, C. and Hagen, I., eds.), Stockton, New York, pp. 47–70.
15. Foung, S. K. H. and Perkins, S. (1985) Cloning by limiting dilution, in *Human Hybridomas and Monoclonal Antibodies* (Engleman, E. G., Foung, S. K. H., Larrick, J., and Raubitschek, A., eds.), Plenum, New York, p. 476.
16. Larrick, J., Raubitschek, A., and Senyk, G. (1985) Soft agar cloning protocol, in *Human Hybridomas and Monoclonal Antibodies* (Engleman, E. G., Foung, S. K. H., Larrick, J., and Raubitschek, A., eds.), Plenum, New York, pp. 474–475.
17. Schmitt, J. J. and Zimmermann, U. (1989) Enhanced hybridoma production by electrofusion in strongly hypo-osmolar solutions. *Biochemica Et Biophysica Acta* **983,** 42–50.
18. Bradshaw, P. A., Perkins, S., Lenette, E. T., Rowe, J., and Foung, S. K. H. (1988) Generation and applications of human monoclonal antibodies to herpes viruses, in *Proceedings of the International Symposium on Clinical Applications of Human Monoclonal Antibodies* (Hubbard, R. and Marks, V., eds.), Plenum, New York, pp. 149–158.
19. Bradshaw, P. A., Duran, M., Lee, E., Young, L., Reyes, G. R., Perkins, S., Pande, H., and Foung, S. K. H. (1991) Anti-CMV human monoclonal antibodies epitope identification and immune response analysis, in *Progress in Cytomegalovirus Research* (Landini, M. P., ed.) Elsevier, Amsterdam, The Netherlands, pp. 157–160.
20. Hadlock, K., Perkins, S., Rehman, S. M. M., Chan, L., Lipka, J., Reyes, G. R., and Foung, S. K. H. (1992) Production, characterization and epitope mapping of a human monoclonal antibody to HTLV-I p19 gag protein. Fifth International Conference on Human Retrovirology: HTLV. Kumamoto, Japan.

CHAPTER 25

Electrofusion of Mammalian Oocytes and Embryonic Cells

Josef Fulka, Jr., Robert M. Moor, and Josef Fulka

1. Introduction

The methods of induced cell fusion are very useful procedures in reproductive and developmental biology. They are used to answer basic questions associated with cell-cycle regulation in mammalian oocytes and embryos *(1,2)*, to produce tetraploid embryos *(3)*, and of particular interest, these procedures are central to the construction of clones, i.e., by allowing nuclear transplantation in mammals (*see* Chapter 26; *4,5*). Various techniques can be used for the induction of fusion. The most commonly used in mammalian embryology are the techniques of Sendai virus-induced fusion, polyethylene glycol (PEG)-induced fusion, and the presently widely used technique of electrofusion *(6,7)*.

Electrofusion of oocytes and embryos is a very straightforward procedure. Simply stated, closely apposed cells are placed between the electrodes, and the necessary fusion pulses are applied. An important factor is the means by which the embryonic cells are prepared. In this chapter, we describe the basic techniques necessary for the preparation of high-quality mammalian oocytes and embryonic cells; it cannot be overemphasized that viability after fusion is only possible with high-quality starting material.

2. Materials
2.1. Oocytes and Embryos

It is beyond the scope of this chapter to describe either the collection and culture of oocytes, their fertilization in vitro and subsequent culture to the appropriate developmental stage, or the collection of embryos from the genital tracts (for details, see refs. *8–10*). We therefore assume that appropriate embryos for fusion are available. We also cannot describe the many possible methods that now exist for the preparation of cells for fusion. Only those methods considered optimal and routinely used in our laboratory are described. We stress that cells destined for fusion must be fully and normally developed. After each manipulation step, their morphology must be assessed to ensure normality. Requisite criteria are round and a consistently homogenous cytoplasm.

2.2. Handling and Culture Media

1. Manipulation solution: Phosphate-buffered saline (PBS, Sigma, St. Louis, MO), supplemented with 4 mg/mL bovine serum albumin (BSA; fraction V, Sigma) or 10–20% fetal calf serum (FCS, Sigma) (*see* Note 1).
2. Culture medium: Medium M-199 (Sigma) supplemented with sodium pyruvate (100 µg/mL, Sigma), Gentamicin (25 µg/mL, Sigma), and either BSA-V (4 mg/mL, Sigma) or FCS (10–20%, Sigma) (*see* Note 2). Prepare fresh media each week.

2.3. Preparation of Cells for Fusion

1. Cumulus removal solution: Manipulation medium (PBS) with 1 mg/mL hyaluronidase (Sigma).
2. Zonae pellucidae removal solutions: (a) PBS (Sigma) supplemented with pronase (Sigma). Dissolve pronase (0.1–0.5%) in PBS supplemented with 10 mg/mL polyvinylpyrrolidone (PVP). After mixing for 5–10 min, centrifuge at $1000g$ for 5 min, filter-sterilize, aliquot (0.25 mL), and store at –20°C; (b) Acidified saline solution: Tyrode's solution, acidic (Sigma) (*see* Note 3).
3. Cell agglutination solution: PBS with 10 mg/mL PVP is supplemented with 200–300 µg/mL phytohemagglutinin (PHA-P; Sigma). Store at 4°C for up to 2 wk.
4. Dissociation media: (a) Ca^{2+}- and Mg^{2+}-free Dulbecco's PBS (Sigma) supplemented with 1 mg/mL BSA; (b) Manipulation medium (PBS) with cytochalasin B (CCB, 5 µg/mL; Sigma). Dissolve CCB in DMSO (1 mg/mL), and store at –20°C (*see* Note 4).

5. Fusion media: 0.3M mannitol, 100 μM CaCl$_2$, 100 μM MgCl$_2$ (*see* Note 5).
6. Enucleation solution: 50 mg/mL etoposide (ETO; Sigma) dissolved in DMSO, aliquoted and stored at $-20°C$, and 10 mg/mL cycloheximide (CHXM; Sigma) dissolved in saline and stored at $-20°C$. Prepare both ETO and CHXM fresh every month (*see* Note 6).
7. Culture equipment and conditions: Disposable culture dishes or multiwell plates are convenient for egg culture. However, for manipulation, glass dishes are preferred because the cells do not adhere to the glass, provided that the dishes are absolutely clean and detergent free. During culture, the medium can be covered with paraffin oil. This is not recommended during manipulation (*see* Note 7). Cells are cultured in an atmosphere of 5% CO$_2$ in air at 37°C for mouse and 38–39°C for domestic animals.
8. Pipets: Made from hematocrit capillary tubes (capillary tubes for microhematocrits, 75-mm length, plain; Drummond Scientific Company, Broomall, PA).

3. Methods

3.1. Preparation of Micropipets

Consistency in the preparation of these important pieces of equipment is vital, and may require a period of training and practice for satisfactory results. Each pipet is pulled to the required diameter using an ethanol burner and the tip is broken to give a clean, sharp end for egg manipulation. Score the finely drawn part of the pipet with ethanol-sterilized sandpaper, and snap the tip with fine forceps. A sharp, clean break is required; pipets whose internal diameter is too small (<3/4 of manipulated cell diameter) can be rebroken. Pipets designated for use during cell agglutination require heat polishing on a microforge. Position the pipet at a safe distance from the filament, move it slowly toward the heated filament, and under microscopic control, complete the process of polishing such that the tip diameter is not changed. The pressure inside the pipet can be regulated by digital pressure through an attached rubber tube constructed from a 1–2 cm tube of appropriate diameter sealed at one end and connected at the other to the pipet. Soft tubing is ideal for cell-transfer purposes, whereas more rigid tubing gives more precise control and is recommended for use during the agglutination process.

3.2. Preparation of Oocytes for Fusion

1. Before fusion, the cumulus cells and zonae pellucidae must be removed. Cumulus from immature (germinal vesicle; GV) and metaphase I oocytes

is not mucified, and can therefore be removed easily by gentle and repeated pipeting using a cleanly broken, but unpolished pipet. The diameter of the pipet end should be the same diameter or marginally smaller than that of the zona-enclosed oocyte. Ovulated and cultured metaphase II oocytes are enclosed by cumuli that are very sticky. Preincubate these with hyaluronidase (1 mg/mL in PBS) for 3–5 min, and then remove the cumulus by the standard pipeting procedure outlined above.

2. Remove the zona pellucida by incubating the oocytes in a droplet (0.1 mL) of pronase (0.1–0.5%) at 37°C for 5–10 min. Wash once or twice in manipulation medium, and remove the zona remnants by rapid pipeting using a pipet with a tip having an inside diameter equivalent to the diameter of the oocyte cytoplasm. Oocyte viability is not compromised even if the zonae are dissolved completely by pronase (see Note 8).

Tyrode's solution gives basically the same results. Transfer the oocytes in a minimal volume of PBS (1–3 µL) into Tyrode's solution at 37°C; after a few seconds, the zonae will dissolve. Rapidly transfer the oocytes to PBS containing BSA to avoid destruction caused by adhesion of the zona-free oocytes to the dish. Store the oocytes in manipulation medium until the agglutination.

3. To agglutinate the oocytes to be fused, transfer oocytes into PBS (0.1–0.2 mL) with PHA-P. Using the tip of a fine polished pipet, push them gently against each other to establish weak contact. To increase the contact area, draw the apposed cell pairs very slowly into the manipulation pipet, which must be polished and have a diameter approx 25% less than that of the combined diameter of the cell pairs. Work on a glass surface only. When contact is established, transfer the oocytes in small numbers into manipulation medium to prevent adhesion between pairs (see Note 9).

3.3. Electrofusion of Oocytes

1. Using stereomicroscopic control, transfer the oocyte pairs in a nonelectrolyte solution into an electrofusion chamber (see Note 10). The precise positioning of the paired cells relative to the electrodes varies according to the design of the chamber, but we recommend positioning the pairs between the electrodes of a chamber with a large electrode gap (~500 µm). Any direct contact between the electrode and vitelline membrane often leads to membrane disruption and cell lysis.

2. Use ac current for cell orientation with a field strength of 3–6 V and 600 kHz. A perpendicular orientation of the paired cells between electrodes is slowly induced and can be controlled using a micropipet by gentle nudging until the appropriate cell position is obtained.

3. When all pairs are perpendicularly oriented, switch on dc pulses. Optimal parameters must be defined empirically, and differ for different types of cells or combination of cells. In general 1–2 pulses of 1 kV each of 50-µs duration are sufficiently strong to induce fusion. The pulses are applied at room temperature.
4. Transfer the cell pairs into culture medium, wash twice using culture medium, and incubate at 37–39°C, depending on species of origin. Morphological signs of fusion are observed after about 15–20 min, but definitive fusion success can only be accurately determined after about 45–60 min (*see* Note 10).

3.4. Fusion of Embryonic Cells

1. Remove zona pellucida as described in Section 3.2., step 2.
2. Disaggregate the embryonic cells by incubating zona-free embryos for 5–10 min in manipulation medium supplemented with CCB, and pipet them vigorously using a heat-polished pipet with inner diameter about half that of the embryo.
3. Regardless of the combination to be fused (oocyte-blastomere, cytoplast-blastomere, blastomere-blastomere), agglutinate, fuse, and culture the cells as described above for oocyte heterokaryons.

3.5. Production of Cytoplasts

This is an especially important part of the embryo cloning procedure (*see* Note 6). In the mouse, a novel and very simple enucleation procedure has been published recently *(11)*.

1. Collect the oocytes from large antral follicles of pregnant mare serum gonadotropin, 5–10 IU (PMSG) stimulated females, and then remove the cumulus cells by pipeting (Section 3.2., step 1). Select those oocytes of the same size with intact GV.
2. Culture them in M-199 (37°C, 5% CO_2 in air) for 90 min before further selection. Those that have undergone GV breakdown are cultured for an additional 5 h before being incubated for 2 h in M-199 supplemented with ETO (50 µg/mL).
3. Transfer the oocytes into M-199 with ETO and CHXM (50 µg/mL). After about 12 h, they will start to extrude the polar bodies containing the entire chromosome complement.
4. After an additional 4-h culture in medium 199 with ETO and CHXM, wash the oocytes in ETO- and CHXM-free medium 199. Culture and maintain them in this medium until required. Virtually all oocytes are successfully enucleated using this approach and can thereafter be used for fusion *(12)* (*see* Note 11).

4. Notes

1. Many different media exist for embryo manipulation. In general, manipulation solutions must be nontoxic, with an osmolality of about 280–300 mosM and a pH of between 7.2 and 7.4 when used under normal atmosphere conditions. These solutions are generally supplemented with BSA or FCS. A list of liquid balanced salt solutions is given in Sigma Cell-Culture Catalogue, 1994, p. 28.
2. Many suitable media are now available for the culture of cells after in vitro manipulation (Sigma Cell-Culture Catalogue, 1994, pp. 22–26). For short-term culture, no special media are necessary beyond the standard requirements of a lack of toxicity, an osmolality between 275 and 300 mosM, and the capacity to maintain a pH of 7.2–7.4 when kept in a culture atmosphere. The most commonly used atmosphere is 5% CO_2 in air. However, for certain purposes, special media are necessary *(8,9)*.
3. Neither zona-removal method has any deleterious effects on oocytes or embryos. They differ only in the speed of reaction.
4. CCB-treated cells are resistant to destruction, making this a method highly suitable for beginners. Alternative treatments are described in manuals mentioned in Section 1. *(8–10)*.
5. A number of different media can be used. Those recommended are non-electrolyte solutions that ensure the proper (perpendicular) orientation of the cells in the ac electric field. Some examples include $0.3M$ mannitol (Sigma) and isotonic glucose solution, 5.5% (Sigma), which are made in double-distilled water and filter-sterilized.
6. The most commonly used approach involving aspiration of the polar body and adjacent cytoplasm is described in refs. *8* and *9*. In addition, the bisection procedure described by Tarkowski *(13)* can be used.
7. We consider that successful manipulations are best carried out in droplets of medium not covered with paraffin oil, since oil often sticks to the pipet wall and destroys the cells. The procedure must therefore be performed rapidly to minimize evaporation. For longer-term manipulations, use a larger volume of the medium.
8. Naturally ovulated bovine oocytes and embryos are resistant to pronase, so Tyrode's solution must be used to remove zonae pellucidae from these cells.
9. Close contact is difficult to achieve in immature (GV×GV) oocytes because of their lack of elasticity. This can be overcome by a 30-min culture period (37°C) prior to agglutination. By contrast, maturing oocytes are sufficiently elastic to ensure a large area of contact and, consequently, are easily agglutinated.

10. During cell fusion, contact between the electrode and the cell membrane must be avoided. A fusion rate of 75% is satisfactory. A lower rate suggests the need to increase dc voltage, length of duration of pulses, or number of pulses. Similarly, if cells are damaged during fusion, reductions in these fusion parameters should be made. The optimal starting material for cell fusion in our view are maturing bovine oocytes because of their ease of manipulation and general resilience. Initial fusion parameters should be low (for example 0.5 kV, 50 µs, 1 pulse). If fusion does not occur, fusion parameters can be increased, and the process repeated. Fusion should be evaluated after about 60 min. Oocytes must be handled with great care, since even extremely small lesions result in lysis after the induction of fusion.
11. We have tested chemical enucleation using different culture media. Neither Basal Eagle Medium nor Minimal Essential Medium was satisfactory; M-199 gave excellent results. CZB (Chattot, Ziomek, Bavister) gives 100% enucleation (R. Procházka, personal communication). The time-course of maturation must be carefully assessed for optimal chemical enucleation. Under our conditions, GV breakdown occurs in most oocytes within 60–90 min and the first polar body is extruded between 8 and 9 h after the start of culture. Oocytes must be transferred to ETO medium about 2 h before the extrusion of the polar body. It is important to maintain a constant temperature during the chemical enucleation process.

References

1. Fulka, J., Jr. and Fulka, J. (1991) Regulation of processes involved in mammalian oocyte maturation. *Bull. Assoc. Anat.* **228,** 59–61.
2. Moor, R. M. (1988) Regulation of the meiotic cycle in oocytes of domestic animals. *Ann. NY Acad. Sci.* **541,** 248–258.
3. Nagy, A. and Rossant, J. (1993) Production of completely ES-derived fetuses, in *Gene Targeting, A Practical Approach* (Joyner, A. L., ed.), IRL, London, pp. 147–179.
4. Robl, J. M., Collas, P., Fissore, R., and Dobrinsky, J. (1992) Electrically induced fusion and activation in nuclear transplant embryos, in *Guide to Electroporation and Electrofusion* (Chang, D. C., Chassy, B. M., Saunders, J. A., and Sowers, A. E., eds.), Academic, New York, pp. 535–551.
5. Smith, L. C. (1992) Production of genetically identical embryos by electrofusion, in *Guide to Electroporation and Electrofusion* (Chang, D. C., Chassy, B. M., Saunders, J. A., and Sowers, A. E., eds.), Academic, New York, pp. 371–391.
6. (1984) *Cell Fusion,* Ciba Foundation Symposium 103, Pitman, London, p. 282.
7. Chang, D. C., Chassy, B. M., Saunders, J. A., and Sowers, A. E., eds. (1992) *Guide to Electroporation and Electrofusion.* Academic, New York, p. 569.
8. Hogan, B., Constantini, F., and Lacy, E., eds. (1986) *Manipulating the Mouse Embryo, A Laboratory Manual.* Cold Spring Harbor Laboratory, Cold Spring Harbor, NY, p. 315.

9. Monk, M., ed. (1987) *Mammalian Development, A Practical Approach.* IRL, London, p. 306.
10. Dziuk, P. and M. Wheeler, W., eds. (1991) *Handbook of Methods for the Study of Reproductive Physiology in Domestic Animals.* University of Illinois Press.
11. Fulka, J., Jr. and Moor, R. M. (1993) Noninvasive chemical enucleation of mouse oocytes. *Mol. Reprod. Dev.* **34,** 427–430.
12. Fulka, J., Jr., Notarianni, E., Passoni, L., and Moor, R. M. (1993) Early changes in embryonic nuclei fused to chemically enucleated mouse oocytes. *Int. J. Dev. Biol.* **7,** 433–439.
13. Tarkowski, A. K. (1977) In vitro development of haploid mouse embryos produced by bisection of one-cell fertilised eggs. *J. Embryol. Exp. Morphol.* **38,** 187–202.

CHAPTER 26

Nuclear Transfer in Bovine Embryos

Akira Iritani and Tasuku Mitani

1. Introduction

In early work to produce identical animals, such as twins, triplets, or quadruplets, blastomeres of two- to four-cell stage sheep embryos were separated *(1)*, and each blastomere was sealed in an empty zona pellucida, embedded in agar gel, and temporarily cultured in vivo in rabbit or sheep oviducts until the blastocyst stage. Developed embryos derived from separated blastomeres were recovered 4–5 d later, and transferred into final recipients. However, when producing more than eight identical animals, the developmental ability of the separated blastomere at the 16- to 32-cell stage was quite low, probably because of a shortage of cytoplasmic material. Later, the technique of nuclear transplantation into enucleated oocytes was developed to compensate for the shortage of cytoplasm. Production of cloned animals by nuclear transfer has been reported in many mammalian species, including the mouse *(2–5)*, rabbit *(6)*, sheep *(7)*, cattle *(8,9)*, and pig *(10)*.

Early embryos, such as those at the 16- to 32-cell stage, are usually used as nuclear donors in domestic animal species. However, other types of cells, such as embryonic stem cells (ES cells), primordial germ cells (PGCs), and spermatocytes/spermatids, have been recently used for nuclear transplantation. Since different cell types fused with enucleated oocyte cytoplasm have different characteristics, each requires a specific protocol. This chapter describes techniques for nuclear transfer by electrofusion of blastomeres and oocyte cytoplasm in cattle.

Table 1
Composition of Dulbecco's Phosphate-Buffered Saline
(D-PBS) (11)

Component	Concentration, g/L
NaCl	8.000
KCl	0.200
$Na_2HPO_4 \cdot 12H_2O$	2.900
KH_2PO_4	0.190
$MgCl_2 \cdot 6H_2O$	0.100
$CaCl_2 \cdot 2H_2O$	0.132
Sodium pyruvate	0.036
Glucose	1.000
BSA	4.000
Penicillin G potassium salt	0.075
Phenol red	0.005

2. Materials
2.1. Media and Solutions

1. Dulbecco's phosphate-buffered saline (D-PBS) (11): Prepare as shown in Table 1, filter-sterilize (pore size: 0.22 μm), and store at 4°C. This can be used up to 1 wk. D-PBS/calf serum, for flushing uteri: Add 1% calf serum instead of BSA. D-PBS/pronase, for removing zona pellucida: Add 2.5% (w/v) pronase, filter-sterilize, and store at –20°C. D-PBS/trypsin, for blastomere separation: Add 0.25% (w/v) trypsin, filter-sterilize, and store at –20°C. D-PBS/collagenase, for removing cumulus cells: Add 0.05% (w/v) collagenase, filter-sterilize, and store at –20°C.
2. BO medium (12): Prepare as shown in Table 2; for washing the collected oocyte-cumulus complexes (OCC). Filter-sterilize and store at 4°C up to 1 wk.
3. m-TCM199 medium: 9.8 g/L Medium 199 (Gibco-BRL), 2.2 g/L NaHCO$_3$, 55 mg/L sodium-pyruvate, 50 mg/L streptomycin sulfate, 100 IU/mL penicillin G potassium salt (13,14).
4. m-TCM199 maturation medium: add 10% fetal calf serum (FCS), 10 μg/mL leutinizing hormone, and 1 μg/mL estradiol to m-TCM199; filter-sterilize. Since the medium contains gonadotropins, store only for 1 d at 4°C.
5. m-TCM199/FCS: Add 10% FCS. Filter-sterilize and store at 4°C (14,15).
6. M2 medium (HEPES buffered): Prepare as shown in Table 3 (16), filter-sterilize, and store at 4°C up to 1 wk.
7. M2/FCS medium for cutting zona pellucida and donor cell injection in the manipulation chamber: Add 10% FCS to M2 medium instead of BSA.

Table 2
Preparation of BO Medium (12)[a]

Stock A component	g/L	Stock B component	g/100 mL	Other components	g/100 mL
NaCl	8.188	NaHCO$_3$	1.552	Glucose	0.250
KCl	0.376	Phenol red	0.001	Na-pyruvate	0.014
CaCl$_2$·2H$_2$O	0.412			Penicillin	0.006
NaH$_2$PO$_4$·H$_2$O	0.142			BSA	0.300
MgCl$_2$·6H$_2$O	0.133				
Phenol red	0.005				

[a]Before use, mix stock A, 80 mL; stock B, 20 mL; and other components.

Table 3
Composition of M2 Medium (16)

Component	Concentration, g/L
NaCl	5.530
KCl	0.360
KH$_2$PO$_4$	0.200
CaCl$_2$·2H$_2$O	0.250
MgSO$_4$·7H$_2$O	0.350
NaHCO$_3$	5.000
HEPES	2.600
Sodium lactate	0.036
Sodium pyruvate	1.000
Glucose	
BSA	4.000
Penicillin G potassium salt	0.075
Streptomycin sulfate	0.050
Phenol red	0.005

8. M2-Hoechst medium for nuclear staining: Add 1 μg/mL Hoechst 33342 to M2 medium and store at 4°C. This solution is light-sensitive.
9. M2 medium for enucleation (CB-CC medium): Add 5 μg/mL cytochalasin B and 0.1 μg/mL colcemid to M2 medium.
10. Fusion medium: (see Table 4 [17–19]).
11. Ca^{2+}-, Mg^{2+}-free PBS: 1.37M NaCl, 26.5 mM KCl, 88.4 mM Na$_2$HPO$_4$, 14.7 mM KH$_2$PO$_4$. Autoclave to sterilize.
12. Physiological saline: 0.85% NaCl.
13. Ca ionophone A23187 (Sigma, St. Louis, MO).
14. CRI aa medium: 114.7 mM NaCl, 3.1 mM KCl, 0.4 mM sodium pyruvate,

Table 4
Formulation of Zimmermann Mammalian Cell-Fusion Medium (19)

Component	Mol wt	g/L	Concentration
Sucrose	342.3	95.84	280 mM
Mg($C_2H_3O_2$)$_2 \cdot 4H_2O$	214.5	0.107	0.5 mM
Ca($C_2H_3O_2$)$_2$	158.2	0.016	0.1 mM
K_2HPO_4(anh)	174.2	0.174	1.0 mM
Glutathione	307.3	0.031	0.1 mM
Bovine serum albumin	25,000	0.01	0.01 mg/mL

^{a}Adjust to pH 7.0.

26.2 mM NaHCO$_3$, 5.0 mM hemicalcium lactate, 100 μ/mL penicillin, 50 μg/mL streptomycin, 10 μg/mL phenol red, 3 mg/mL calf serum albumin, 10 μL/mL MEM essential amino acid solution (Gibco), 10 μL/mL MEM nonessential amino acid solution (Gibco), 20 μL/mL L-glutamic acid (2 mg/mL). Filter-sterilize and store at 4°C (20).

2.2. Preparation of Glass Microtools

1. Holding pipet: Hand-pull a microcapillary microcap (1-mm outside diameter [OD], Drummond Scientific, PA) by heating with a microburner (Fig. 1). Cut the pulled microcap at a point of 80–120 μm OD (depending on the size of oocyte or embryo). Heat the tip of cut surface using a microforge (MF-79, Narishige, Tokyo) to produce a smooth surface and a 10–15 μm inside diameter (ID) (Fig. 2).
2. Microneedle: Place a hand-pulled microcap vertically on the microforge, then heat, and pull the tip down. Cut the tip, and adjust its size (thickness and elasticity) to be suitable for oocytes or embryos from the particular species.
3. Enucleation/injection pipet: Pull a microcap using a horizontal micropuller (PN-3, Narishige, Tokyo). Make a glass bead at the tip of the platinum filament of the microforge. As shown in Fig. 3A, attach the glass bead to the cutting site (with ID depending on the size of blastomere), and cut the micropipet by quickly heating and cooling the filament (Fig. 3B). Slightly heat to produce a smooth surface (Fig. 3C).

3. Methods
3.1. Preparation of Donor Cells

In vivo produced preimplantation embryos are used as a source of donor cells, since the developmental ability after nuclear transplantation is generally higher than reconstituted embryos with in vitro produced donor cells.

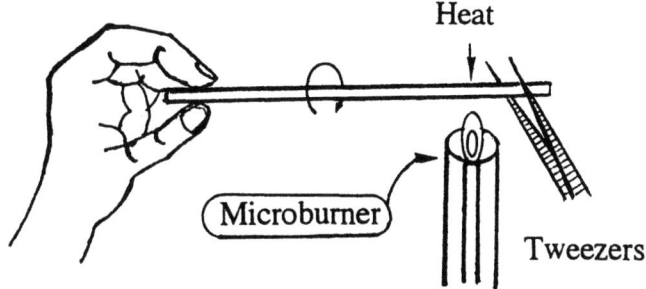

Fig. 1. Preparation of holding pipet.

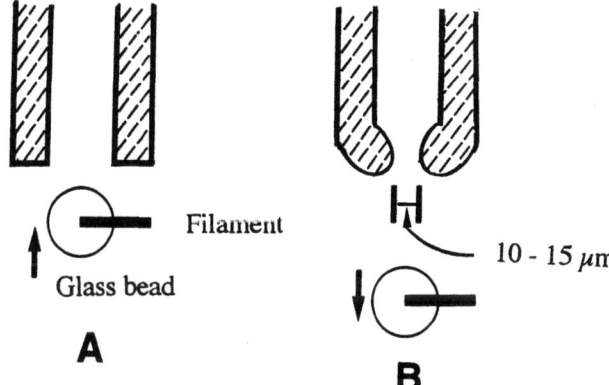

Fig. 2. Shaping the tip of holding pipet.

1. Recover 8- to 32-cell embryos nonsurgically by flushing uteri of donor cows with D-PBS containing 1% calf serum 5–8 d after artificial insemination.
2. Remove the zona pellucida of the embryos by incubating in 1–2 mL of D-PBS/pronase for 5 min, followed by pipeting.
3. To isolate blastomere donor cells, separate 16- to 32-cell embryos by gently pipeting with micropipets in Ca^{2+}-, Mg^{2+}-free PBS. Separate blastomeres of compacted morulae as described above, except use D-PBS/trypsin. The inner cell mass of blastocysts is first exposed by microsurgical cutting, and then separated into single cells as described above. We usually prepare several micropipets having different ID sizes, so that we can change the micropipet to fit the different blastomere sizes (see Note 2). Each set of separated donor cells from one embryo is kept in a microdrop (about 20 µL) of m-TCM199/FCS, scattered to prevent reaggregation.

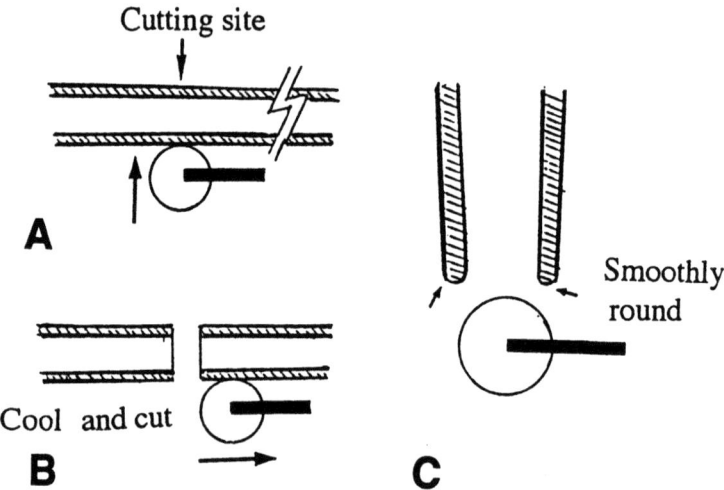

Fig. 3. Preparation of enucleation/injection pipet. (**A**) Put the glass bead on the cutting site, and then turn the filament on and off. (**B**) Cut the pipet by cooling the glass bead. (**C**) Slightly melt the tip of the pipet to be smooth.

3.2. Preparation of In Vitro Matured Oocytes (14,15)

Performing nuclear transplantation using large animal species, such as cattle, requires in vitro production of matured oocytes by culture to supply a large number of recipient cytoplasms.

1. Obtain ovaries at a local slaughterhouse, and transport in physiological saline at 30°C to the laboratory.
2. Aspirate immature oocytes from small follicles 2–5 mm in diameter using a syringe with an 18-gage needle. Wash the collected OCC twice in BO medium.
3. Culture for 24 h at 39°C in an atmosphere of 5% CO_2 and 95% air in m-TCM199 maturation medium *(13,14)*. Approximately 90% of the cultured oocytes will mature to the second metaphase (M-II).

3.3. Preparation of Micromanipulation Chamber

1. Place three small drops (25 µL each) of M2/FCS, CB-CC medium *(13)*, and M2/FCS in the center of a 90-mm diameter Petri dish separated by 5 mm.
2. Cover drops with paraffin oil, and use them for zona cutting (M2/FCS), enucleation (CB-CC), and donor cell injection (M2/FCS).

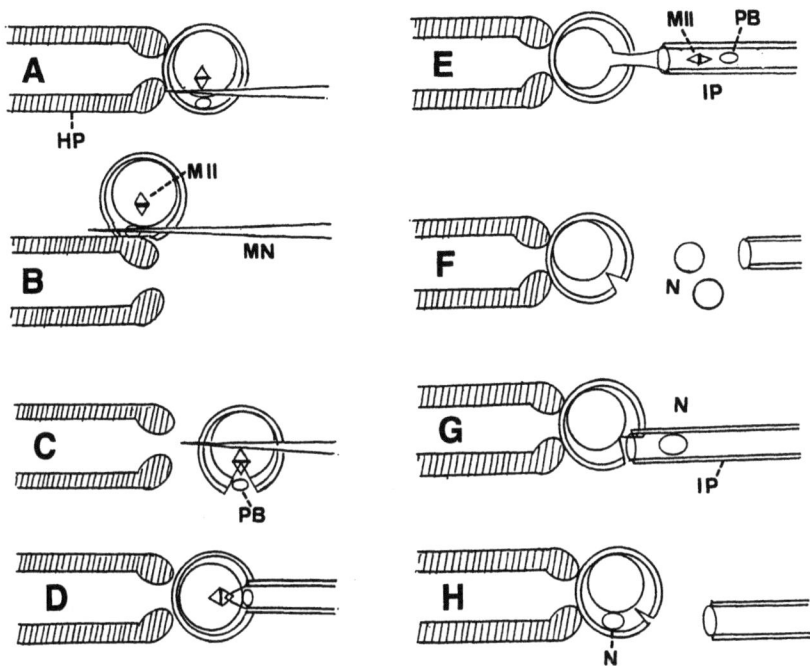

Fig. 4. Procedures for enucleation and blastomere (nucleus) injection. M-II: M-II chromosome plate, PB: first polar body, MN: microneedle, HP: holding pipet, IP: enucleation/injection pipet, OC: enucleated cytoplasm, N: donor nucleus.

3.4. Cutting the Zona Pellucida

1. Incubate cultured and matured M-II oocytes with 1–2 mL of D-PBS/collagenase for a few minutes.
2. Remove cells by gently pipeting with a Pasteur pipet. We usually prepare several pipets having an ID larger or smaller than that of the oocyte.
3. Transfer mature oocytes into the manipulation drop for zona cutting.
4. Hold the oocyte, insert the microneedle into the perivitelline space (Fig. 4A), and cut the zona pellucida by rubbing it against the upper edge of the holding pipet (Fig. 4B). Then press open the cut region of zona pellucida with a microneedle (Fig. 4C).

3.5. Removal of M-II Oocyte Nuclei

Oocytes matured in vivo or in vitro can be used as the recipient oocyte cytoplasm. It is difficult to recognize the nucleus of bovine matured oocyte at the M-II stage under the microscope, because its cytoplasm contains a large amount of lipids. Generally, the nucleus of the M-II oocyte is closely associated with the first polar body (PB-I) and is efficiently enucleated by removing the oocyte cytoplasm around the PB-I (*see* Note 3). The enucleated oocyte should be examined by fluorostaining of the nuclear materials with Hoechst 33342 to confirm the successful enucleation and disappearance of nuclear materials.

1. Connect the enucleation/injection pipet to an air-filled syringe.
2. Incubate five to ten of the zona-cut oocytes in a small drop (50 µL) of CB-CC medium for 15 min at room temperature.
3. Transfer these oocytes into the manipulation drop of CB-CC medium, and insert the enucleation pipet into the perivitelline space (Fig. 4D). Aspirate one-third of the volume of cytoplasm adjacent to the PB-I (Fig. 4E).
4. Wash the enucleated oocytes three times in small drops (50 µL) of M2 medium to reduce the cytotoxicity of cytochalasin B and colcemid.
5. Incubate oocytes in a small drop (20–40 µL) of M2/Hoechst 33342 for 10 min in the dark, and monitor the disappearance of nuclear materials using a fluorescent microscope.
6. Select oocytes that are completely enucleated.
7. Wash the enucleated oocytes three times in small drops (50 µL) of M2 medium, and transfer into drops (50 µL) of storage medium, m-TCM199/FCS.

3.6. Insertion of Donor Cells

1. Transfer donor cells and enucleated oocytes into the manipulation drop of M2/FCS.
2. Aspirate a donor cell into the injection pipet, and insert it into the perivitelline space (Fig. 4F–H). The oocytes inserted with a donor cell derived from the same embryo are kept together in a small drop (50 µL) of m-TCM199/10% FCS until needed for cell fusion.

3.7. Electrofusion

An efficient electrofusion method was developed by Zimmermann et al. (*17–19*). The electrofusion is achieved through a process of electrical breakdown at contacted membranes, phase transition, formation of bridges between bilayers of two membranes, and resealing of the lipid layers (*see* Notes 5 and 6).

There are two types of electric generators: One is equipped with ac and dc pulse generators, and the other is equipped with only a dc pulse generator. We have generally used the latter one. In this case, since the

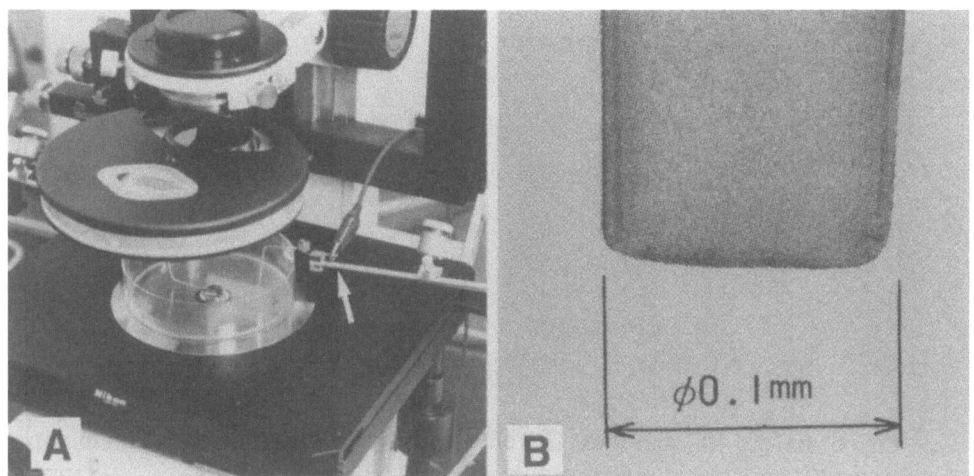

Fig. 5. Electrofusion apparatus (arrow) equipped to microscope (**A**) and tip of the electrode (**B**).

distance between the electrodes is equal to the diameter of oocyte (approx 100 µm), a 1-kV pulse generator is sufficient. In our laboratory, the pulse generator and the polar electrodes shown in Fig. 5 have been used. In nuclear transplantation by this method, about 90% of the blastomeres separated from 32-cell embryos could be successfully fused with the enucleated oocyte cytoplasm in cattle (Fig. 6; Tables 4 and 5).

3.8. Embryo Culture and Transfer

Nuclear transplanted embryos can be cultured in vitro using a coculture system appropriate for in vitro production of bovine embryos *(14,15)*. Blastocyst-stage embryos produced by in vitro cultivation are then transferred into recipient cows (*see* Note 7).

3.9. New Procedure for Nuclear Transfer in Cattle

A procedure involving a combination of aging of enucleated oocyte cytoplasm and double activation improved the developmental ability of the nuclear-transplanted embryos (Aoyagi, Y., personal communication).

1. Activate in vitro matured (at 24-h culture) and enucleated oocytes by treatment with Ca-ionophore A23187 (5 µM) for 5 min.
2. Pulse with 100 V/mm, 90 µs.
3. Incubate in cycloheximide (10 µg/mL) for 5–6 h.

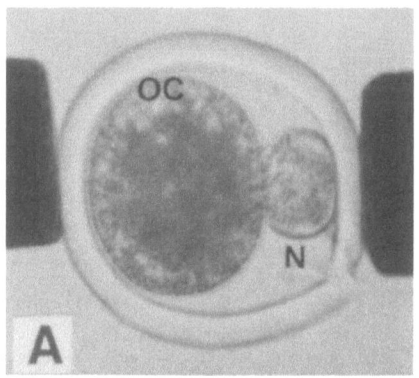 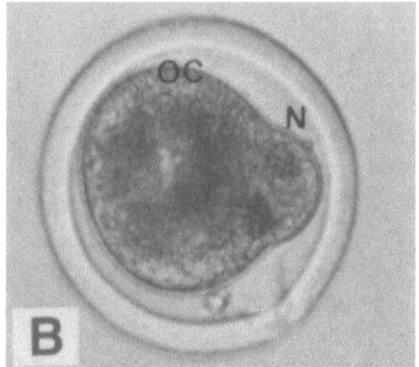

Fig. 6. Before (**A**) and after (**B**) electrofusion of blastomere and recipient oocyte. OC: enucleated oocyte cytoplasm, and N: donor nucleus.

Table 5
Electrofusion of Blastomeres of 8- to 32-Cell In Vivo Bovine Embryos with Enucleated Oocytes Matured In Vivo or In Vitro

Stage of donor embyros	Source of enucleated oocytes	No. (%) of eggs	
		Pulsed	Fused
8-cell	In vivo	19	17 (89)
9- to 16-cell	In vitro	20	17 (85)
17- to 32-cell	In vitro	23	22 (96)

4. Fuse donor blastomeres from in vivo produced day 5 embryo with above prepared recipient cytoplasms in a solution of $0.32 M$ mannitol, 0.1 mM $MgSO_4$ and 0.05 mM $CaCl_2$ using a dc pulse of 100 V/mm, 60-μs duration.
5. Culture for 7 d in a chemically defined medium, CR1aa *(20)* including 5% calf serum (*see* Note 7).

4. Notes

1. Many factors influence the efficiency of nuclear transfer, such as the quality of donor cells and recipient cytoplasm, interactions of donor cells and recipient oocyte cytoplasm, activation of recipient oocyte, and synchronization of cell cycles.
2. Glass microtools necessary for the practice of micromanipulation include holding pipets, microneedles, and enucleation/injection pipets. Skill in the preparation of glass microtools is an important factor to increase the effi-

Table 6
Parthenogenetic Activation of Bovine Oocytes Matured In Vitro by Electrical Stimulation[a]

Condition		No. of oocytes used	No. (%) of activated oocytes with				
Strength, V	Length, μs		Total	1-PN[b] +2PBS	1-PN only	2-PN only	≥3-PN
0	0	35	3 (9)	0	2	1	0
5	50	19	1 (5)	0	1	0	0
	100	20	4 (16)	0	3	1	0
	150	10	5 (50)	1	2	1	1
10	50	26	20 (77)	2	12	6	0
	100	27	21 (78)	5	11	4	1
15	50	34	29 (85)	7	20	2	0
	100	33	27 (82)	3	13	9	2
20	50	10	10 (100)	1	7	2	0
	100	8	8 (100)	0	5	3	0

[a]The oocytes were examined 18 h after electrical stimulation.
[b]PN: pronucleus(ei). 2PBs: two polar bodies.

ciency of nuclear transplantation. The shape of the microneedle differs slightly depending on the species that oocytes are derived from; for example, the zona pellucida of mouse and rat oocytes is thin and soft, so a thin and pointed needle is suitable for these species, but for sheep and cattle oocytes, a less pointed needle is suitable, since the zona pellucida from these animals is thick and hard.

3. After aspirating about one-third of the cytoplasm adjacent to the first polar body, 51 of 63 oocytes (81%) matured in vitro were successfully enucleated *(13)*.

4. Various methods are available for inducing parthenogenetic activation, and two methods, electrical stimulation and ethanol treatment, are the methods of choice because of their high efficiency and stability of inducing activation; they yield a high proportion of viable oocytes, and they are simple to perform. Electric stimulation may be used to induce activation and to effect cell fusion simultaneously. Table 6 shows the conditions for electric stimulation and the results of activation *(13)*. After incubating oocytes in HEPES-buffered solution (HeBS) containing 7% ethanol (v/v) for 7 min at room temperature, 89% of the oocytes were successfully activated *(13)*.

5. The fusion medium consists of nonelectrolytic solutions (0.3M mannitol or 0.3M sucrose) supplemented with 0.1 mM Ca^{2+} and 0.05 mM Mg^{2+}, or Zimmermann cell fusion medium (ZCFM) as shown in Table 4 *(17)*.
6. Factors influencing the electrofusion:

   The important parameters for the efficient electrofusion are the intensity of the dc pulse and the pulse duration for the membrane breakdown. It is important to monitor the viability of reconstituted embryos. In the electrofusion process, the extracellular Ca^{2+} in the fusion medium increases the fusion efficiency, and it also increases cell viability. At least 0.05–1 mM Ca^{2+} should be included in the fusion medium. Mg^{2+} also influences fusion efficiency: The optimum concentration of Mg^{2+} is approx 1 mM.

   Temperature influences the fusion efficiency, since the process of electrofusion involves membrane breakdown and resealing. Too low a temperature will induce membrane breakdown, but simultaneously result in cytotoxicity by inhibiting membrane resealing. Too high a temperature will increase the fusion efficiency, but reduce the cell viability by exhausting metabolic activity of the cell. Joule heat produced during the process of cell alignment by an ac pulse raises the temperature by several degrees, so the fusion treatment should be done at room temperature.
7. Rates at which cows became pregnant after transfer of blastocyst stage embryos were 42% (50/119) and 53% (8/15) (Aoyagi, Y., et al., personal communication).

References

1. Willadsen, S. M. (1979) A method for culture of micromanipulated sheep embryos and its use to produce monozygotic twins. *Nature* **277,** 298–300.
2. McGrath, J. and Solter, D. (1983) Nuclear transplantation in the mouse embryos by microsurgery and cell fusion. *Nature* **220,** 1300–1302.
3. McGrath, J. and Solter, D. (1983) Nuclear transplantation in mouse embryos. *J. Exp. Zool.* **228,** 355–362.
4. McGrath, J. and Solter, D. (1984) Inability of mouse blastomere nuclei transferred to enucleated zygotes to support development in vitro. *Science* **226,** 1317–1319.
5. Tsunoda, Y., Yasui, T., Shioda, Y., Nakamura, K., Uchida, T., and Sugie, T. (1987) Full term development of mouse blastomere nuclei transplanted into enucleated two-cell embryos. *J. Exp. Zool.* **242,** 147–151.
6. Stice, S. L. and Robl, J. M. (1988) Nuclear reprogramming in nuclear transplant rabbit embryos. *Biol. Reprod.* **39,** 657–664.
7. Willadsen, S. M. (1986) Nuclear transplantation in sheep embryos. *Nature* **320,** 63–65.
8. Robl, J. M., Prather, R., Barnes, F., Eyestone, W., Northey, D., Gilligan, B., and First, N. L. (1987) Nuclear transplantation in bovine embryos. *J. Anim. Sci.* **64,** 642–647.

9. Prather, R. S., Barnes, F. L., Sims, M. M., Robl, J. M., Eyestone. W. H., and First, N. L. (1987) Nuclear transplantation in the bovine embryo; assessment of donor nuclei and recipient oocyte. *Biol. Reprod.* **37,** 859–866.
10. Prather, R. S., Sims, M. M., and First, N. L. (1989) Nuclear transplantation in early porcine embryos. *Biol. Reprod.* **41,** 414–418.
11. Dulbecco, R. and Vogt, M. (1954) Plaque formation and isolation of pure lines with poliomyelitis viruses. *J. Exp. Med.* **99,** 167–174.
12. Brackett, B. G. and Oliphant, G. (1975) Capacitation of rabbit spermatozoa in vitro. *Biol. Reprod.* **12,** 260–274.
13. Mitani, T., Utsumi, K., and Iritani, A. (1993) Developmental ability of enucleated bovine oocytes matured in vitro after fusion with single blastomeres of eight-cell embryos matured and fertilized in vitro. *Mol. Reprod. Dev.* **34,** 314–322.
14. Kato, H and Iritani, A. (1993) In vitro fertilization in cattle. *Mol. Reprod. Dev.* **36,** 229–231.
15. Goto, K., Iwai, N., Takuma, Y., and Nakanishi, Y. (1994) Viability of one-cell bovine embryos cultured in vitro: comparison of cell-free culture with co-culture. *J. Reprod. Fertil.* **100,** 239–243.
16. Quinn, P., Barros, C., and Whittingham, D. G. (1982) Preservation of hamster oocytes to assay the fertilizing capacity of human spermatozoa. *J. Reprod. Fertil.* **66,** 161–168.
17. Zimmermann, U. and Vienken, J. (1982) Electric field-induced cell-to-cell fusion. *J. Membrane Biol.* **67,** 165–182.
18. Zimmermann, U. (1982) Electric field-mediated fusion and related electrical phenomena. *Biochem. Biophys. Acta* **694,** 227–277.
19. Wolfe, B. A. and Kraemer, D. C. (1992) Methods in bovine nuclear transfer. *Theriogenology* **37,** 5–15.
20. Rosenkrans, C. F. and First, N. L. (1994) Effect of free amino acids and vitamins on cleavage and developmental rate of bovine zygotes in vitro. *J. Anim. Sci.* **72,** 434–437.

CHAPTER 27

Electrofusion of Mouse Embryos to Produce Tetraploids

Ulrich Petzoldt

1. Introduction

It is possible to generate tetraploidy in mammalian embryos by chemical or physical suppression of a cleavage division causing endoreduplication of the genome *(1,2)*, or by using techniques to fuse karyoblasts or cells with nucleated or enucleated eggs and blastomcrcs, which results in transfer of nuclei or production of tetraploid embryos. Treatments with inactivated Sendai virus or with polyethylene glycol are effective in this regard *(3–5)*, but the use of electrofusion and its specific modifications for mammalian embryos has greatly facilitated the procedure *(6–9)*.

The degree of complexity and difficulty of electrofusion procedures used to produce tetraploid mouse embryos depends on the objective of the project. The easiest method is to fuse two blastomeres of the same embryo inside the zona pellucida. However, when the aim of the experiments requires the fusion of eggs or blastomeres of different ages or genetic backgrounds, the procedure is more time-consuming, because it is necessary to remove the zonae, isolate the blastomeres, and agglutinate the cells to be fused to each other. These additional steps of chemical treatment and handling, as well as the subsequent development without the zona pellucida, may result in lower viability of the fusion products. In general, such tetraploid embryos offer a useful tool for studying a variety of embryological questions, e.g., nucleocytoplasmic ratio *(10,11)*, cell cycle *(12)*, stage-specific gene expression *(13–15)*, or developmental capacities *(16,17)*.

2. Materials
2.1. Animals

Fusion experiments can be performed with eggs and embryos of any mouse strain (*see* Note 1). For the combination of defined genomes, it is necessary to use animals from well-characterized inbred strains *(18)*.

2.2. Media

Most of the media for collection, treatment, and culture of mouse embryos are generally used and prepared according to standard manuals *(18)*. The chemicals for the solutions should be of tissue-culture quality.

1. Mouse embryo culture medium M16 *(19)*: Weigh out into a clean 100-mL bottle: 552.8 mg NaCl, 35.6 mg KCl, 25.1 mg $CaCl_2 \cdot 2H_2O$, 16.2 mg KH_2PO_4, 29.3 mg $MgSO_4 \cdot 7H_2O$, 211.0 mg $NaHCO_3$, 261.0 mg sodium lactate, 3.6 mg sodium pyruvate, 100.2 mg glucose, 7.5 mg penicillin-G potassium salt, 5.0 mg streptomycin sulfate, and 3.7 mg EDTA (Titriplex III, Merck, Darmstadt, Germany). Add double-distilled H_2O (ddH_2O) to a final volume of 100 mL, and dissolve by careful shaking. Add 0.1 mL of phenol red (13 mg/mL in $0.154M$ $NaHCO_3$). Gas the medium by bubbling 5% CO_2, 5% O_2, and 90% N_2 through it for 10 min (5% CO_2 in air also may be used). Add 400.0 mg of bovine serum albumin (BSA, Fraction V, Sigma, St. Louis, MO) on top of the medium, allow to dissolve slowly, and mix gently. Filter through Millipore filter into a sterile 100-mL bottle, discarding the first few milliliters through the filter. Gas the surface with the gas mixture for 5 min, and store at 4°C for up to 1 wk. Regas the surface after every use.
2. Mouse embryo culture medium M2 *(20)*: Weigh out into a clean 100-mL bottle: 552.8 mg NaCl, 35.6 mg KCl, 25.1 mg $CaCl_2 \cdot 2H_2O$, 16.2 mg KH_2PO_4, 29.3 mg $MgSO_4 \cdot 7H_2O$, 34.9 mg $NaHCO_3$, 261.0 mg sodium lactate, 3.6 mg sodium pyruvate, 100.2 mg glucose, 7.5 mg penicillin-G potassium salt, 5.0 mg streptomycin sulfate, 3.7 mg EDTA. Dissolve in 80 mL of ddH_2O, and add 0.1 mL of phenol red. Weigh 496.9 mg of HEPES into a clean flask, and dissolve in 5 mL of ddH_2O. Adjust to pH 7.4 with $0.2N$ NaOH. Add HEPES solution to the other mixture, and add ddH_2O to a final volume of 100 mL. Add 400.0 mg of BSA on top of the medium, allow to dissolve, and mix gently. Filter-sterilize as described before, and store without gasing at 4°C for up to 2 wk. M2 without BSA is used as fusion buffer, and is prepared and stored the same way, but omitting the BSA.

3. Mouse embryo culture medium PB1 without Ca^{2+} and Mg^{2+} *(21)*: Weigh out into a clean 100-mL bottle: 624.2 mg NaCl, 21.0 mg KCl, 299.1 mg $Na_2HPO_4 \cdot 12H_2O$, 20.0 mg KH_2PO_4, 3.8 mg sodium pyruvate, 7.8 mg penicillin-G potassium salt, 104.0 mg glucose, 208.0 mg polyvinylpyrrolidone (PVP, 25,000–30,000, Merck (Darmstadt, Germany), lyophilized against ddH_2O). Add ddH_2O to a final volume of 104 mL and 0.1 mL of phenol red. Filter-sterilize as described, and store at 4°C for up to 2 wk.
4. Embryo culture medium MEM *(22)*: Pipet into a clean tube: 6.62 mL ddH_2O, 0.8 mL 10X MEM (Flow Laboratories, UK), 0.08 mL glutamine (20 mM, Flow Laboratories), 0.1 mL penicillin-G potassium salt solution (6.27 mg/mL in 0.9% NaCl solution), 0.1 mL streptomycin sulfate solution (5 mg/mL in 0.9% NaCl solution), 2 mL fetal calf serum (FCS, Flow Laboratories), 0.3 mL $NaHCO_3$ (7.5% solution, Flow Laboratories). Filter-sterilize into a tube or bottle, and gas surface as described. Store at 4°C for up to 1 wk, and regas the surface after every use.
5. Hyaluronidase solution: Dissolve hyaluronidase (Sigma) to a final concentration of 150 IU/mL in embryo culture medium M2, filter-sterilize as described, divide into 1-mL aliquots, and store at –20°C for up to 6 mo.
6. Pronase solution: Dissolve 89.3 mg of pronase (Calbiochem, Frankfurt, Germany, SA 70.000 PUK) and 250 mg of dialyzed PVP in 25 mL of embryo culture medium M2 without BSA, and incubate at 37°C for 2 h. Dialyze the solution in tubing (Viking, exclusion limit 8–15 kDa, Serva, Heidelberg, Germany) against 500 mL of M2 without BSA at 4°C overnight, filter-sterilize, and store in 1-mL aliquots at –20°C for up to 6 mo.
7. Phytohemagglutinin solution *(23)*: Dissolve 1 mg of phytohemagglutinin (PHA-P, Sigma) in 1 mL of embryo culture medium M2. Store in 10-µL aliquots in small tubes at –20°C for up to 6 mo. Before use, dilute to 1 mL with embryo culture medium M16 (final concentration = 10 µg/mL).
8. Culture oil: Add 80 mL of embryo culture medium M16 (without BSA and phenol red) to 500 mL of paraffin oil (British Drug Houses, Poole, UK) in a sterile dark bottle, and stir slowly for 24 h. Incubate at 37°C for 1 h. Remove the medium, and replace it with a fresh 80-mL aliquot. Repeat the complete procedure, and store the oil in sterile dark bottles at room temperature.

2.3. Equipment

1. Sterile watch glasses with lids.
2. Sterile mouth pipets (fine-drawn 50-µL Duran Ringcaps, Hirschmann, Germany) connected with a mouth piece through a rubber tubing equipped with a cotton filter.
3. Tissue-culture dishes 60 × 15 mm (Falcon, Becton Dickinson, Oxnard, CA).

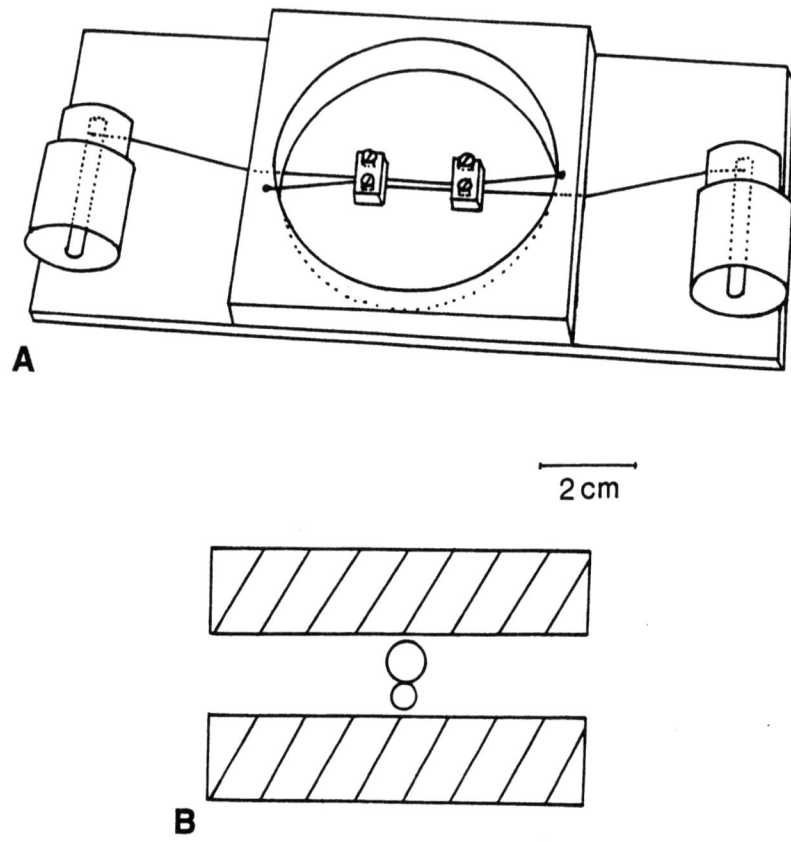

Fig. 1. Electrofusion chamber. (**A**) The fusion chamber consists of transparent Plexiglas™ with a container for the fusion buffer and platinum electrodes fixed at the bottom of the container and separate connections to electric plugs. (**B**) Enlarged view of the electrodes (200-μm diameter each) and a cell pair (fertilized egg and two-cell blastomere) in the 160-μm space between the electrodes.

4. Petri dishes, 60 × 15 mm (Greiner, Nürtingen, Germany); these prevent cells from attaching to the bottom of the dish.
5. Dissection microscope (in a laminar flow hood, if possible).
6. Heating plates or tables to keep material and media at 37°C.
7. Fusion chamber: Consisting of transparent Plexiglas™ with a container for the fusion buffer and platinum wire electrodes at the bottom of the container (Fig. 1A). The platinum wires (200-μm diameter) are fixed at a distance of 160 μm apart and each separately connected to an electric plug.

3. Methods

All media, culture dishes, and equipment should be prepared and warmed to 37°C in advance (and during the experiment), and be ready for use according to standard procedures *(18)*. As far as possible, the embryos should be kept on heating plates or in the incubator when not being handled for experiments.

3.1. Preparation of Eggs and Embryos

1. Collect eggs and embryos of the chosen developmental stages in watch glasses *(18)*. Release eggs, still surrounded by cumulus cells, directly into hyaluronidase solution, and observe until the follicle cells are dispersed. Wash three times in medium M2 in watch glasses. If not used immediately, transfer eggs and embryos into droplets of M16 under oil in Falcon culture dishes, wash through three droplets of the medium under oil, and store in an incubator (*see* Note 1).
2. To remove the zona pellucida, transfer eggs and embryos into pronase solution in watch glasses at 37°C. After a few minutes (usually about 5 min), the zonae start to dissolve. Then wash the embryos three times in M2 in watch glasses (BSA is required to inactivate the remaining pronase) (*see* Note 2).
3. Allow embryos to recover. Since the pronase treatment might also affect membrane surface molecules of the egg and blastomere, it is advisable to let them recover in droplets of M16 under oil in Falcon dishes at 37°C, 100% humidity, and 5% CO_2 in air for about 1 h.
4. To separate blastomeres, transfer cleavage embryos into watch glasses with Ca^{2+}- and Mg^{2+}-free PB1 (singly or in groups, when embryos are at the identical cell stages). While observing under a microscope, repeatedly bring the embryos into a mouth pipet, and blow them out gently. When all blastomeres are separated (*see* Note 3), wash through three droplets of M16 under oil in Greiner dishes and store singly or in small groups, but do not allow them to contact each other.
5. To form cell pairs for fusion, generate a close surface contact between the cells to be fused by treating cells with PHA. To simplify the procedure, several rows with medium droplets under oil can be formed in the same culture dish (Greiner). Place a row of PHA in the middle, and one or more rows of M16 at each side containing the different cell types to be fused. Transfer one cell of each cell type into the same droplet of PHA, and gently push them together using a mouth pipet. When about five droplets are filled with cell pairs, transfer the dish to the incubator. The agglutination of cells is observed at intervals of about 5 min and facilitated by gently

pushing them together again if necessary. After 5–15 min, the cells are usually well attached to each other and are then washed three times through droplets of M2 under oil (Greiner dishes).

3.2. Electrofusion

1. First prepare the fusion chamber: Mount the chamber on the platform of a dissection microscope. Cells are handled under the microscope using transmission light through the transparent bottom of the chamber. (Clean the chamber after each use with ddH$_2$O, and sterilize with absolute ethanol.)
2. Transfer attached pairs of cells into the fusion chamber, filled with M2 without BSA (see Note 4), using a mouth pipet; orient cell pairs between electrodes as shown in Fig. 1B. It is usually possible to arrange several cell pairs along the electrodes and to treat them together. The cells should not touch the electrodes during the fusion process. Subject the cells to two square pulses of 1 kV/cm, with a duration of 0.3 ms each and with an interval of about 1 s *(7,8,24)*. With the distance between the electrodes fixed at 160 µm, the intensity of the square pulses should be 16 V (see Note 5). Replace the fusion buffer in the fusion chamber with fresh and warmed (37°C) solution after several fusions (depending on the working speed and the degree of contamination with oil from the cell transfers).
3. After the fusion pulses, transfer the cell pairs individually back into the droplets of M2. Wash through three droplets of M16 under oil in Falcon dishes, and keep there separately in culture. The success of the fusion should be monitored after 5–15 min and after 1–2 h, if necessary. Signs of fusion are the disappearance of the cell borders between the two cells and the progressive rounding of the fusion product. As an additional control, the number of nuclei in potential tetraploid cells can be verified (two pronuclei coming from fertilized eggs, one nucleus from blastomeres; *see* Note 2). Usually, the first signs of cell fusion are visible after a few minutes under the dissection microscope. Fusion to a round cell should be finished after about 10 min, but may be delayed owing to cooling during handling.

The tetraploid fusion product can be cultured up to the blastocyst stage in M16 in droplets under oil *(13)*. For culture through implantation, transfer the blastocysts to droplets of embryo culture medium MEM in Falcon dishes *(14,22)*. Alternatively, transfer embryos to a pseudopregnant foster mother *(16)*.

The technique outlined here allows the fusion of two fertilized eggs or blastomeres from two different embryos of the same age to generate synchronous tetraploid fusion embryos or to achieve asynchronous tetraploid embryos by fusing a fertilized egg with a blastomere or two blastomeres of different ages (*see* Note 6). It is technically much easier to obtain a synchronous tetraploid embryo by using a normal two-cell embryo, still surrounded by the zona pellucida, by arranging it between the electrodes and fusing the blastomeres to a tetraploid egg *(16)*. Also, such embryos are easier to handle than fusion products without zona, and they usually develop better in vivo and in vitro. This technique is also effective with embryos of other mammals (*see* Note 7), e.g., rabbit, rat, and cattle *(8,24,25)*.

4. Notes

1. Fertilized eggs of some mouse strains, especially when randomly bred, do not develop well in vitro and arrest at the two-cell stage ("two-cell block"). If they are collected from the oviduct after this period, they can be cultured without difficulty. When planning experiments with fertilized eggs as fusion partners, it is advisable to use mouse strains that do not show a two-cell block in vitro or to determine whether eggs of the selected strain can be cultured in vitro from the one-cell stage.
2. After removing the zona pellucida from fertilized eggs, the second polar body is usually still attached to the egg's surface. It may be removed before the egg is transferred to the fusion chamber, but under the conditions used, we have not observed fusion between the second polar body, as was also shown by Sun and Moor *(26)*.
3. When separating blastomeres in Ca^{2+}- and Mg^{2+}-free medium, cells often become sticky and attach to the bottom of the watch glasses. It is helpful to use Greiner dishes, or to cover the bottom with 1% agar or agarose in the separation medium.
4. Even though nonelectrolyte solutions, such as $0.3M$ mannitol or glucose, are often used as a fusion buffer, we selected the electrolyte solution M2 without BSA for our experiments. Such solutions were shown to be equally effective for mouse blastomere fusion *(7)*, but may be replaced if necessary.
5. The 160-µm distance between the two electrodes in our fusion chamber is just sufficient to place a pair of mouse eggs or smaller cells between them. When working with slightly larger eggs from other species, a greater distance may be required.
6. When generating asynchronous tetraploid fusion embryos, there is a general problem that the two cells are usually at different stages of the cell cycle. Nuclear transfer experiments show that such asynchrony between the donor nucleus and recipient cytoplasm greatly influences the developmental capacity of the resulting embryo *(27,28)*. A similar effect was seen in asynchronous tetraploid embryos *(13,29)*. These problems cannot be

completely circumvented owing to varying cell-cycle lengths during early cleavage, but by careful observation of the timing of superovulation, using in vitro fertilization, and by observing the cleaving behavior of the embryos *(30)*, it is possible to adjust the cell cycles of the two fusion partners to a certain degree.
7. Mouse embryos can be easily cultured in vitro after removal of the zona pellucida, and they may also implant when transferred into a foster mother *(18)*. Development of embryos from other mammalian species, however, is quite dependent on the presence of a zona pellucida, and fusions are usually performed inside the zona *(8,24,25)*, thus generating synchronous tetraploid embryos. When the objective of the experiment requires the combination of cells from two different embryos, and incubation for further development or transfer to a foster mother, it might be necessary either to arrange the two cells inside a zona pellucida before fusion or to transfer the fusion product back into an empty zona pellucida. For such operations, micromanipulators and an inverted microscope are indispensable.
8. With practice, electrofusion of embryonic mouse cells is relatively easy, and success rates of 90% are common when suitable fusion conditions are selected *(26)*. The subsequent development of tetraploid fusion eggs is limited by the combination of cells used. When the fusion embryos are synchronous tetraploid, they can develop in vitro up to the blastocyst stage and beyond. When they are asynchronous tetraploid, the developmental capacity in vitro is more restricted *(13,29)*. After generation of synchronous tetraploid embryos by fusion of blastomeres inside the zona pellucida and transfer to a foster mother, they often reach postimplantation stages *(11,16,17)*. Development to term, however, has been reported only in tetraploid embryos generated by endoreduplication of the genome after cytochalasin treatment *(31)*.

References

1. Beatty, R. A. and Fischberg, M. (1952) Heteroploidy in mammals. III. Induction of tetraploidy in pre-implantation mouse eggs. *J. Genet.* **50,** 471–479.
2. Snow, M. H. L. (1973) Tetraploid mouse embryos produced by cytochalasin B during cleavage. *Nature* **244,** 513–515.
3. Graham, C. F. (1971) Virus assisted fusion of embryonic cells. *Acta. Endocrinol. (Suppl.)* **153,** 154–167.
4. Eglitis, M. A. (1980) Formation of tetraploid mouse blastocysts following blastomere fusion with polyethylene glycol. *J. Exp. Zool.* **213,** 309–313.
5. McGrath, J. and Solter, D. (1983) Nuclear transplantation in the mouse embryo by micro-surgery and cell fusion. *Science* **220,** 1300–1302.
6. Berg, H. (1982) Biological implications of electric field effects. Part V. Fusion of blastomeres and blastocysts of mouse embryos. *Bioelectrochem. Bioenerget.* **9,** 223–228.

7. Kubiak, J. Z. and Tarkowski, A. K. (1985) Electrofusion of mouse blastomeres. *Exp. Cell Res.* **157,** 561–566.
8. Ozil, J.-P. and Modlinski, J. A. (1986) Effects of electric field on fusion rate and survival of 2-cell rabbit embryos. *J. Embryol. Exp. Morph.* **96,** 211–228.
9. Willadsen, S. M. (1986) Nuclear transplantation in the sheep embryos. *Nature* **320,** 63–65.
10. Wiley, L. M. and Obasaju, M. F. (1988) Induction of cytoplasmic polarity in heterokaryons of mouse 4-cell-stage blastomeres fused with 8-cell- and 16-cell-stage blastomeres. *Dev. Biol.* **130,** 276–284.
11. Henery, C. C. and Kaufman, M. H. (1992) Relationship between cell size and nuclear volume in nucleated red blood cells of developmentally matched diploid and tetraploid mouse embryos. *J. Exp. Zool.* **261,** 472–478.
12. Henery, C. C. and Kaufman, M. H. (1991) Cleavage rates of diploid and tetraploid mouse embryos during the preimplantation period. *J. Exp. Zool.* **259,** 371–378.
13. Eid, R. and Petzoldt, U. (1993) Stage-specific gene expression in asynchronous tetraploid mouse embryos formed by fusion of blastomeres and fertilized eggs. *Roux's Arch. Dev. Biol.* **202,** 198–203.
14. Petzoldt, U. (1991) Developmental profile of glucose phosphate isomerase allozymes in parthenogenetic and tetraploid mouse embryos. *Development* **112,** 471–476.
15. Eglitis, M. A. and Wiley, L. M. (1981) Tetraploidy and early development: effects on developmental timing and embryonic metabolism. *J. Embryol. Exp. Morph.* **66,** 91–108.
16. Kaufman, M. H. and Webb, S. (1990) Postimplantation development of tetraploid mouse embryos produced by electrofusion. *Development* **110,** 1121–1132.
17. Henery, C. C., Bard, J. B. L., and Kaufman, M. H. (1992) Tetraploidy in mice, embryonic cell number, and the grain of the developmental map. *Dev. Biol.* **152,** 233–241.
18. Hogan, B., Constantini, F., and Lacy, F. (1986) *Manipulating the Mouse Embryo. A Laboratory Manual.* Cold Spring Harbor Laboratory, Cold Spring Harbor, NY.
19. Whittingham, D. G. (1971) Culture of mouse ova. *J. Reprod. Fertil.* **14,** 7–14.
20. Quinn, P., Barros, C., and Whittingham, D. G. (1982) Preservation of hamster oocytes to assay the fertilizing capacity of human spermatozoa. *J. Reprod. Fertil.* **66,** 161–168.
21. Whittingham, D. G. (1974) Embryo banks in the future of developmental genetics. *Genetics* **78,** 395–402.
22. Monk, M. and Ansell, J. (1976) Patterns of lactic dehydrogenase isozymes in mouse embryos over the implantation period in vivo and in vitro. *J. Embryol. Exp. Morph.* **36,** 653–662.
23. Mintz, B., Gearhart, J. D., and Guymont, A. O. (1973) Phytohemagglutinin-mediated blastomere aggregation and development of allophenic mice. *Dev. Biol.* **13,** 195–199.
24. Kurischko, A. and Berg, H. (1986) Electrofusion of rat and mouse blastomeres. *Bioelectrochem. Bioenerget.* **15,** 513–519.
25. Iwasaki, S., Kono, T., Fukatsu, H., and Nakahara, T. (1989) Production of bovine tetraploid embryos by electrofusion and their developmental capability in vitro. *Gamete Res.* **24,** 261–267.

26. Sun, F. Z. and Moor, R. M. (1989) Factors controlling the electrofusion of murine embryonic cells. *Bioelectrochem. Bioenerget.* **21,** 149–160.
27. Smith, L. C., Wilmut, I., and Hunter, R. H. F. (1988) Influence of cell cycle stage at nuclear transplantation on the development in vitro of mouse embryos. *J. Reprod. Fertil.* **84,** 619–624.
28. Kono, T., Kwon, O. Y., Watanabe, T., and Nakahara, T. (1992) Development of mouse enucleated oocytes receiving a nucleus from different stages of the second cell cycle. *J. Reprod. Fertil.* **94,** 481–487.
29. Taniguchi, T., Cheong, H. T., and Kanagawa, H. (1991) Fusion and developmental rates of single blastomere pairs of mouse two- and four-cell embryos using the electrofusion method. *Theriogenology* **36,** 645–654.
30. Smith, R. K. W. and Johnson, M. H. (1986) Analysis of the third and fourth cell cycles of mouse early development. *J. Reprod. Fertil.* **67,** 393–399.
31. Snow, M. H. L. (1975) Embryonic development of tetraploid mice during the second half of gestation. *J. Embryol. Exp. Morph.* **34,** 707–721.

CHAPTER 28

Spectrofluorometric Assay for Cell-Tissue Electrofusion

Richard Heller

1. Introduction

Electrofusion is a process by which fusion between cell membranes can be induced by exposure to electrical fields *(1,2)*. Many practical applications of electrofusion have been demonstrated, such as the formation of hybridomas *(3–6)*, the production of monoclonal antibodies (MAb) *(7–9)*, studying membrane fusion mechanisms *(10–14)*, and examining cytosolic events *(15–17)*. In addition, liposome-cell fusion has been utilized for the introduction of material into cells *(18)*. Cell-tissue electrofusion (CTE) represents another electrofusion area of interest. In this process, individual cells are incorporated into intact tissue *(19)*. CTE has been performed in vivo and has been shown to be useful for the interspecies transfer of membrane-surface components *(20)*.

CTE employs electromechanical processes by which individual animal cells can be electrofused directly to histologically intact tissues in vitro or in vivo in anesthetized animals *(19,20)*. CTE is achieved by applying electrical fields that result in the coalescence of juxtaposed plasma membranes of the individual cells and the cells within the tissues. To assess future biological applications of CTE, it is important to be able to evaluate the degree of cell fusion quantitatively. This chapter describes an assay developed to determine fluorometrically the number of cells fused to intact tissue.

The fluorometric assay for quantitating electrofusion (FAQE; *21*) employs a variety of procedures to quantitate the number of cells fused. The first step is to label the cells effectively. A method to produce fluorescently labeled cells while maintaining cell viability is described. An important aspect of cell fusion is to maintain contact between the membranes to be fused. To facilitate this, a procedure is described in which cells are layered on electrostatically charged membranes, which allows cells to be held in place and in contact with the tissue during the fusion process. The final two methods are the quantitation procedure and the development of standard curves to determine the number of cells fused.

2. Materials

2.1. Cell Lines

Many cell types, including adherent and nonadherent cells, are suitable for CTE protocols (*see* Note 1). Cells should be maintained in cell-culture media until needed for fusion.

2.2. Animal Tissue

1. Obtain tissue from anesthetized or euthanized animal (e.g., rabbit, mouse, and so forth).
2. Handle tissue aseptically, and place immediately into storage media.
3. Store tissue in appropriate media, and use within 48 h (*see* Note 2).

2.3. Chemicals and Miscellaneous Materials

1. Vital fluorescent dye, hydroethidine (HE, Polysciences, Warrington, PA): Prepare by adding 7 mg of HE to 1 mL of *N,N*-dimethyl-formamide (Aldrich, Milwaukee, WI) to obtain a 2.2-mM solution.
2. PBS: phosphate-buffered saline (Mediatech, Fisher Scientific; Pittsburgh, PA).
3. Sodium dodecyl sulfate (SDS): 0.2% solution prepared in PBS.
4. Zeta-Probe nylon blotting membranes, normally used in molecular biology *(22,23)*, are used because they contain a positive electrostatic charge on their surface that has the ability to hold negatively charged cells in place during the fusion procedure.

2.4. Electrofusion Equipment

An appropriate electrofusion chamber is insulated and consists of one to six 10.5-mm diameter wells. Each well has a 10-mm diameter flat platinum electrode at the bottom. The chamber is equipped with a water-

Assay for Cell-Tissue Electrofusion 343

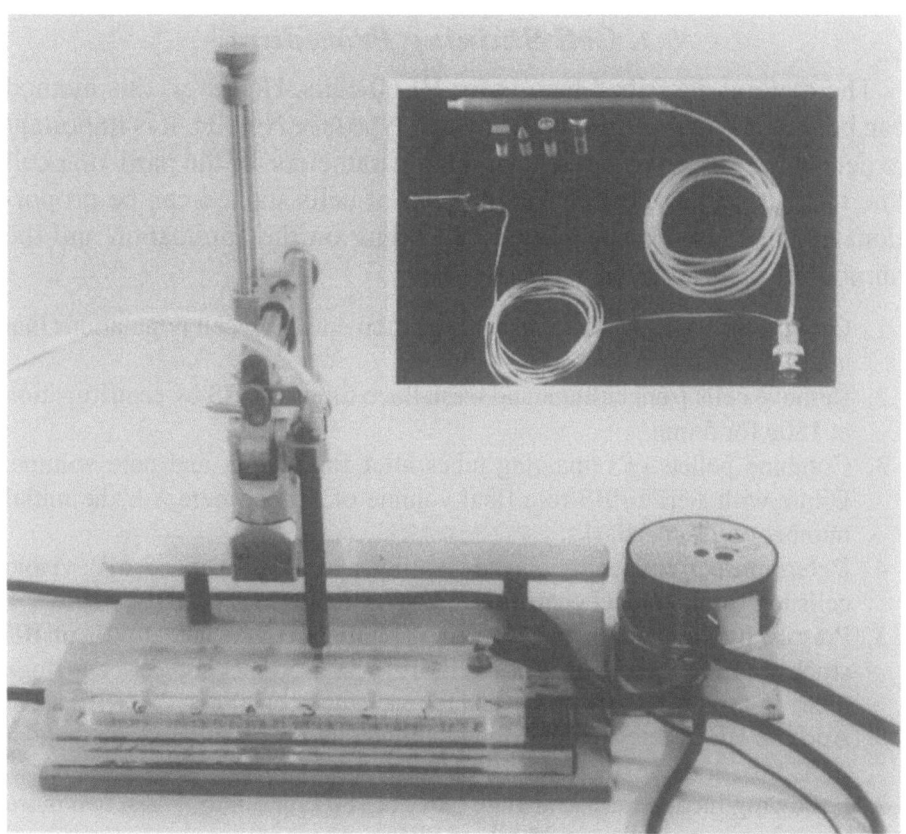

Fig. 1. CTE fusion chamber. Six-well fusion chamber is shown in lower portion of figure. Attached air pump circulates temperature-regulating water. Insert shows prototype electrode holder with a variety of shapes and sizes of electrodes.

flow system to regulate the well temperature. An electrode holder allows for insertion or removal of electrodes of various shapes and sizes (Fig. 1). Connect the disk electrode in the chamber and the other electrodes to the power sources (*see* Note 3). Pulse generators for electrofusion experiments can be either specially designed and constructed or obtained from commercial sources (e.g., Vivofuser, constructed at U.S.F., Tampa, FL; T800 Transfector, BTX, Inc., San Diego, CA).

3. Methods

3.1. Cell Staining Procedure

The method described here is for HL60 cells. However, this method can be modified to be used with any cell type (*see* Note 1). It is important to determine the most efficient staining parameters for the particular cell line to be used. In addition, the number of cells stained can be proportionately increased or decreased depending on the application and the number of cells required.

1. Grow cells under standard growth requirements. Use cell populations that are >90% viable.
2. Remove cells from culture, and wash three times in PBS by centrifugation at 180g for 5 min.
3. Combine pellets of remaining tubes after final wash, and note volume. Dilute with sterile PBS to a final volume of X mL where X = the initial number of 75-cm^2 flasks.
4. Determine cell concentration and viability, and add approx 2.5×10^7 viable cells to 50-mL conical centrifuge tubes (as many as needed).
5. Prepare hydroethidine (HE) working solution (WS): Place 90 µL of HE stock solution into 45 mL of sterile PBS in a 50-mL conical centrifuge tube; 45 mL of HE WS are needed for each tube containing 2.5×10^7 cells.
6. Add the HE WS to the tubes containing cells, mix well, and incubate at 37°C for 1 h in the dark.
7. Following incubation, centrifuge tubes at 180g for 5 min. Wash two more times by resuspending in 25 mL of PBS and centrifuging.
8. Combine cells from all tubes, determine total volume, and dilute to a final volume of $2X$ mL with sterile PBS, where X is equivalent to number of tubes used to stain cells. Determine both the total cell concentration and fraction of stained cells using a hemocytometer and fluorescence microscope. Dilute cells to desired concentration with PBS.

3.2. Layering of HE Stained Cells on Zeta-Probe Membranes

Zeta-Probe (ZP) membranes are utilized to hold individual cells in place, and assure contact between cells and tissue during the electrofusion process. The quantitation of CTE is performed using HE-stained cells. However, unstained cells can be used if quantitation is not necessary. This method provides a simple way to remove cells reproducibly from suspension by collecting them on ZP blotting membrane disks *(24)*.

These disks can be utilized to control better the number of cells intended for contact with intact tissue in CTE experiments (*see* Note 4).

1. Cut ZP blotting membranes into 10-mm diameter disks that have a handle (*see* Note 5).
2. Aseptically dilute cell suspension with sterile PBS to a final concentration that can range from 2–8×10^6 cells/mL, depending on the number of cells required on the membranes (*see* Note 6).
3. Pick up ZP disks, prepared in step 1, with sterile forceps, and place into the wells of a 48-well tissue-culture plate. The number of disks used will depend on the experimental design.
4. Aseptically place 500 µL of cell suspension into each of the wells containing a ZP disk, and place plate in a centrifuge plate carrier.
5. Centrifuge at 180*g* for 5 min using a plate-spinning rotor.
6. Remove plates from centrifuge, and remove disk with sterile forceps by grasping the handle of the disk to lift out of the well. Place disk on a flat dry Kimwipe for 1–2 min to dry the disk partially.
7. Perform CTE protocols by placing disk, cell side down, onto the tissue.

3.3. CTE Procedure

The procedure described here is for the fusion of human HL60 cells to rabbit corneal epithelium. The procedure can be modified to fuse other cell types or tissues (*see* Note 7). In addition, the procedure can be performed with stained or unstained cells. If quantitation is the goal of the procedure, then cells should be stained with HE as described in Section 3.1.

1. Obtain animal tissue as described in Section 2.2. Process the corneal tissue further by excising 10-mm circular buttons. Store in PBS or other appropriate media at 4°C until needed, and then place in the fusion chamber (Fig. 1).
2. Centrifuge cells to be fused to tissue onto ZP disks, and place disk cell side down onto corneas.
3. Place flat platinum electrode in holder (Fig. 1) and lower by a micromanipulator to apply appropriate mechanical force (*see* Note 8) to juxtapose cells on the ZP disk and cells in tissue.
4. Apply one to three 20-µs square-wave pulses of 0.6 kV/cm, 1 s apart via a dc generator to initiate the fusion process.
5. Raise the electrode, remove the tissue (cornea), and wash in PBS. Wash by holding an edge of the tissue (cornea) with a forceps, and shake vigorously with a back and forth motion a total of 20 times (*see* Note 9).
6. Quantitate the number of cells fused by following Section 3.4., or alternatively, fix the histologically modified tissue (cornea) and examine by scanning electron microscopy.

3.4. Quantitative Assay for CTE

1. Stain cells with HE as described in Section 3.1. Wash stained cells by centrifugation to remove extracellular dye.
2. Prepare tissue as described in Section 3.3., CTE procedure. Cut tissue into 10-mm diameter buttons, and place in the fusion chamber.
3. Layer HE-stained cells on to ZP disks as described in Section 3.2., and perform electrofusion procedure.
4. Remove unfused stained cells, following fusion, by washing the tissue. Place tissue into the wells of a 48-well tissue-culture plate. Lyse the epithelial layer containing the stained fused cells by adding 0.5 mL of 0.2% SDS, which releases the dye.
5. Remove the remaining corneal tissue, transfer the supernatant to a 5-mL centrifuge tube, and add 1.5 mL of PBS. Measure the fluorescent intensities in the supernatant fluids with a spectrofluorometer (Perkin-Elmer Model LS3B, Norwalk, CT) at an excitation wavelength of 480 nm and an emission wavelength of 615 nm.
6. Estimate the number of cells electrofused from standard curves by linear regression (*see* Notes 10 and 11).

3.5. Generation of Standard Curves

1. Lyse cell suspensions containing 6.0×10^6 HE-stained cells in 0.5 mL of PBS in each of two tubes by the addition of 0.5 mL of 0.2% SDS.
2. Combine the supernatant fluids of the two tubes, and perform twofold serial dilutions to yield cell concentrations ranging from 7.2×10^2 to 6.0×10^6 cells/mL.
3. Add 1 mL of sterile PBS to each tube for a final volume of 2 mL and final concentration range of 3.7×10^2 to 3.0×10^6 cells/mL.
4. Measure the supernatants in each tube with a spectrofluorometer, at an excitation wavelength of 480 nm and an emission wavelength of 615 nm, and plot the cell density vs measured fluorescence on a log–log scale (*see* Note 10).

4. Notes

1. The staining procedure and the procedure for layering cells on ZP disks can be utilized with both substrate-adherent and suspension cultured cells. The methods reported here have been used with the nonadherent human HL60, human U937, and murine WEHI-3 lymphoma cell lines, and the adherent human ME-180 epidermal and human Chang conjunctival cell lines. There was no significant difference in the number or viability of the layered cells among these cell lines. The specific example of CTE described utilized human HL60 promyelocytic leukemia cells (ATCC CCL

240; American Type Culture Collection, Rockville, MD). The nonadherent HL60 cells were grown as suspension cultures in plastic tissue-culture flasks (Costar, Cambridge, MA) containing Dulbecco's Modified Eagle's Medium (DMEM; Mediatech, Washington, DC) supplemented with 100 µg/mL gentamicin (Gibco, Grand Island, NY) and 10% heat-inactivated (56°C, 45 min) fetal bovine serum (Hyclone, Logan, UT) at 37°C in an atmosphere of humidified air containing 5% CO_2.
2. A variety of tissues can be used in CTE protocols. The majority of the work performed in the author's laboratory was done with rabbit corneas and was handled as follows. Tissue was obtained from adult New Zealand white rabbits (3–4 kg) that were housed, fed, watered, and handled in compliance with NIH regulations (25). Rabbits were humanely euthanized by iv administrations of pentobarbital overdoses. Animal corneas were excised by standard procedures employed routinely for collecting corneas from human donors. Briefly, using iris forceps and tenotomy scissors, the conjunctiva was cut at the limbus. A #15 Bard-Parker blade was used to make an incision through the sclera about 2 mm from the limbus. Castroviejo corneal scissors were used to complete the incision around the cornea. Suturing forceps were used to lift the cornea, and the final cut was made with the corneal scissors at the scleral spur. Corneas excised from rabbits were used immediately after collection or placed at 4°C, for up to 3 d prior to use. The medium used for storage was either modified McCarey-Kaufman (M-K) corneal storage medium (Aurora Biologicals, Buffalo, NY) or DMEM, neither of which was supplemented with serum, gentamicin, or phenol red. Prior to use, 10-mm diameter corneal buttons were excised from the tissue.
3. The studies described in this chapter were performed using custom-built equipment. However, CTE can also be accomplished using commercially available dc pulse generators or fusion instruments. Currently, there are no readily available fusion chambers and electrodes that can accommodate CTE procedures. The author would recommend that investigators design the chamber and electrodes to fit the intended application.
4. An important aspect of the electrofusion procedure is maintaining contact between the individual cells and the tissue. This is accomplished by layering the cells onto ZP blotting membranes. As described in Section 3.2., the cells are centrifuged onto the ZP disks in a 48-well plate. The variation between the number of cells layered onto ZP disks was less when the cells were centrifuged compared to when they were allowed to settle by gravity, and the optimal centrifugation force was found to be 180g. HE-stained cells at various concentrations were placed onto ZP disks to determine if different concentrations of cells could be reproducibly layered onto ZP

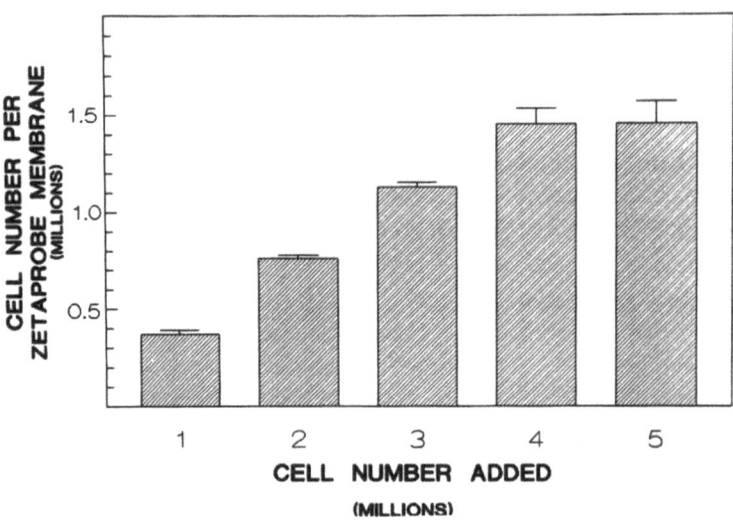

Fig. 2. Effect of cell concentration on membrane loading. Human HL60 cells suspended in 0.5 mL PBS were placed in wells containing ZP disks. Cell density ranged from 1×10^6 to 5.0×10^6 cells/0.5 mL. Cells were layered onto disks by centrifugation at 180g. Cell number per ZP membrane was determined spectrofluorometrically. At each density, except 5.0×10^6, approx 37% of the cells added were layered onto the disks; $n = 6$, error bars represent standard deviations. From Heller, R. and Grasso, R. J. (24). Copyright © 1991 by the Tissue Culture Association. Reprinted by permission of the copyright owner.

disks. The maximum number of cells that were reproducibly layered onto ZP disks was obtained by adding 4.0×10^6 cells/0.5 mL. The minimum number of cells that could be added to obtain a reproducible number of cells was 1.0×10^6 cells/0.5 mL. At all concentrations tested except 5.0×10^6, approx 37% of the cells added to the well were found to be adhered to the ZP disks (Fig. 2). The percent error was approx 5%, except at the highest concentration used. From these results, it was clear that the number of cells placed onto ZP membranes could be controlled to obtain a particular cell density for CTE experiments.

5. The ZP blotting membrane is packaged in sheets. For the layering procedure, the membranes are cut into 10-mm diameter disks with a #5 cork borer. To facilitate the removal of the disk from the plate, preparation of the ZP disks should include a handle, which extends at least 5–6 mm above the disk (Fig. 3). To accomplish this, a notch was etched on the side of the

Assay for Cell-Tissue Electrofusion 349

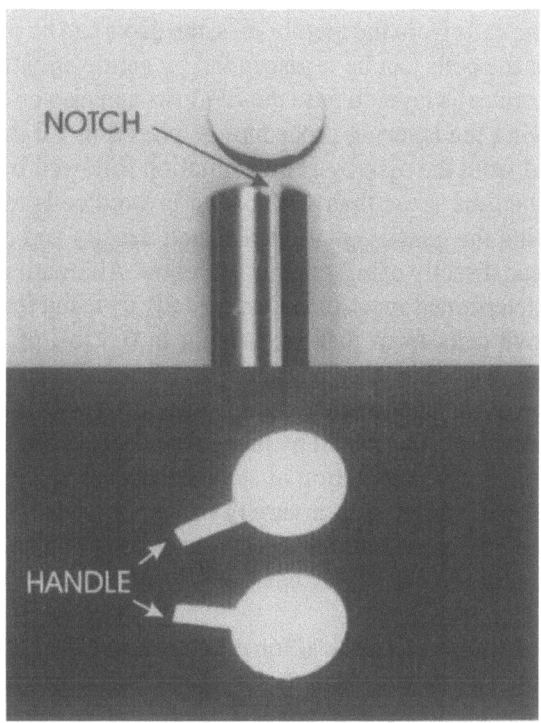

Fig. 3. ZP disks for CTE. The ZP disks are prepared by cutting a 10-mm diameter circle with a #5 cork borer. The borer contains a 1–2 mm "notch" (upper panel) that leaves a small portion of the disk attached to the ZP sheet. From this point of attachment, an extension that measures 5–6 mm long by 1–2 mm wide is cut with a pair of scissors, and the disk is removed from the sheet. The extension serves as the "handle" (lower panel), which facilitates the handling of the ZP disk. From Heller, R. and Grasso, R. J. *(24)*. Copyright © 1991 by the Tissue Culture Association. Reprinted by permission of the copyright owner.

borer. The notch is placed in an area of the ZP sheet that will yield a "handle" approx 5–6 mm in length and 1–2 mm wide (Fig. 3). This last section of the membrane is removed from the sheet by cutting with scissors. ZP disks are sterilized by wrapping disks in suitable material, i.e., aluminum foil, and autoclaving. The disks are easily moved without the loss of the cells, even when inverted. In addition, the disks, before removal from the plate, can remain at room temperature for 3–5 h without a significant change in the number of layered cells. Less than a 5% variation in cell

number from one sample to the next was found. However, if it is necessary to remove the cells from the membranes, the disks can be placed into media or PBS, and the cells can be resuspended by gentle agitation.
6. The number of cells layered onto the ZP disks can also be quantitated. The disk, following the layering procedure, is placed in 1.0 mL of PBS. Cells are removed from the disk by gentle agitation followed by removal of the disk from the tube. Less than 10% of the layered cells remain on the ZP disk following the gentle agitation. The cell density and cell viability can be determined directly using a hemocytometer. Alternatively, the cell density can be determined spectrofluorometrically by using fluorescent stained cells. Remove cells from disk by placing in 0.5 mL of PBS and gently agitating the disk. The disk is removed, 1.5 mL PBS is added, and the fluorescence of the supernatant fluid is measured with a spectrofluorometer. Cell density is determined by linear regression using a standard curve (*see* Section 3.5. for preparation of standard curve).
7. CTE can be performed with a variety of tissue types. In addition to the rabbit cornea, the procedure has been successfully performed with bovine and guinea pig corneas. Cells have also been fused to mouse skin and mouse liver.
8. CTE procedures use mechanical force to juxtapose the cells prior to initiation of the fusion process. With the development of the quantitative assay, this parameter could be evaluated. The amount of force applied to the system was measured in g/electrode diameter by placing the fusion chamber on a torsion balance. ZP disks containing HL60 cells were placed cell side down onto rabbit corneas. Using a 2-mm diameter electrode, between 25 and 400 g of force were applied to the cornea through the ZP disk. A single 20-μs, 0.6 kV/cm square-wave pulse was administered. The corneas were washed, fixed in 2.5% glutaraldehyde, and examined by SEM. The results of these experiments indicate that optimum fusion is reached at 100 g. Although a higher yield of fused cells was obtained at 400g, there was also a large amount of tissue damage. By using FAQE, a set of standard parameters to obtain reproducible CTE was established as follows: Cells on a ZP disk are placed on the tissue and a single 20-μs, 0.6 kV/cm square-wave pulse is administered through a flat, 2-mm diameter, circular electrode while applying 100 g of force *(21,26)*.
9. A critical portion of the assay is the wash procedure following electrofusion. It is important to establish with unfused control cells the amount and severity of washes necessary to remove unfused cells. An additional consideration is the amount of time between the fusion procedure and the wash procedure. A kinetic study of CTE *(27)* has shown that the individual cells are incorporated into the epithelial cells within 20 min and that the

Table 1
Quantitation of CTE[a]

Experiment no.	Measured fluorescence		No. of cells fused, SD
	Controls, SD	Experimental, SD	
1	230.2	258.0	4,523[b]
	(15.0)	(1.0)	(154)
2	191.9	217.3	3,845[b]
	(15.7)	(6.7)	(999)
3	199.1	273.0	6,369[b]
	(23.3)	(5.0)	(463)
4	215.7	270.0	4,693[b]
	(41.1)	(2.6)	(250)
5	249.3	318.3	7,119[b]
	(49.9)	(2.1)	(236)
Mean	217.4	267.3	5,309[b]
(S.E.)	(2.9)	(9.3)	(514)

[a] To determine the number of cells fused, the measured fluorescence of the controls was subtracted from the measured fluorescence of the experimental and compared to the standards by linear regression. The controls represent the mean of three groups of corneas ($n = 9$). Each experimental group represents the mean of three corneas.

[b] = $p \leq 0.05$ by Student's t-test when comparing the experimental group with the control groups. Fluorescence was measured in arbitrary units (A.U.). From Heller, *(21)*. Copyright © 1992 by Academic Press, Inc., San Diego, CA. Reprinted by permission of the copyright owner.

cells appeared to be stably fused after 6–9 min. Therefore, it is best to wait at least 6 min before vigorously washing the tissue to remove unfused cells.

10. While performing FAQE, several control groups should be included in the experimental protocol:
    a. Untreated tissue—no cells, no pulses;
    b. Stained cells placed on tissue under the proper mechanical pressure, no pulses; and
    c. Tissue placed under the proper mechanical pressure in the absence of cells, proper pulsing conditions.

The combination of these three groups will establish the background fluorescence of the tissue and the procedure. The epithelial layers of these control groups are lysed with the same procedure as the fused samples, and the resulting supernatant fluids are measured in a spectrofluorometer. Table 1 shows the results from a series of quantitations performed with the following fusion parameters: a single 20 μs dc pulse, 0.6 kV/cm field strength, and 100 g mechanical force. The fusion procedure was performed at three sites on each cornea *(21)*.

11. A limitation of this quantitation procedure is that it is necessary to destroy the sample to determine the number of fused cells. However, once the criteria are established for reproducible fusion, then the procedure can be performed with unstained cells and the histologically modified tissue can be used for a variety of biological applications. FAQE is primarily used to evaluate and optimize fusion parameters for particular cell and tissue types. An additional factor to consider is the potential for background fluorescence of the tissue. Although it is important to lyse the epithelial layer containing the fused cells completely, it is important to perform this lysis in a gentle manner. Sonication was attempted as an alternative method to the use of SDS, but it was found that this process greatly increased the background fluorescence. Other detergents or agents can be used for the lysis process. However, the release of natural fluorochromes and potential pH changes should be considered, since both contribute to the background fluorescence of the samples.

References

1. Bates, G., Saunders, J., and Sowers, A. E. (1987) Electrofusion: principles and applications, in *Cell Fusion* (Sowers, A. E., ed.), Plenum, New York, pp. 367–395.
2. Zimmermann, U. and Vienken, J. (1982) Electric field induced cell-to-cell fusion. *J. Membrane Biol.* **67,** 165–182.
3. Glassy, M. (1988) Creating hybridomas by electrofusion. *Nature* **333,** 579,580.
4. Glassy, M. C. and Pratt, M. (1989) Generation of human hybridomas by electrofusion, in *Electroporation and Electrofusion in Cell Biology* (Neumann, E., Sowers, A. E., and Jordan, C. A., eds.), Plenum, New York, pp. 271–282.
5. Hewish, D. R. and Werkmeister, J. A. (1989) The use of an electroporation apparatus for the production of murine hybridomas. *J. Immunol. Methods* **120,** 285–289.
6. White, J., Blackman, M., Bill, J., Kappler, J., Marrack, P., Gold, D. P., and Born, W. (1989) Two better cell lines for making hybridomas expressing specific T cell receptors. *J. Immunol.* **143,** 1822–1825.
7. Foung, S. H. K. and Perkins, S. (1989) Electric field induced cell fusion and human monoclonal antibodies. *J. Immunol. Methods* **116,** 117–122.
8. Lo, M. M. S., Tsong, T. Y., Conrad, M. K., Strittmatter, S. M., Hester, L. D., and Snyder, S. H. (1984) Monoclonal antibody production by receptor mediated electrically induced cell fusion. *Nature* **310,** 792–794.
9. Lo, M. M. S. and Tsong, T. Y. (1989) Producing monoclonal antibodies by electrofusion, in *Electroporation and Electrofusion in Cell Biology* (Neumann, E., Sowers, A. E., and Jordan, C. A., eds.), Plenum, New York, pp. 259–270.
10. Sowers, A. E (1987) The long lived fusogenic state induced in erythrocyte ghosts by electric pulses is not laterally mobile. *Biophys. J.* **52,** 1015–1020.

11. Sowers, A. E. (1988) Fusion events and nonfusion contents mixing events induced in erythrocyte ghosts by an electric pulse. *Biophys. J.* **54**, 619–626.
12. Sowers, A. E. (1989) The study of membrane fusion and electroporation mechanisms, in Charge and Field Effects in Biosystems—II (Allen, M. J., Cleary, S. F., and Hawkridge, F. M., eds.), Plenum, New York, pp. 315–337.
13. Sowers, A. E. (1989) Electrofusion of dissimilar membrane fusion partners depends on additive contributions from each of the two different membranes. *Biochim. Biophys. Acta* **985**, 339–342.
14. Abidor, I. G. and Sowers, A. E. (1992) Kinetics and mechanism of cell membrane electrofusion. *Biophys. J.* **61**, 1557–1569.
15. Chakrabarti, R., Wylie, D. E., and Schuster, S. M. (1989) Transfer of monoclonal antibodies into mammalian cells by electroporation. *J. Biolog. Chem.* **264**, 15,494–15,500.
16. Mir, L. M., Banoun, H., and Paoletti, C. (1988) Introduction of definite amounts of nonpermeant molecules into living cells after electropermeabilization: direct access to the cytosol. *Exp. Cell Res.* **175**, 15–25.
17. Ozawa, K., Hosoi, T., Tsao, C. J., Urabe, A., Uchida, T., and Takaku, F. (1985) Microinjection of macromolecules into leukemic cells by cell fusion technique: search for intracellular growth-suppressor factors. *Biochem. Biophys. Res. Commun.* **130**, 257–263.
18. Chernomordik, L. V., Sokolov, A. V., and Budker, V. G. (1990) Electrostimulated uptake of DNA by liposomes. *Biochim. Biophys. Acta* **1024**, 179–183.
19. Grasso, R. J., Heller, R., Cooley, J. C., and Haller, E. M. (1989) Electrofusion of individual animal cells directly to intact corneal epithelial tissue. *Biochim. Biophys. Acta* **980**, 9–14.
20. Heller, R. and Grasso, R. J. (1990) Transfer of human membrane surface components by incorporating human cells into intact animal tissue by cell-tissue electrofusion. *Biochim. Biophys. Acta* **1024**, 185–188.
21. Heller, R. (1992) Spectrofluorometric assay for the quantitation of cell-tissue electrofusion. *Anal. Biochem.* **202**, 286–292.
22. Church, G. M. and Gilbert, W. (1984) Genomic sequencing. *Proc. Natl. Acad. Sci. USA* **81**, 1991–1995.
23. Reed, K. C. and Mann, D. A. (1985) Rapid transfer of DNA from agarose gels to nylon membranes. *Nucleic Acids Res.* **13**, 7207–7221.
24. Heller, R., and Grasso, R. J. (1991) Reproducible layering of tissue culture cells onto electrostatically charged membranes. *J. Tissue Culture Methods* **13**, 25–30.
25. (1985) *Guide for the Care and Use of Laboratory Animals* NIH Publication No. 85-23, Washington, DC.
26. Heller, R. and Gilbert, R. (1992) Development of cell-tissue electrofusion for biological applications, in *Guide to Electroporation and Electrofusion* (Chang, D. C., Chassy, B. M., Saunders, J. A., and Sowers, A. E., eds.), Academic, San Diego, CA, pp. 393–410.
27. Heller, R. (1993) Incorporation of individual cells into intact tissue by electrofusion, in *Electricity and Magnetism in Biology and Medicine* (Blank, M., ed.), San Francisco Press, San Francisco, CA, pp. 115–118.

CHAPTER 29

Cytometric Detection and Quantitation of Cell–Cell Electrofusion Products

Mark J. Jaroszeski, Richard Gilbert, and Richard Heller

1. Introduction

Cell–cell electrofusion (CCE) is a process that involves forcing cells into close juxtaposition and then inducing fusion by delivering electric pulses to the cells. CCE has proven to have many practical applications. It has been used for monoclonal antibody (MAb) production (1,2), hybridoma production (3–5), and to transfer membrane-surface markers (6). Many other applications are described in this volume. In addition, the study of membrane fusion mechanisms has been the focus of some researchers (7–12). Electrofusion techniques seldom result in 100% yield between fusion partners. Therefore, a major aspect of CCE applications is the ability to detect and quantitate fusion products.

Traditional methods for rating the success or failure of fusion are visual. Light and fluorescence microscopy have been implemented for this purpose. One popular light microscopic technique uses a polynucleation index (12–14). Fluorescence methods include using a variety of different internal dyes (15–17). Microscopic methods have been used extensively because they are simple and use a one-cell-at-a-time approach to analysis. However, they also have several disadvantages. First, they are very time-consuming. Manual enumerations can take from 5–15 min/ sample depending on the application. Time becomes particularly important if large sets of fusion samples are analyzed. Second, it is only practical to determine fusion success for a small number of cells. Microscopic methods typically use visual determinations from several hundred cells as an analytic basis. Most fusion methods utilize several thousand to several million cells per fusion sample. Therefore, manual determinations are normally based on a small fraction of the cells in a fusion sample. Third, microscopic methods are subject to human bias and error. Alternative assays for quantitating fusion include analyzing for the secretion of a specific product *(4)*, flow cytometric methods *(18,19)*, and spectrofluorometric methods *(20)*.

Flow cytometry offers several advantages over microscopic methods while retaining a one-cell-at-a-time approach to analysis. The time required to analyze a fusion sample cytometrically is on the order of seconds. Therefore, cytometric analysis offers a time savings that is of practical significance. The number of cells analyzed is very large compared to microscopic methods. Common cell numbers for cytometric samples range from several thousand to >40,000. Another advantage of cytometry is that human bias and error are removed from the detection and quantitation method. Finally, many cytometers have sorting capability. Therefore, it is possible to separate hybrid cells from a sample.

The premise of the cytometric detection and quantitation method given in this chapter is dual fluorescence (Fig. 1). Opposing fusion partners are first stained with different fluorescent dyes. One cell type is loaded with 5-chloromethylfluorescein diacetate (CMFDA) and the other is loaded with 5-(and 6)-{[(4-chloromethyl)benzoyl]amino} tetramethylrhodamine (CMTMR). After loading, CCE is performed using the stained cells. After the fusion process is complete, three different cell types will be present. Cells that exhibit the fluorescence of both dyes (dual fluorescence) are hybrid cells. The other two cell types will exhibit single fluorescence of the respective stains; these are unfused cells. Flow cytometry is utilized to detect and quantitate the number of hybrids and unfused cells based on fluorescence.

This chapter describes the method used to stain cells prior to fusion. It also addresses the cytometric protocol. The detection and quantitation scheme has been used with a variety of cell types. Some of these are human promyelocytic leukemia cells (ATCC CCL 240), murine lymphoma cells (ATCC TIB 53), rat Sertoli cells from primary culture, human fibroblasts from primary culture, human breast carcinoma cells (ATCC

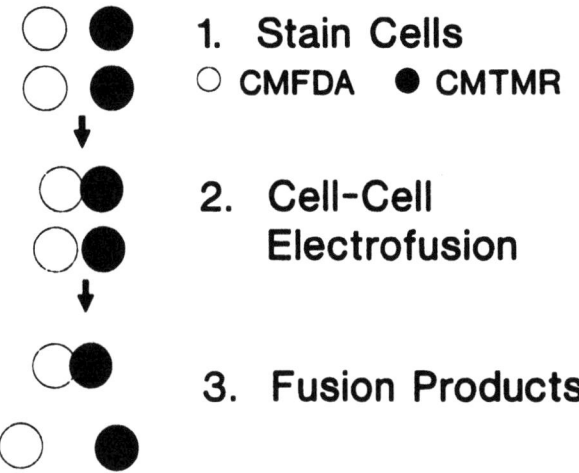

Fig. 1. The premise of cytometric detection and quantitation. (1) Opposing fusion partners are first stained with different fluorescent dyes, CMFDA and CMTMR. (2) CCE is performed on the stained cells. (3) Fusion products will be composed of three different cell types. Fused cells will be dual-fluorescing. Unfused cells will retain the single fluorescence of their respective stained fusion partners.

HTB 132), human bladder carcinoma cells (ATCC HTB 9 5637), rat Sertoli cells (ATCC CRL 1715), and hamster β-cells (ATCC CRL 1777). The method detailed below has yielded consistent results irrespective of the cell types used for fusion indicating that it is applicable to many different cell–cell systems.

2. Materials

1. Two different cell types to be fused: Cells from primary sources and lines can be used. Cells of the same type can also be used as opposing fusion partners.
2. Growth media used to maintain the cell type(s) used for fusion.
3. Stock solution of CMFDA (Molecular Probes, Eugene, OR): 5 mM solution (*see* Note 1) in dimethyl sulfoxide (DMSO; Sigma, St. Louis, MO).
4. Stock solution of CMTMR (Molecular Probes, Eugene, OR): 5 mM solution (*see* Note 1) in DMSO.
5. A flow cytometer that is capable of acquiring fluorescent data from two different bandwidths of light. Cytometers that are set up to detect cells that are labeled with fluorescein and rhodamine compounds should provide

excellent results. For example, a Becton Dickinson FACStar Plus (Becton Dickinson, San Jose, CA) is used to detect fusion products from a variety of different cell–cell systems. Excitation is provided by an 80 mW argon laser tuned to a wavelength of 488 nm. CMFDA emission is detected in the FL1 (green) channel; CMTMR emission is detected in the FL2 (red) channel. Detection wavelengths for FL1 and FL2 channels are 530 ± 15 and 585 ± 21 nm, respectively. Forward light scatter (FSC) and side-angle light scatter (SSC) data are also collected. A Becton Dickinson FACScan instrument has also been used with equivalent results.
6. Dulbecco's Phosphate-Buffered Saline (PBS) Solution (Mediatech, Washington, DC).
7. Conical-bottom centrifuge tubes with a 15- or 45-mL capacity. Tube size will depend on the number of cells stained with the fluorescent dyes.

3. Methods
3.1. Cell Staining
1. Mix separate suspensions of each fusion partner in the media used for growth. Cell densities of approx 1.5×10^6 cells/mL yield excellent results. Mix the suspensions in conical-bottom centrifuge tubes to facilitate subsequent washing. Note the total volume in each tube; this will be referred to as one staining volume. If adherent cells are under investigation, detach cells in order to make the appropriate suspension(s).
2. Stain one cell type by adding CMFDA stock solution to the appropriate suspension so that the dye concentration is 0.5 μM. Stain the other cell type by adding CMTMR stock solution to make the final dye concentration 6.0 μM.
3. Incubate both suspensions for 30 min at 37°C. Shield the mixtures of stain and cells from light during incubation.
4. Wash each suspension twice by centrifugation. Replace the liquid between washes with one staining volume of growth media. After the second wash, suspend the cells in one staining volume of growth media.
5. Incubate the stained cell suspensions for 60 min at 37°C. The suspensions should be shielded from light (*see* Note 2).
6. Wash each suspension three times by centrifugation. Use one staining volume of PBS to replace the liquid after each wash.
7. Cytometrically analyze a mixture of both stained fusion partners. The mixture should contain approximately equal fractions of CMFDA- and CMTMR-stained cells. A plot of FL1 (green fluorescence) vs FL2 (red fluorescence) should be made in order to determine if each fusion partner has a suitable fluorescent magnitude. The resulting plot should appear similar to Fig. 2. Note that the CMFDA- and CMTMR-stained cell popula-

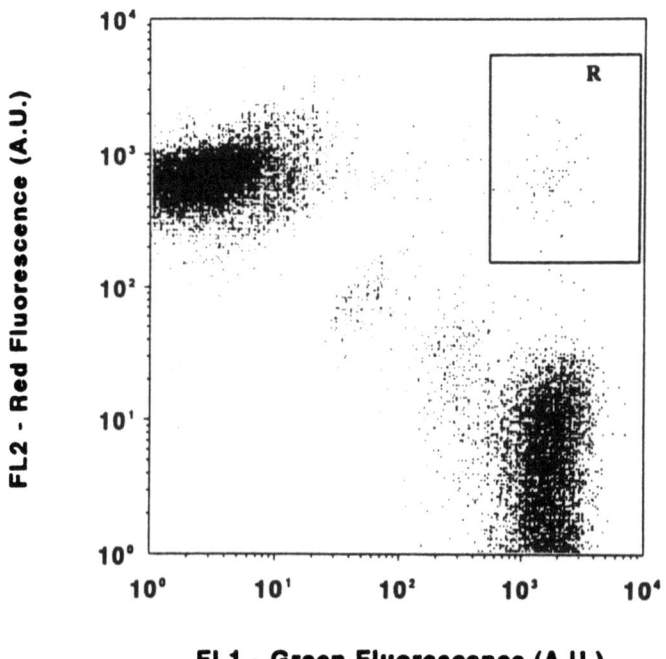

Fig. 2. A typical fluorescent dot plot from a sample that contained a mixture of two stained fusion partners. Cells stained with CMFDA appear as a population near the FL1 axis. Cells stained with CMTMR populate plots near the FL2 axis. Note that the stained fusion partners both appear as separate and distinct populations on the plot. This facilitates detection and quantitation of fusion products that will be present in region R after fusion. A background level of aggregated cells is typically present in region R for samples that have not been electrically treated. Fluorescence was measured in arbitrary units (A.U.). Reprinted with permission *(19)*.

tions in the figure are separate and distinct. This characteristic is critical to the detection and quantitation method. Fusion products will appear as a dual-fluorescing population with FL1 and FL2 magnitudes that are approximately equal to those of the stained fusion partners. This region is labeled R in Fig. 2. If the populations are not separate and distinct, graphical resolution of the hybrid population will be impossible. Dye concentrations used for staining should be increased if the fusion partner populations cannot be distinguished (*see* Notes 3 and 4).

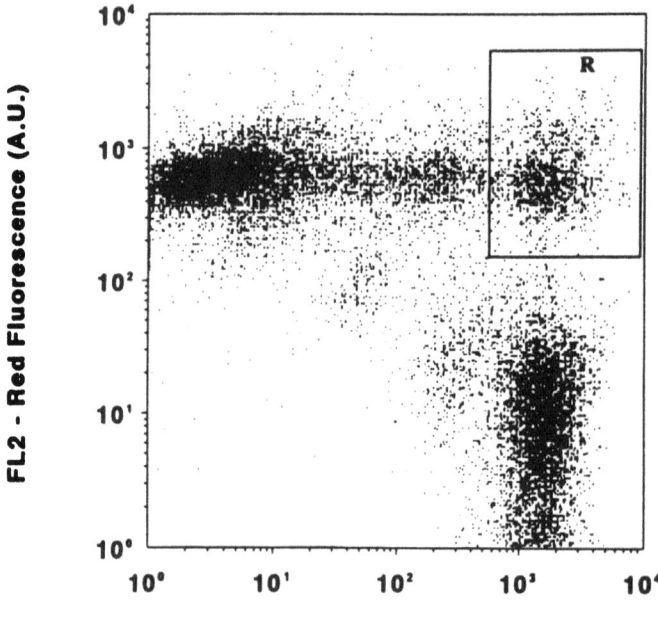

Fig. 3. A typical fluorescent dot plot from a fusion sample. The population within region R contains hybrid cells. The hybrids within the region represent 6.9% of the 20,000 cells represented by the plot. Unfused fusion partners retain their respective positions near the FL1 and FL2 axes. Cells located between the hybrid and unfused cell populations are typically damaged cells or debris that result from electrofusion. Fluorescence was measured in arbitrary units (A.U.). Reprinted with permission (19).

8. Perform fusion after establishing that the staining procedure will yield a graphically resolvable hybrid population (see Notes 5 and 6).

3.2. Detection and Quantitation of Fusion Products

1. Cytometrically analyze cells that remain after fusion. Acquire FL1, FL2, FSC, and SSC data.
2. Examine the FL1 vs FL2 plot from each fusion sample. A typical dot plot from a fusion sample is given in Fig. 3 (see Notes 7 and 8). Note that the plot shown contains three major populations. The population within the region labeled R contains fused cells. The populations near each of the axes are unfused CMFDA- and CMTMR-stained cells.

3. Quantitation of fusion products is conducted using flow cytometer software. A region similar to R in Fig. 3 should be drawn around the hybrid population. Flow cytometer software then tabulates the number of cells within the region. Most software packages express the population within any region as a percentage of the cells represented on the plot and as the number of cells within the region. If multiple samples are fused, the same region is applicable to all samples.

4. Notes

1. Stock solutions of CMFDA and CMTMR should be stored frozen (−10°C) and shielded from light. Consistent staining is achieved if the stock solutions are quickly defrosted prior to use and then refrozen as soon as possible. If used in this manner, the stock solutions will provide reproducible staining if used within several months.
2. According to the dye manufacturer (Molecular Probes), CMFDA and CMTMR pass freely through the membranes into the cytosol. Once they pass through the membranes, both dyes undergo what is thought to be a glutathione S-transferase-mediated reaction, which renders the dyes membrane impermeable. Therefore, the purpose of incubating cells in step 5 of the cell-staining procedure is to allow time for this reaction to occur.
3. The CMFDA- and CMTMR-staining concentrations that yield distinct and separate populations on FL1 vs FL2 cytometric plots vary with cell type. Fluorescent dye concentrations ranging from 0.25–0.8 μM CMFDA and 4.0–12.0 μM CMTMR have been utilized for several different cell lines.
4. Quenching does not appear to be a complicating factor for CMFDA or CMTMR. Stained cell samples exposed to room light for several hours did not show decreased fluorescent magnitudes with time.
5. Several different cell lines containing CMFDA and CMTMR have been placed back into culture. Fluorescence is retained for several days with no marked cytotoxic affect.
6. The staining method yields cells and hybrids that can be visualized using fluorescence microscopy. Standard filters for viewing fluorescein isothyocyanate allow CMFDA-stained cells to be viewed, and filters normally used for phycoerythrin allow CMTMR-stained cells to be seen. We prefer to view emission from both dyes simultaneously using a Leitz Orthoplan 2 microscope (Leica, West Germany). Filters are designed for simultaneous excitation by 490 ± 20 and 575 ± 30 nm. In addition, the filters allow wavelengths of 525 ± 40 and 633 ± 20 nm to be simultaneously viewed.
7. Electrofusion can produce small cellular debris. This debris may hinder interpretation of FL1 vs FL2 plots. Examination of an FSC vs SSC plot is useful for detecting this type of debris. It is generally present as particles

that are smaller in FSC magnitude than the major population of FSC vs SSC plots. The small particles can be gated out of sample data prior to making FL1 vs FL2 plots.

8. In principle, this method can be used with other dye combinations as long as both stains can be detected in the FL1 (530 ± 14 nm) and FL2 (535 ± 15 nm) cytometer channels. Many stain combinations were investigated during development of the detection and quantitation method. The combination of CMFDA and CMTMR provides consistent and reproducible results. A common problem with fluorescent dye combinations other than CMFDA and CMTMR (i.e., combinations of fluorescein isothiocyanate [Sigma]; [5 and 6] sulfofluorescein diacetate [Molecular Probes]; hydroethidine [Polysciences, Warrington, PA]; and Calcium Crimson [Molecular Probes]) was that mixtures containing cells stained with both dyes migrated off their respective axes in time. Both stained cell populations moved toward the dual-fluorescing region of FL1 vs FL2 plots over the course of minutes to hours. This complicates interpretation because mixtures of stained cells are present during and after fusion procedures. One possible explanation for this is that dye may leak out of a stained cell type and then diffuse into the other stained cell type. This problem can seriously hinder detection and quantitation because cytometric populations will not occupy the same positions on plots over a normal experimental time frame. Also, migration toward the dual-fluorescing region would increase the likelihood of false-positive results. Mixtures of CMFDA- and CMTMR-stained cells showed no population migration in time; therefore, they were judged suitable for use. If the methods presented in this chapter are applied using different dyes, it is important to verify that stained cell populations do not migrate in time.

References

1. Lo, M. M. S., Tsong, T. Y., Conrad, M. K., Strittmatter, S. M., Hester, L. D., and Snyder, S. (1984) Monoclonal antibody production by receptor-mediated electrically induced cell fusion. *Nature* **310,** 792–794.
2. Foung, S. K. H. and Perkins, S. (1989) Electric field-induced cell fusion and human monoclonal antibodies. *J. Immunol. Methods* **116,** 117–122.
3. Glassy, M. (1988) Creating hybridomas by electrofusion. *Nature* **333,** 579,580.
4. Hewish, D. R. and Werkmeister, J. A. (1989) The use of an electroporation apparatus for the production of murine hybridomas. *J. Immunol. Methods* **120,** 285–289.
5. Wojchowski, D. M. and Sytkowski, A. J. (1986) Hybridoma production by simplified avidin-mediated electrofusion. *J. Immunol. Methods* **90,** 173–177.
6. Grasso, R. J., Heller, R., Cooley, J. C., and Haller, E. M. (1989) Electrofusion of individual animal cells directly to intact corneal epithelial tissue. *Biochim. Biophys. Acta* **980,** 9–14.

7. Abidor, I. G. and Sowers, A. E. (1992) Kinetics and mechanism of cell membrane electrofusion. *Biophys. J.* **61,** 1557–1569.
8. Zimmerman, U., Pilwat, B., and Riemann, F. (1974) Dielectric breakdown of cell membranes. *Biophys. J.* **14,** 881–899.
9. Dimitrov, D. S., Apostolova, M. A., and Sowers, A. E. (1990) Attraction, deformation, and contact of membranes induced by low frequency electric fields. *Biochim. Biophys. Acta* **1023,** 389–397.
10. Sowers, A. E. (1989) The mechanism of electroporation and electrofusion in erythrocyte membranes, in *Electroporation and Electrofusion in Cell Biology* (Neumann, E., Sowers, A. E., and Jordan, C. A., eds.), Plenum, New York, pp. 229–256.
11. Rols, M. P. and Teissie, J. (1989) Ionic-strength modulation of electrically induced permeabilization and associated fusion of mamalian cells. *Eur. J. Biochem.* **179,** 109–115.
12. Teissie, J. and Blangero, C. (1984) Direct experimental evidence of the vectorial character of the interaction between electric pulses and cells in cell electrofusion. *Biochim. Biophys. Acta* **775,** 446–448.
13. Teissie, J. and Rols, M. P. (1986) Fusion of mammalian cells in culture is obtained by creating the contact between cells after their electropermeabilization. *Biochem. Biophys. Res. Comm.* **140-1,** 258–266.
14. Sukharev, S. I., Bandrina, I. N., Barbul, A. I., Fedorova, L. I., Abidor, I. G., and Zelenin, A. V. (1990) Electrofusion of fibroblasts on the porous membrane. *Biochim. Biophys. Acta* **1034,** 125–131.
15. Sowers, A. E. (1988) Fusion events and nonfusion contents mixing events induced in erythrocyte ghosts by an electric pulse. *Biophys. J.* **54,** 619–626.
16. Sowers, A. E. (1986) A long-lived fusogenic state is induced in erythrocyte ghosts by electric pulses. *J. Cell Biol.* **102,** 1358–1362.
17. Bakker Schut, T. C., Kraan, Y. M., Barlag, W., de Leij, L., de Grooth, B. G., and Greve, J. (1993) Selective electrofusion of conjugated cells in flow. *Biophys. J.* **65,** 568–572.
18. Shi, T., Eaton, A. M., and Ring, D. B. (1991) Selection of hybrid hybridomas by flow cytometry using a new combination of fluorescent vital stains. *J. Immunol. Methods* **141,** 165–175.
19. Jaroszeski, M. J., Gilbert, R., and Heller, R. (1994) Detection and quantitation of cell-cell electrofusion products by flow cytometry. *Anal. Biochem.* **216,** 271–275.
20. Heller, R. (1992) Spectrofluorometric assay for the quantitation of cell-tissue electrofusion. *Anal. Biochem.* **202,** 286–292.

Index

A

Aminoglycoside transferase, *see* Neomycin phosphotransferase
Antibodies, 83–92, 232, 235, 284, 341, 355
  assays for incorporation into cells, 90, 91
  hybridoma formation, 295–307, 341, 355
  inhibiting cell division, 84
  inhibiting cellular enzymes, 84
  introduction into adherent cells, 95, 104–110
  microinjection, 108
  purification, 85
Antigen presenting cells (APC), 73–81
Antisense RNA, 123

B

B-lymphocytes, 289, 295–307
  activating with Epstein Barr virus, 295, 297, 303
Basilen blue, 110
β-galactosidase, 64, 128, 143, 146, 192–196, 212, 214, 217, 219, 221, 222, 234
Blastomeres,
  electrofusion to oocytes, 284, 331–340
  tetraploid, 331
  use in nuclear transfer, 317–329
Bone marrow cells, 274, 278
  preparation, 276
Bovine,
  blastomeres, 284, 289, 317–329
  embryonic cells, 314, 317–329
  oocytes, 314, 315, 317–329
  spermatozoa, 161–166
  fertilization of oocytes, 164
  preparation, 163

C

Calcium ion, cellular effects, 186, 254
Calcium phosphate coprecipitation, 115, 119, 141, 151
Capacitance, 32, 46, 47, 104, 188, 207
Cardiac cells, 253–271
  electric field effects on single cells, 260, 262
Carrier DNA, 119, 191, 212, 213, 220, 229
Cell stress, 22
Cell viability, 22, 33, 38, 65, 68, 88–91, 101, 120, 127, 186, 188, 207, 292, 305, 327, 328
Chicken cells (cardiac), 258–260
Chloramphenicol acetyltransferase (CAT), 128, 142, 143, 145, 234, 246
Chromosome aberrations, 67, 68
Coelectroporation, *see* cotransfection
Cotransfection, 119, 134, 137, 157, 158, 200, 208
Cuvets, see electroporation chambers
Cytochalasin B (CCB), 319, 324, 338
Cytotoxic T-lymphocytes (CTL), 73–81
  priming in vitro, 76
  priming in vivo, 76

365

D

DEAE dextran transfection, 115, 141, 212
Dexamethasone (DEX), 116, 118, 146
Dielectrophoresis, 41, 284
Dihydrofolate reductase *(dhfr)*,
  gene amplification with methotrexate, 119, 225–237
  selectable marker, 119, 152
Diphtheria toxin A, 171
DNA, *see* Carrier DNA, Plasmid DNA
Drosophila embryos, 239-243
  aldehyde oxidase gene, 239, 240, 242
  ribozyme expression, 242

E

Electric field,
  alternating current (ac), 39, 284, 287–289, 297
  exponential pulse, 43, 46, 65
  generators, 44–50
  Joule heat, 260, 328
  length, 47–49
    effect on electroporation, 29–40
  multiple pulses, 39, 254
  radio frequency, 50
  square pulse, 43
  strength, effect on electroporation, 29–40
Electrofusion,
  ac alignment pulse, 287, 288, 291, 297, 312
  cell-tissue (CTE), 341-353
  chambers, 285, 298, 305, 312, 334, 342, 343, 347
  dc fusion pulse, 288, 297, 305, 313, 336
  fusion media, 286, 299, 303–306, 314, 327, 337
  hybridoma formation, 295–307, 341, 355

cell lines, 298
mammalian cells, 283–293
microscope, 286
nuclear transfer into oocytes, 317–329
oocytes, 289, 309–316, 331–340
quantitation, 355–363
temperature-dependence, 291, 328
Electroporation,
  adherent cells, 93–113
  antibodies, 68, 83–92, 95, 104–110
  antigen, 73–81
  pathological effects, 253
  protein, 63–71
  stable high-copy-number, 225–237
  synchronized cells, 275
  temperature-dependence, 64, 68, 165, 194, 205, 207, 249, 278
Electroporation chambers, 50–59
  concentric electrodes, 54
  flow through, 54
  meander, 52
  microelectrodes, 255, 264
  microslides, 52, 95–100, 103
  parallel electrodes, 43, 51, 53, 54
  Petri dish electrodes, 53, 54
  reusing, 155
Embryonic stem (ES) cells, 167–184
  cell morphology, 181
  culture, 173–175
  DNA isolation, 177
  injection into blastocysts, 180
  nuclear donors, 317
  storage, 174, 177
Embryos (embryonic cells),
  as nuclear donors, 317–329
  cell preparation, 320, 321, 335
  electrofusion, 309–316
  tetraploid, 331–340
Enucleation, 313, 315

F

Fibroblast cells,
  normal human, 133–140

Index

Fish cells, 245–251
  goldfish cells, 246, 247
  medaka embryos, 246, 248, 249
  plasmid vectors, 246, 247
Flow cytometry, 90, 91, 355–362
  cell staining, 358, 359, 361
Fluorescein isothiocyanate (FITC), 90, 254, 256, 362

G

G418-resistance (G418^r) marker, see Neomycin phoshotransferase
Gene amplification, 119, 225–237
  vectors, 228, 233
Gene targeting, see Homologous recombination
Gene therapy, 273–280
  factor IX gene, 277
  selectable markers, 274
Generators,
  electrofusion, 49, 50
  electroporation, 44–49
  exponential wave, 46
  square wave, 49
Geneticin resistance marker, see Neomycin phosphotransferase
Goldfish cells, 246, 247

H

Hamster cells, 63–71, 115–121, 167, 194, 213, 226, 235, 357
  inducible expression, 118
  selectable markers, 116–119
Hematopoietic cell lines, 274, 276
Hepatocytes, 141–150
  cell extracts, 145
  preparation from rats, 144
  reporter genes, 143, 145, 146–149
Herpes simplex virus thymidine kinase (HSV-tk), 171, 199, 200, 207, 208
Histidinol resistance marker, 207
Homologous recombination, 156–158, 167, 168
  detection by polymerase chain reaction, 182, 275
  vectors, 168, 171, 178, 181
Human cells,
  B-lymphocytes, 289, 295–307
  bladder carcinoma, 357
  bone marrow, 274, 278
  breast carcinoma, 356
  fibroblasts, 133–140, 356
    initiating a strain, 135, 136
    proliferative potential, 133, 135, 138
    selectable markers, 134, 138
  leukemia cells, 346, 356
  lymphoblastoid cells, 151–160
    culture, 153
    G418 selection, 154–157
    selectable markers, 152, 157
  lymphoma, 346
  stromal cells, 277, 278
Hybridomas, 284, 295–307, 341, 355
  feeding schedule, 302
  measuring hybrid yield, 303
Hygromycin resistance marker, 110, 116, 117, 169–171

I

"Intracellular" buffer, 185–197

L

Lac Z, see β-galactosidase
Liposome fusion, 83, 141, 151, 212, 341
Lucifer yellow, 94, 101, 103, 104, 107–111
Luciferase, 128, 149, 212–214, 217, 219–222
Lymphoblastoid cells, 151–160
  culture, 153
  G418 selection, 154–157
  selectable markers, 152, 157

M

Medaka (fish) embryos, 246, 248, 249
Membrane,
  barrier, 5, 6, 8

electrical breakdown,
  irreversible, 10, 33
  reversible (REB), 4, 10, 12–17, 21, 22, 29–32, 37
fusion, 341
permeability, 5, 9, 21, 29–31, 37, 64, 108
pores, 3, 8, 11, 29–33, 37, 254, 291
  conducting, 15
  diminishing, 255
  distribution of sizes, 12
  metastable, 17
  transient aqueous, 8, 9, 20, 24
recovery, 22
rupture, 13–15, 24
Methotrexate (MTX), 119, 152, 226, 231–236
Microinjection, 83, 108, 141, 161, 239, 245
Molecular transport, 19, 21, 30–32, 64
Monoclonal antibodies (MAb), *see* antibodies
Mouse cells, 102, 104, 119, 167–184, *see also* Embryonic stem cells
  blastomeres, 284, 289
  bone marrow, 278
  DNA isolation, 177
  lymphoma, 346, 356
  negative selection markers, 171, 176
  selectable markers, 169–171, 176, 181, 182
Mouse mammary tumor virus (MMTV) promoter, 118, 128
Murine leukemia inhibitory factor (LIF), 169, 174, 176
Mycophenolic acid resistance marker, *see* Xanthine-guanine phosphoribosyltransferase

N

Negative selection markers, *see* Herpes simplex virus thymine kinase, Diphtheria toxin A

Neomycin phosphotransferase *(neo)* 116, 117, 152, 157, 169–172, 181, 182, 200, 204, 207, 274, 277
Nuclear transfer, 317–329
  electrofusion, 324, 325
  microtools, 320, 327
  parthenogenic activation, 327

O

Oocytes, 161–166
  electrofusion, 289, 309–316, 331–340
    cell preparation, 311, 312
  enucleation, 313, 315, 317, 322
  fertilization, 164

P

Pituitary cells (rat),
  culture, 124
  hormone secretion, 123, 126, 127
  reporter genes, 128
Plasmid DNA,
  effect of concentration, 119, 137, 138, 155, 194, 205, 213, 226, 241, 249
  effect of form (linear, circular), 119, 125, 127, 137, 192, 205, 213, 233, 234, 278
  effect of size, 205
  preparation, 124, 134, 136, 142, 153, 192, 202, 215, 220, 233, 240, 249, 275
Polyethylene glycol (PEG), 283, 309, 331
Polymerase chain reaction, assays for homologous recombination, 182, 275
Porcine cells, 102, 110, 284, 289
Pores, see Membrane, Pores
Promoters,
  actin, 246, 247
  adenovirus major late, 228
  antifreeze protein, 247
  β-globin, 128

Index

c-fos, 143
carbomoyl phosphate synthetase, 190, 191
cytomegalovirus (CMV), 246, 247
heat shock, 246
Herpes simplex virus thymidine kinase, 249
metallothionein, 128, 246, 247
Miw (chicken), 246
Moloney murine leukemia virus, 110
mouse mammary tumor virus (MMTV), 118, 128
mouse TSHβ, 128
phoshoglycerate kinase (PGK), 169
rat growth hormone, 128
rat prolactin, 128
Rous sarcoma virus (RSV), 128, 138, 143, 148, 190, 191, 212, 214, 219, 220, 246, 247
Spi, 143, 146
SV40, 117, 118, 128, 190, 200, 246
thymidine kinase, 143, 146, 191
X47 *(Xiphophorus)*, 246
Propidium iodide, 90, 91

R

Rabbit cells, 167
  cornea, use in cell-tissue electrofusion, 347–352
Rat cells, 102, 104–109, 123–131, 356, 357
  colon cells, 188
  fibroblasts, 188, 196
  hepatocytes, 141–150
    cell extracts, 145
    preparation, 144
    selectable markers, 143
  hepatoma cells, 188–196
    reporter genes, 190–196
  kidney cells, 188
  pituitary cells,
    culture, 124
    hormone secretion, 123, 126, 127
    reporter genes, 128

Reporter genes, *see* Chloramphenicol acetyltransferase, luciferase, and β-galactosidase
Resistance (electrical), 32, 43, 46, 47
Restriction endonucleases, 63–71
Retrovirus vectors, 141, 225, 273

S

Satellite DNA, 199–210
Sendai virus, 63, 283, 309, 331
Sodium butyrate treatment, 138, 212
Spermatozoa, 161–166, 289
  nuclear donors, 317

T

Temperature-dependence, *see* Electroporation, Electrofusion
Tetraploid cells, 331–340
Thymidine kinase,
  mutant cells, 119, 201
  promoter, 143, 146
  selectable marker, 119
Time constant, 32, 47, *see also* Electric field, Length, Capacitance
Tissue electroporation, 23, 258
Transgenic fish, 245–251
Transgenic mice, 161, 167–184
Transient gene expression, 127, 128, 142–149, 185–197, 211–225, 241, 242, 278
Transmembrane potential (voltage), 4, 9, 29–31
  electro-optical dyes, 256
  measuring, 255, 256, 259, 265
Trypan blue, 78, 88, 89, 108, 127, 276, 292

V

Viability, *see* Cell viability
Viral vectors, *see* Retrovirus vectors

X

Xanthine-guanine phosphotransferase *(gpt)*, 116, 117

Methods in Molecular Biology™ Series

Methods in Molecular Biology™ manuals are available at all medical bookstores. You may also order copies directly from Humana by filling in and mailing or faxing this form to: Humana Press, 999 Riverview Drive, Suite 208, Totowa, NJ 07512 USA, Phone: 201-256-1699/Fax: 201-256-8341.

☐ 55. **Plant Cell Electroporation and Electrofusion Protocols**, edited by Jac A. Nickoloff, 1995 • 0-89603-328-7 • Comb $49.50 (T)
☐ 54. **YAC Protocols**, edited by David Markie, 1995 • 0-89603-313-9 • Comb $69.50 (T)
☐ 53. **Yeast Protocols: Methods in Cell and Molecular Biology**, edited by Ivor H. Evans, 1995 • 0-89603-319-8 • Comb $69.50 (T)
☐ 52. **Capillary Electrophoresis: Principles, Instrumentation, and Applications**, edited by Kevin D. Altria, 1995 • 0-89603-315-5 • Comb $64.50 (T)
☐ 51. **Antibody Engineering Protocols**, edited by Sudhir Paul, 1995 • 0-89603-275-2 • Comb $69.50
☐ 50. **Species Diagnostics Protocols: PCR and Other Nucleic Acid Methods**, edited by Justin P. Clapp, 1995 • 0-89603-323-6 • Comb $69.50 (T)
☐ 49. **Plant Gene Transfer and Expression Protocols**, edited by Heddwyn Jones, 1995 • 0-89603-321-X • Comb $69.50 (T)
☐ 48. **Animal Cell Electroporation and Electrofusion Protocols**, edited by Jac A. Nickoloff, 1995 • 0-89603-304-X • Comb $64.50 (T)
☐ 47. **Electroporation Protocols for Microorganisms**, edited by Jac A. Nickoloff, 1995 • 0-89603-310-4 • Comb $69.50
☐ 46. **Diagnostic Bacteriology Protocols**, edited by Jenny Howard and David M. Whitcombe, 1995 • 0-89603-297-3 • Comb $69.50
☐ 45. **Monoclonal Antibody Protocols**, edited by William C. Davis, 1995 • 0-89603-308-2 • Comb $64.50
☐ 44. **Agrobacterium Protocols**, edited by Kevan M. A. Gartland and Michael R. Davey, 1995 • 0-89603-302-3 • Comb $69.50
☐ 43. **In Vitro Toxicity Testing Protocols**, edited by Sheila O'Hare and Chris K. Atterwill, 1995 • 0-89603-282-5 • Comb $69.50
☐ 42. **ELISA: Theory and Practice**, by John R. Crowther, 1995 • 0-89603-279-5 • Comb $59.50
☐ 41. **Signal Transduction Protocols**, edited by David A. Kendall and Stephen J. Hill, 1995 • 0-89603-298-1 • Comb $64.50
☐ 40. **Protein Stability and Folding: Theory and Practice**, edited by Bret A. Shirley, 1995 • 0-89603-301-5 • Comb $69.50
☐ 39. **Baculovirus Expression Protocols**, edited by Christopher D. Richardson, 1995 • 0-89603-272-8 • Comb $69.50
☐ 38. **Cryopreservation and Freeze-Drying Protocols**, edited by John G. Day and Mark R. McLellan, 1995 • 0-89603-296-5 • Comb $79.50
☐ 37. **In Vitro Transcription and Translation Protocols**, edited by Martin J. Tymms, 1995 • 0-89603-288-4 • Comb $69.50
☐ 36. **Peptide Analysis Protocols**, edited by Ben M. Dunn and Michael W. Pennington, 1994 • 0-89603-274-4 • Comb $64.50
☐ 35. **Peptide Synthesis Protocols**, edited by Michael W. Pennington and Ben M. Dunn, 1994 • 0-89603-273-6 • Comb $64.50
☐ 34. **Immunocytochemical Methods and Protocols**, edited by Lorette C. Javois, 1994 • 0-89603-285-X • Comb $64.50
☐ 33. **In Situ Hybridization Protocols**, edited by K. H. Andy Choo, 1994 • 0-89603-280-9 • Comb $69.50
☐ 32. **Basic Protein and Peptide Protocols**, edited by John M. Walker, 1994 • 0-89603-269-8 • Comb $59.50 • 0-89603-268-X • Hardcover $89.50
☐ 31. **Protocols for Gene Analysis**, edited by Adrian J. Harwood, 1994 • 0-89603-258-2 • Comb $69.50
☐ 30. **DNA–Protein Interactions**, edited by G. Geoff Kneale, 1994 • 0-89603-256-6 • Paper $64.50
☐ 29. **Chromosome Analysis Protocols**, edited by John R. Gosden, 1994 • 0-89603-243-4 • Comb $69.50 • 0-89603-289-2 • Hardcover $94.50

☐ 28. **Protocols for Nucleic Acid Analysis by Nonradioactive Probes**, edited by Peter G. Isaac, 1994 • 0-89603-254-X • Comb $59.50
☐ 27. **Biomembrane Protocols: II. Architecture and Function**, edited by John M. Graham and Joan A. Higgins, 1994 • 0-89603-250-7 • Comb $64.50
☐ 26. **Protocols for Oligonucleotide Conjugates: Synthesis and Analytical Techniques**, edited by Sudhir Agrawal, 1994 • 0-89603-252-3 • Comb $64.50
☐ 25. **Computer Analysis of Sequence Data: Part II**, edited by Annette M. Griffin and Hugh G. Griffin, 1994 • 0-89603-276-0 • Comb $59.50
☐ 24. **Computer Analysis of Sequence Data: Part I**, edited by Annette M. Griffin and Hugh G. Griffin, 1994 • 0-89603-246-9 • Comb $59.50
☐ 23. **DNA Sequencing Protocols**, edited by Hugh G. Griffin and Annette M. Griffin, 1993 • 0-89603-248-5 • Comb $59.50
☐ 22. **Microscopy, Optical Spectroscopy, and Macroscopic Techniques**, edited by Christopher Jones, Barbara Mulloy, and Adrian H. Thomas, 1993 • 0-89603-232-9 • Comb $69.50
☐ 21. **Protocols in Molecular Parasitology**, edited by John E. Hyde, 1993 • 0-89603-239-6 • Comb $69.50
☐ 20. **Protocols for Oligonucleotides and Analogs: Synthesis and Properties**, edited by Sudhir Agrawal, 1993 • 0-89603-247-7 • Comb $69.50 • 0-89603-281-7 • Hardcover $89.50
☐ 19. **Biomembrane Protocols: I. Isolation and Analysis**, edited by John M. Graham and Joan A. Higgins, 1993 • 0-89603-236-1 • Comb $64.50
☐ 18. **Transgenesis Techniques: Principles and Protocols**, edited by David Murphy and David A. Carter, 1993 • 0-89603-245-0 • Comb $69.50
☐ 17. **Spectroscopic Methods and Analyses: NMR, Mass Spectrometry, and Metalloprotein Techniques**, edited by Christopher Jones, Barbara Mulloy, and Adrian H. Thomas, 1993 • 0-89603-215-9 • Comb $69.50
☐ 16. **Enzymes of Molecular Biology**, edited by Michael M. Burrell, 1993 • 0-89603-322-8 • Paper $59.50
☐ 15. **PCR Protocols: Current Methods and Applications**, edited by Bruce A. White, 1993 • 0-89603-244-2 • Paper $54.50
☐ 14. **Glycoprotein Analysis in Biomedicine**, edited by Elizabeth F. Hounsell, 1993 • 0-89603-226-4 • Comb $64.50
☐ 13. **Protocols in Molecular Neurobiology**, edited by Alan Longstaff and Patricia Revest, 1992 • 0-89603-199-3 • Comb $59.50
☐ 12. **Pulsed-Field Gel Electrophoresis: Protocols, Methods, and Theories**, edited by Margit Burmeister and Levy Ulanovsky, 1992 • 0-89603-229-9 • Hardcover $69.50
☐ 11. **Practical Protein Chromatography**, edited by Andrew Kenney and Susan Fowell, 1992 • 0-89603-213-2 • Hardcover $59.50
☐ 10. **Immunochemical Protocols**, edited by Margaret M. Manson, 1992 • 0-89603-270-1 • Comb $69.50
☐ 9. **Protocols in Human Molecular Genetics**, edited by Christopher G. Mathew, 1991 • 0-89603-205-1 • Hardcover $69.50
☐ 8. **Practical Molecular Virology: Viral Vectors for Gene Expression**, edited by Mary K. L. Collins, 1991 • 0-89603-191-8 • Paper $54.50
☐ 7. **Gene Transfer and Expression Protocols**, edited by Edward J. Murray, 1991 • 0-89603-178-0 • Hardcover $79.50
☐ 6. **Plant Cell and Tissue Culture**, edited by Jeffrey W. Pollard and John M. Walker, 1990 • 0-89603-161-6 • Comb $69.50
☐ 5. **Animal Cell Culture**, edited by Jeffrey W. Pollard and John M. Walker, 1990 • 0-89603-150-0 • Comb $69.50

Name _____
Department _____
Institution _____
Address _____
City/State/Zip _____
Country _____
Phone # _____ Fax # _____

Postage & Handling: *USA Prepaid (UPS):* Add $4.00 for the first book and $1.00 for each additional book. *Outside USA (Surface):* Add $5.00 for the first book and $1.50 for each additional book.

☐ My check for $ _____ is enclosed
(Drawn on US funds from a US bank).
☐ Visa   ☐ MasterCard   ☐ American Express

Card # _____
Exp. date _____
Signature _____

"T" denotes a tentative price. Prices listed are Humana Press prices, current as of June 1995, and do not reflect the prices at which books will be sold to you by suppliers other than Humana Press. All prices subject to change without notice.
UK, Europe, Middle East, and Africa: Order directly from Chapman & Hall by faxing to: +44-171-522-9623.

If you have any concerns about our products,
you can contact us on
ProductSafety@springernature.com

In case Publisher is established outside the EU,
the EU authorized representative is:
**Springer Nature Customer Service Center GmbH
Europaplatz 3, 69115 Heidelberg, Germany**

Printed by Libri Plureos GmbH
in Hamburg, Germany